Djamel A. Zighed, Shusaku Tsumoto, Zbigniew W. Ras,
and Hakim Hacid (Eds.)

Mining Complex Data

Studies in Computational Intelligence, Volume 165

Editor-in-Chief

Prof. Janusz Kacprzyk
Systems Research Institute
Polish Academy of Sciences
ul. Newelska 6
01-447 Warsaw
Poland
E-mail: kacprzyk@ibspan.waw.pl

Djamel A. Zighed
Shusaku Tsumoto
Zbigniew W. Ras
Hakim Hacid
(Eds.)

Mining Complex Data

 Springer

Djamel A. Zighed
ERIC Laboratory
University of Lyon 2
5 avenue Pierre Mendès-France
69676 Bron cedex
France
Email: abdelkader.zighed@univ-lyon2.fr

Zbigniew W. Ras
Department of Computer Science
University of North Carolina
9201 University City Blvd.
Charlotte, N.C. 28223
USA
Email: ras@uncc.edu

Shusaku Tsumoto
Shimane University
School of Medicine
89-1 Enya-cho
Izumo 693-8501
Japan
Email: tsumoto@med.shimane-u.ac.jp

Hakim Hacid
ERIC Laboratory
University of Lyon 2
5 avenue Pierre Mendès-France
69676 Bron cedex
France
Email: hhacid@eric.univ-lyon2.fr

ISBN 978-3-540-88066-0   e-ISBN 978-3-540-88067-7

DOI 10.1007/978-3-540-88067-7

Studies in Computational Intelligence   ISSN 1860949X

Library of Congress Control Number: 2008935498

© 2009 Springer-Verlag Berlin Heidelberg

Typeset & *Cover Design:* Scientific Publishing Services Pvt. Ltd., Chennai, India.

Printed in acid-free paper

9 8 7 6 5 4 3 2 1

springer.com

Preface

Data mining and knowledge discovery can today be considered as stable fields with numerous efficient methods and studies that have been proposed to extract knowledge from data. Nevertheless, the famous golden nugget is still challenging. Actually, the context evolved since the first definition of the KDD process and knowledge has now to be extracted from data getting more and more complex. The structure of the data, for instance, doesn't match the attribute-value format when considering the web, texts or videos.

In the framework of Data Mining, many software solutions have been developed for the extraction of knowledge from tabular data (which are typically obtained from relational databases). Methodological extensions have been proposed to deal with data initially obtained from other sources, like in the context of natural language (text mining) and image (image mining). KDD has thus evolved following a unimodal scheme instantiated according to the type of the underlying data (tabular data, text, images, etc), which, at the end, always leads to working on the classical double entry tabular format.

However, in a large number of application domains, this unimodal approach appears to be too restrictive. Consider for instance a corpus of medical files. Each file can contain tabular data such as results of biological analyzes, textual data coming from clinical reports, image data such as radiographies, echograms, or electrocardiograms. In a decision making framework, treating each type of information separately has serious drawbacks. It appears therefore more and more necessary to consider these different data simultaneously, thereby encompassing all their complexity. Many examples of complex data can thus be found in potential knowledge extraction processes. These data can be:

- Semi-structured or unstructured;
- Sensor data such as scientific or medical data;
- Representing the same information at different periods;
- Grouping different kinds of information (images, text, ontologies, etc.).

Hence, a natural question arises: how could one combine information of different nature and associate them with a same semantic unit, which is for instance the patient? On a methodological level, one could also wonder how to compare such complex units via similarity measures. The classical approach consists in aggregating partial dissimilarities computed on components of the same type. However, this approach tends to make superposed layers of information. It considers that the whole entity is the sum of its components. By analogy with the analysis of complex systems, it appears that knowledge discovery in complex data can not simply consist of the concatenation of the partial information obtained from each part of the object. The aim would rather be to discover more « global » knowledge giving a meaning to the components and associating them with the semantic unit. This fundamental information cannot be extracted by the currently considered approaches and the available tools.

The new data mining strategies shall take into account the specificities of complex objects (units with which are associated the complex data). These specificities are summarized hereafter:

- *Different kind*. The data associated with an object are of different types. Besides classical numerical, categorical or symbolic descriptors, text, image or audio/video data are often available.
- *Diversity of the sources*. The data come from different sources. As shown in the context of medical files, the collected data can come from surveys filled in by doctors, textual reports, measures acquired from medical equipment, radiographies, echograms, etc.
- *Evolving and distributed*. It often happens that the same object is described according to the same characteristics at different times or different places. For instance, a patient may often consult several doctors, each one of them producing specific information. These different data are associated with the same subject.
- *Linked to expert knowledge*. Intelligent data mining should also take into account external information, also called expert knowledge, which could be taken into account by means of ontology. In the framework of oncology for instance, the expert knowledge is organized under the form of decision trees and is made available under the form of "best practice guides" called Standard Option Recommendations (SOR).
- *Dimensionality of the data*. The association of different data sources at different moments multiplies the points of view and therefore the number of potential descriptors. The resulting high dimensionality is the cause of both algorithmic and methodological difficulties.

The difficulty of Knowledge Discovery in complex data lies in all these specificities.

The aim of this book is to gather the most recent works that address issues related to the concept of mining complex data. The whole knowledge discovery process being involved, our goal is to provide researchers dealing with each step of this process by key entries. Actually, managing complex data within the KDD process implies to work on every step, starting from the pre-processing (e.g. structuring and organizing) to the visualization and interpretation (e.g. sorting or filtering) of the results, via the data mining methods themselves (e.g. classification, clustering, frequent patterns extraction, etc.). The papers presented here are selected from the workshop papers held yearly since 2006.

The book is composed of four parts and a total of sixteen chapters. Part I gives a general view of complex data mining by illustrating some situations and the related complexity. It contains five chapters. Chapter 1 illustrates the problem of analyzing the scientific literature. The chapter gives some background to the various techniques in this area, explains the necessary pre-processing steps involved, and presents two case studies, one from image mining and one from table identification.

In Chapter 2, the authors use genetic algorithms for imbalanced datasets classification. The chapter presents a new methodology to extract a Fuzzy System by using Genetic Algorithms for the classification of imbalanced datasets when the intelligibility of the Fuzzy Rules is an issue. A method for fuzzy variable construction, based on modifying the set of fuzzy variables obtained by the DDA/RecBF clustering algorithm is proposed. Afterwards, these variables are recombined to obtain Fuzzy Rules by means of a Genetic Algorithm. The method has been developed for the prenatal Down's syndrome detection during the second-trimester of pregnancy.

Chapter 3 deals with another problem which adds a complexity to mining processes, i.e. the noisy data. In this chapter, the authors propose new approach and modifications carried out on AdaBoost to handle the noise in data. They demonstrate that it is possible to improve the performance of the Boosting by exploiting assumptions generated with the former iterations to correct the weights of the examples.

Missing values can be seen as a possible type of noise in the data. In Chapter 4, authors use probabilistic decision trees to deal with that problem. The proposed approach is explained and the comparative tests using several real databases are discussed.

Chapter 5 deals with Kernel-based algorithms on interval data and their visualisation. The chapter aims at extending kernel methods to interval data mining and using graphical methods to explain the obtained results. Interval data type can be an interesting way to aggregate large datasets into smaller ones or to represent data with uncertainty. No algorithmic changes are required from the usual case of continuous data other than the modification of the Radial Basis Kernel Function evaluation. Thus, kernel-based algorithms can deal easily with interval data. Interactive graphical decision tree algorithms and visualization techniques to give an insight into support vector machines results are used.

One of the main goals in Knowledge Discovery is to find interesting associations between values of attributes, those that are meaningful in a domain of interest. The objective of Part II of this book is to illustrate the complexity in terms of data and mining processes related to the general area of rules extraction. Chapter 6 presents evaluation models based on objective indices. To enhance the adaptability of rule evaluation models, authors introduced a constructive meta-learning system for the construction of appropriate learning algorithms.

Chapter 7 is related to the statistical association rules in the specific context of medical images. Authors discuss how to take advantage of association rules mining to promote feature selection from low-level image features. Feature selection can significantly improve the precision of content-based queries in image databases by removing noisy and redundant features. A new algorithm named StARMiner is presented. StARMiner aims at finding association rules relating low-level image features to high-level knowledge about the images. Such rules are employed to select the most relevant features.

Chapter 8 deals with sequence mining. The first part of the chapter proposes a new algorithm for mining frequent sequences. This algorithm processes only one scan of the database because of an indexed structure associated to a bit map representation. Beyond mining plain sequences, taking into account multidimensional information associated with sequential data is of a great interest to many applications. In the second part, author proposes a characterization based multidimensional sequential patterns mining.

The last chapter, Chapter 9, of Part II discusses a specific case of rules, i.e. E-Action rules. E-Action rules can be used not only for automatic analysis of discovered classification rules but also for hints of how to reclassify some objects in a data set from one state into another more desired one.

Part 3 deals with the general problem of graph mining and its applications. This part contains five chapters. Chapter 10 discusses the problem of indexing graph structured data. An own design of an indexing structure for general graph structured data called ρ-index that allows an effective processing of special path queries is presented. These special queries represent for example a search for all paths lying between two arbitrary vertices limited to a certain path length.

Chapter 11 deals with frequent sub-graph mining. Mining graph databases for frequent sub-graphs has recently developed into an area of intensive research. Its main goals are to reduce the execution time of the existing basic algorithms and to enhance their capability to find meaningful graph fragments. Here authors present a method to achieve the former, namely an improvement of what they called "perfect extension pruning" in an earlier work. With this method the number of generated fragments and visited search tree nodes can be reduced, often considerably, thus accelerating the search.

In Chapter 12, a parallel algorithm Peamc (Parallel Enumeration of All Maximal Cliques) is discussed which exploits several new and effective techniques to enumerate all maximal cliques in large-scale complex networks. Experimental results on true-life networks with up to 20 million vertices and 50 million edges show that Peamc can find all the maximal cliques with high efficiency and scalability.

Applications of graph mining on community finding in large networks are discussed in the next two chapters. In Chapter 13, authors introduce the topology of the network to evaluate the feasibility and correctness of a community finding algorithm. A relationship between the rough number of communities and the magnitude of the number of hub nodes in the network is given in detail firstly. Then, an algorithm based on Laplace matrix spectral decomposition is proposed and its key technology, threshold selection of Euclidean distance between nodes, is discussed. Based on the scale-free topology of complex network, the evaluation criterion of community finding algorithm including three conditions is obtained.

Chapter 14 deals with the same problem as the previous chapter but using another approach, i.e. k-dense method. An efficient algorithm for extracting k-dense communities is proposed. The method is applied on the three different types of networks assembled from real data, namely, from blog trackbacks, word associations and Wikipedia references, and demonstrated that the k-dense method could extract communities almost as efficiently as the k-core method, while the qualities of the extracted communities are comparable to those obtained by the k-clique method.

The last part of this book contains two chapters related to the problem of data clustering. Chapter 15 deals with clustering of orders. A new method called k-o'means is developed. This algorithm successfully extracts grouping structures in orders, and was computationally efficient with respect to the number of orders. However, it was not efficient in cases where there are too many possible objects yet. A new method (k-o'means-EBC), grounded on a theory of order statistics is proposed. Furthermore, several techniques to analyze acquired clusters of orders are proposed.

Chapter 16 deals with validity indices and their usage in the context of textual data. Two main contributions are discussed: firstly, since validity indices have been mostly studied in a two or three-dimensional datasets, authors have chosen to evaluate them in a real-world applications, document and word clustering. Secondly, a new context-aware method that aims at enhancing the validity indices usage as stopping criteria in agglomerative algorithms is proposed.

Djamel A. Zighed,
ERIC Laboratory, University of Lyon 2,
France.

Shusaku Tsumoto,
Shimane University, School of Medicine,
Japan.

Zbigniew W. Ras,
Department of Computer Science, University of North Carolina,
Charlotte, USA.

Hakim Hacid,
ERIC Laboratory University of Lyon 2
France.

Contents

Part III: Graph Data Mining

Part IV: Data Clustering

Part I

General Aspects of Complex Data

1

Using Layout Data for the Analysis of Scientific Literature

Brigitte Mathiak, Andreas Kupfer, and Silke Eckstein

Technische Universität Braunschweig, Germany
Institute of Information Systems
mathiak@gmail.com, {kupfer, eckstein}@ifis.cs.tu-bs.de

Summary. It is said that the world knowledge is in the Internet. Scientific knowledge is in the books, journals and conference proceedings. Yet both repositories are too large to skim through manually. We need clever algorithms to cope with the huge amount of information. To filter, sort and ultimately mine the information available it is vital to use every source of information we have. A common technique is to mine the text from the publications, but they are more complex than the text they include. The position of the words gives us clues about their meaning. Additional images either supplement the text or offer proof to a proposition. Tables cannot be understood before deciphering the rows and columns. To deal with the additional information, classic text mining techniques have to be coupled with spatial data and image data. In this chapter, we will give some background to the various techniques, explain the necessary pre-processing steps involved and present two case studies, one from image mining and one from table identification.

1.1 Introduction

A scientific document is more complex than it seems. While readers can easily deduce structure and semantics of the different characters and pictures on a page, most of this structure information is not stored and available when automatically accessing the publication. Most text mining applications from scientific literature are trying to find facts. In biology, these are facts like gene-to-gene relationships [1, 2] or gene expression profiling [3], mostly in abstracts [4], as these are most easily available. It is shown that these techniques can give biologically significant results [5]. Yet, more information can be obtained by searching through full text paper [6]. In [7] it was shown that different kinds of information are stored in different kinds of sections. Although the abstract has the highest information density, other sections contain viable information as well. In [8], it has been observed that analyzing the figure caption is of great value. Classifying the different sections of a paper to analyse them separately like in [9] has been successfully attempted, but curiously, the figure captions and tables have not been examined.

D.A. Zighed et al. (Eds.): Mining Complex Data, SCI 165, pp. 3–22.
springerlink.com   © Springer-Verlag Berlin Heidelberg 2009

In [7], the tested sections did not include tables, figure captions or table captions. Figures have been dealt with, by [10] and our own [11]. However, table contents are a completely new topic and not been reseatched yet, to our knowledge.

1.2 Background

The basic problem of handling PDF documents is that the text information is not freely available. While an HTML file stripped of its tags usually delivers legible text, even the simple task of text extraction from a PDF is rather complicated. Down to the basics, PDF is foremost a visual medium, describing for each glyph (= character or picture) where it should be printed on the page [12]. Most PDF converters simply emulate this glyph-by-glyph positioning in ASCII [13].

Still, since the position of all glyphs is known, the original layout can be deduced and the semantic connection can be restored. For HTML, the layout information has successfully been used to improve the classification of web pages [14]. We extract the same layout information out of PDF documents. This offers us a multitude of possibilities, from the restoration of the original reading order over table recognition up to finding the images in the paper.

Usually, image retrieval in biomedical context is not used for literature retrieval, but for image retrieval on large databases. The general PicHunter approach [15] is an example for such a content-based image retrieval system. With updated Bayesian formulas the framework of PicHunter has been adapted to refine the results of a query by predicting the users action. This approach addresses the image retrieval in general and does not discuss image retrieval within the biological context.

The IRMA-concept (Image Retrieval in Medical Applications) [16, 17] has been developed to handle primitive and semantic queries and to browse medical images with respect to medical applications. The approach is able to support content understanding and highly differentiated queries on an abstract information level. In order to compensate for the different smaller structures of a typical medical image, local representations are used to categorize the entire image. Those local features are then compared with a k-nearest neighbour algorithm.

Tables pose a completely different problem. Green and Krishnamoorthy [18] developed a method that is capable of analyzing model-based tables. To make this approach work, a model or template of the tables in question has to be provided. Zuyev [19] introduced an algorithm for table image segmentation that uses table grids. Tables that use table lines can be identified well, but the approach is not able to find tables without these separators.

Both methods have the problem that they rely much on table separators, like lines or connectors, like dots between the values. Unfortunately, these marks do not come out clearly in vector-based documents. All we have there are the characters and the position of the characters. Also, not all tables use these separators constantly. The T-Recs system [20] follows a bottom-up approach to identify and segment tables in electronic or paper documents. The method is independent from any table separators because only the words are considered. This trait makes it also perfect for adaptation for vector-based documents.

1.3 Overview

In order to extract the layout and being able to process it, several steps are necessary. First, we need the full text literature, which is often a problem due to copyright issues. Since the full text paper is usually available in PDF format, we need to extract the text from there. The real problem in this step is: we need to know where the text is on the page; otherwise we could just use a normal text extractor. Also, the images have to be extracted, again with the positional information intact. This information is handed over from the layout analyzer to the next step (cf. Fig. 1) by storing them in an XML format we call ELL (Enriched Layout Language).

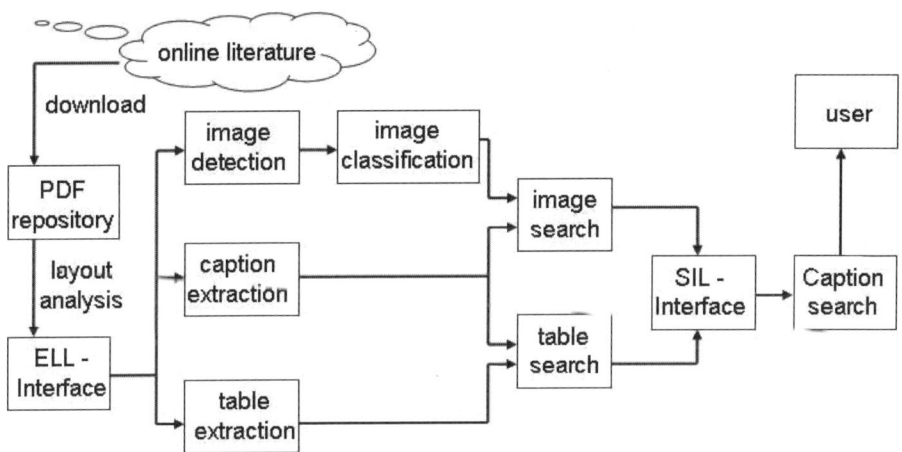

Fig. 1.1. Workflow of the layout analysis

The next steps are modularized. The ELL files are regrouped internally to provide the correct reading order. Also, image or table extraction algorithms are used. To provide a semantic annotation their captions are found by using another algorithm. The results of these algorithms are indexed to be presented on a web-based search engine. In order to have a standardized interface, we created the XML exchange format SIL (Search Interface Language). The SIL files are used by the web platform CaptionSearch to provide a search interface to the user. CaptionSearch allows multiple corpora to be handled separately. Also, different users may be registered to different sets of corpora.

1.3.1   Finding the Literature

The first step in pre-processing is to build a corpus. In order to find a suitable corpus for a specific topic, several open search engines and databases, such as Google, Citeseer or PubMed can be used. Still, a number of papers are not free and can only be downloaded from the publisher website. When downloading from the original publisher website, it is often quite difficult to find the specific paper, as they often have different structures for storing their publications.

The corporae for our case studies were either provided directly by our collaboration partners or had to be acquired using said databases. To build a corpus from scratch automatically, we developed a downloading tool. It only needs a broad search query to describe the topic. That query is fed into suitable open databases and then cross-referenced with the available publisher websites. From there, we should be able to find a link to the full paper. The page is crawled and every link weighed according to whether they contain interesting keywords, like "pdf", "reprint", volume number or issue number. Also, we give penalties for less interesting terms like "abstract", "guide", "faq", and so on. The links are followed in the order of their weights; negative weight links are never followed.

The first link directed to a PDF document is downloaded. In order to prevent the downloading of unwanted documents, we also compiled a blacklist of terms not allowed to be part of a followed link, like "manual", "adobe.com" and so on. When we do not find a suitable link to follow, we assume that we do not have permission to download the paper. The downloader is able to handle multiple proxy configurations, so it is possible to use different licenses simultaneously. So, when we do not find a document, we try to open an alternative connection using another license, or an alternative link given by the database.

The process of downloading is relatively slow, because we want to avoid overloading the source databases. It takes roughly 30 seconds per document, most of that being generous timeouts due to the local rules of the databases involved. The downloader is quite successful. In a test with the query "gene expression microarray rat", there were 1244 full paper links given by PubMed, the database specialising in biomedical literature. By using a regular library license, we achieved to download 598 of them, plus only 34 false positives. Searching for the same documents by just using the free text links provided by PubMed, there were only 485.

The system's precision can be improved through usage, as we can backtrack where the false positives came from and add more terms to the blacklist or the penalty list. An improvement of recall is difficult, as the "missed" paper is rarely noticed.

1.3.2   Extracting Layout Information from the PDF Format

A typical vector-based format like PDF or Postscript does not directly give the position of the text. The description is not pixel-based, but a description of lines and curves that form the text. Like that, the information about the document stays more authentic, because scaling does not matter and a lot of the semantic information prevails, like text. The disadvantage though, is that layout identification becomes more complicated, since more objects are involved.

By convention, the parameters of a command are written before the command and all commands are abbreviated to two letters. All text-related commands are between a BT (line 1) and an ET (line 8) which stands for Begin Text and End Text, respectively. The Tj and TJ commands include the text actually written. The other commands are describing where and how the text is written.

```
1 BT
2 8 0 0 8 52 757.35 Tm
3 /F2 1 Tf
4 0 -1.706 TD
5 (page 354)Tj
6 T*
7 [ (J) -27 (OURN) 27 (AL) -378 (1) ]TJ
8 ET
```

For a correct layout analysis and indexing, it is vital to identify correct word borders; otherwise words may be glued together or torn apart. Also, without spaces the text is undistinguishable from a large table. Unfortunately, the spaces are at times not given directly, but instead the characters are just a little more apart from each other than usual. The problem sharpens as theoretically all characters can be written in any kind of order by jumping around with explicitly set coordinates.

In order to identify the spaces anyway, our first run through the text stream just extracts the characters one by one and calculates their bounding boxes. Then the difference vector $\mathbf{x}_{diff}$ between two adjacent characters is calculated and rotated in writing direction $\mathbf{R}$.

$$rotationmatrix \ \ \mathbf{R} = \begin{pmatrix} x_{old,right} - x_{old,left} & y_{old,right} - y_{old,left} \\ -y_{old,right} + y_{old,left} & x_{old,right} - x_{old,left} \end{pmatrix}$$

$$\mathbf{x}_{diff} = \left(\mathbf{x}_{new,left} - \mathbf{x}_{old,right}\right)\frac{\mathbf{R}}{|\mathbf{R}|}$$

The resulting vector is compared to the current modified font size to determine whether this is a space, no space, carriage return or a new block of text. Next, the blocks are sorted and go through a similar procedure. This way the initial information about the order is conserved best.

The words bounding boxes are extracted, as are all changes in fonts or font size. Additional problems which arise are: text overlaps, when e.g. a special font is used to write the accent over à that overlaps the original "a" and the overall handling of non-identifiable fonts and fonts that give wrong bounding boxes. The results of these calculations are stored together with text in the ELL file format.

1.3.3   Structure in ELL

To get a better overview, we modelled the data structure in UML (cf. Fig. 2). The UML version is then transferred to XML Schema, so all the XML files can

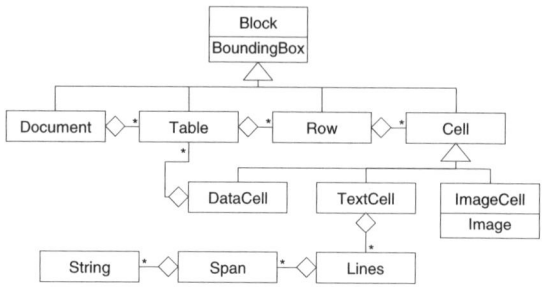

Fig. 1.2. UML model of the ELL format

be checked against it. The various implemented algorithms also use classes in a similar structure.

The model has to fulfil multiple functions at once. First, it represents all the meta-information we gain from the text conversion. Also, there is room for further structuring. Algorithms can nest the text in a recursive data structure and can include additional meta-information on the objects.

The XML document structure is HTML-like so the results can be easily verified with a web browser. The original plain text conversion is preserved, so by simply deleting all tags from the document, the text can be restored at all steps. All metadata is stored in the attributes.

1.3.4   Data Cleansing

The conversion process from PDF to XML documents is not fool-proof. Although the PDF specification is publicly available, many PDFs do not adhere to the recommendations given there, but instead rely on a visual correct appearance in the viewers. To cope with wrong input data, we add a data cleansing step to filter out faulty documents.

The most common errors in PDF conversion are caused by wrong encoding information for a specific font. Usually, each font provides a dictionary to match each character in a font with the Unicode character it represents. Should this be missing, we try to guess by assuming the encoding is similar to ASCII. To catch wrong interpretation though, all of our tests inspect the document text by splitting it into the fonts used. Each text from each font in each document must pass three tests or is removed.

First, it is checked for non-ASCII symbols. These occur with wrong font information and are critical, as they disable the document for XML validation later. The second test removes files which do not contain at least 1000 bytes of document text. This case occurs in documents which are only a series of scanned images without text information or documents which are just an abstract or a form. Most scientific papers contain at least 20 kb of text.

The third test is a cumulative quota of suspicious characters over all characters in the text. If this quota is above 10 percent, the document is rejected. We mark characters as suspicious, if they do not seem to be English text. We started with the language classifier Lingua::Identify by Jose Castro from CPAN. But the results were not very stable, because the library is not designed to identify text not written in a natural language and also scientific papers include a lot of words which do not seem to be English. Therefore, we implemented four simple heuristics to detect conversion flaws. Each one either can either mark a font as suspicious or a continuous piece of text.

The first heuristic makes sure that the text in a font contains at least one whitespace. If there is not a single space in the text written in this font, it is marked as suspicious. As omitted spaces are a common conversion problem, the second heuristic is similar by searching for words with more than 20 letters, with the explicit exception of sequence data. The third marks possible encoding errors, where a character in PDF without a proper encoding reference is converted as letter and number code, like M150. To avoid special names, this pattern must occur in at least 2 consecutive characters. Therefore, single characters, especially mathematical operators, remain undetected. Finally, the fourth heuristic marks all characters which are outside the 7-bit ASCII alphabet.

The data cleansing step is quite fast and can process more than a document per second. In a quite modern test data set 19 out of 605 documents (3.1%) were filtered out. 10 of the documents did not contain enough text, 8 did fail the quota and one included an improper symbol. All of these papers were unintelligible from a text extraction point of view. In older documents this rate can be higher, as older converters tended to emphasis more on shortness than legibility.

1.3.5 CaptionSearch : The Web Application

As a simple demonstration, how the layout information can be used, a web based search engine is used. It is called CaptionSearch, since we are technically not searching for images or tables, but for the caption text beneath them.

After an image or table extraction has been done, the results are written to an XML file in the SIL language. Each publication is represented by one file containing a PDF element (line 2-5), which may contain many image elements (like the one in line 3-5), which have the link to the picture as an attribute (line 3) and the caption as a further element (line 4-5).

```
1 <?xml version="1.0" encoding="iso-8859-2"?>
2 <pdf src="10094677.pdf">
3    <img src="pics/10094677.Im4.jpg">
4       <caption>FIG. 4. DNase I footprint analysis of ...
5 </caption></img></pdf>
```

Alternatively, there may be a text to show, if the image is not available. That way we can also use SIL as a regular search engine. For the indexing itself, we use the Lucene package [21], which offers fast, Java-based indexing, but also some additional functionality, like a built-in query parser and several so-called analyser that allow us to vary how exactly the captions are indexed and what defines a term.

In order to bring the information on the web, we set up a Tomcat web server [22], using Java servlets [23] to produce the website and to present the query results. All queries are executed by a servlet that uses Lucene to fetch the results from the index files and builds a new web page to display the results according to the pre-selected schema.

1.4   Image Extraction

While the location and size of an image in a PDF document are clearly described, there are many possibilities to store the actual image data. The images can be encoded in JPEG, in a BitMap-like format, as a postscript description, even in a fax format and many more. It is very bothersome to implement every single one of those possibilities, especially as third-party products (e.g. www.pdfgrabber.de) are available. Still, those products do not reveal the position of the pictures on the page, so a matching of the resulting pictures to the layout data we already collected is needed. Also, there are graphics not caught by image extractors as they are derived from graphical commands in the PDF script language, which is very similar to Postscript.

1.4.1   Images in PDF

There are two ways to represent images in PDF documents. The textstream can include graphical commands to draw lines or other geometrical objects. Alternatively, they can be stored as external object called XObject. From the text stream an XObject can be called by using the command Do (execute the named XObject).

This example object represents an image that can be called by entering /Im3 Do into the text stream. What happens then is that the object called Im3 is identified and executed. From the object dictionary, we can gain some information like width (line 5) and height (line 6), although this information might not be accurate. The true height and width are calculated and give, together with the current position, the bounding box of the picture. To actually extract the picture, we need the filter (as given in line 10), in this case DCTDecode, which is the PDF name for Jpeg encoding [24].

```
1  22 0 obj
2  << /Type /XObject
3  /Subtype /Image
4  /Name /Im3
5  /Width 580
6  /Height 651
7  /BitsPerComponent 8
8  /ColorSpace /DeviceGray
9  /Length 31853
10 /Filter /DCTDecode>>
11 stream ... endstream endobj
```

Vector based drawings are inserted in the textstream as commands like m for move or l for line. Converting them to pixel pictures is done by using a blank canvas and recreating all commands. To avoid inserting small decoration objects like footnote lines or table separators, we demand a certain density of commands in order to accept it as a full fledged image.

1.4.2   Caption Identification

By convention, images or figures in scientific literature are accompanied by captions to explain their meaning. This is very useful for the mining process, as the figures can be searched for by a simple term search. When identifying the captions, the first step is to look for paragraphs starting with "Fig" in any kind of writing. This feature is very distinctive. So far, we found only one paragraph starting with "Fig" that was not an actual figure caption.

If there is only one caption candidate on the same page, the choice for the picture is clear. Even several pictures for one caption candidate is not a problem, actually this happens quite often, when the figure is composed of more then one picture (e.g. before and after). Two problems may occur though: first, there might be more then one caption candidate (because, for instance, there is more than one figure). Second, there might be no caption candidate, either because the caption does not start with "Fig" or because there is none (it might be a logo or other non-captioned image).

We solve both problems the same way. By using the layout information, gained during the extraction process, we look for likely positions of a caption. The general goal is to pick the candidate closest to the image, preferably below. In order to do that, we give scores for different kinds of proximity.

We considered finding the parameters for this algorithm by machine learning procedures, but it is hard to find a good training set, as we need scientific paper with two or preferably more figures on at least one page, already annotated and not uniformly produced. We found 68 so far, which all performed well with the initial parameters, except for two cases, which were both rather special and hard to interpret even for specialists able to deduce the connection through visually matching caption with image. So, it seems doubtful, if machine learning would not produce an algorithm overfitting the training set.

Candidate figure captions that could not be connected to an image are instead linked to the whole page picture, if existing, or the first picture on the page, if existing, or to a dummy image, so they can be indexed by the search engine and then tracked by looking directly at the PDF. This occurs very rarely, though.

1.4.3   Case Studies for Image Extraction

To evaluate the usefulness of this new search method, we set up a case study with colleagues from biology. There were two things to be observed: how interesting is this method for the scientists and how well does it works in terms of efficiency.

The PRODORIC database [25] contains very special data like DNA binding sites of prokaryotic transcriptional regulators. This data is generated via specific experiments like DNAse I footprints or ElectroMobility gel Shift Assays (EMSA). Unfortunately, the specific experiments conducted are rarely mentioned in the abstract or title of the documents in question. The search for general key words classifying this comprehensive field like "gene regulation", "promoter" or "binding site" results in over 150,000 hits, and even with additional refinement only 10-20% contain appropriate data. Therefore it is necessary to screen all the hits manually to obtain literature references suitable for the database annotation. Of these, those are especially valuable that contain pictures of the DNAse I footprint or EMSA assay, because they represent verified information of high quality. This quality assessment can be important on further exploration of the subject.

In this case study, the corpus included 188 papers that were known to contain information about DNA binding sites (from the PRODORIC database). We extracted 1430 pictures, about one quarter of them pictures of whole pages. In data cleansing, we found that 13% of them were completely unreadable (the oldest ones), due to text conversion errors. The extracted images showed scanned pages of the paper. Another 10% did show fairly good text recognition, but had the pictures not included separately, but as part of a whole page picture. Another 8% showed minor errors, like too short captions, not recognized figure blocks due to text conversion errors, and so on, in some of the captions. All in all, for 80% of the papers the captions could be indexed properly. The rest were set aside for manual inspection.

To find DNAseI footprints the keywords "footprint", "footprinting" and "DNAse" were used to find the appropriate figures in CaptionSearch. Overall, 184 hits were scored of which 163 actually showed experimental data. As a by-product, the thumbnails, presented by the engine, mostly sufficed to make a fast quality assessment. Another positive effect was that the data was much faster available than with the usual method of opening each PDF independently. The search for EMSAs was a bit more difficult, since there is a wide range of naming possibilities. The most significant terms in those names were "shift", "mobility", "EMSA" and "EMS" to catch "EMS assay". We had 91 hits of which 81 were genuine. Recall could not be tested thoroughly, due the sheer numbers of pictures and the limited time of experts, but a random sample did not include interesting pictures that had not also been found by the keywords, which suggests a rather high recall.

The second field study was conducted in collaboration with neurologists. They were interested in finding paper under the topic of "mismatch negativity" and make them searchable through our engine. Technically, the main difference between the two is the variety of age. The binding site corpus is from the years 1995 to 2003, while the mismatch negativity corpus only includes papers from the years 2001 and 2002. And of course the general topic is different, one being from Microbiology and one from Neurology.

In the newer corpus the problems were a little different. It includes 355 papers, all from the years 2001 and 2002, containing 1754 extractable pictures. 2 of the papers had to be omitted in the data cleansing step. We found that 31% of the pictures were logos. Most of these occurred either on the first page, or on the every single page as a copyright reminder. There were much fewer text conversion errors; only 5% of the captions were either too long or too short or otherwise wrongly converted.

While precision in both corpora was reasonably high, often the wrong pictures were quite obvious wrong, from a visual point of view. While so far, only the textual description of the images is considered, the next step is to look at the pictures themselves.

1.5 Image Classification

In the biological data set a significant number (27%) of the images were showing the whole page instead of just the figure. What we are looking for, though, are pictures containing the gel pictures made for the experiments. As gel pictures are visually quite different from other common pictures, a classic image classification seems possible. In the neurological data set the problems are different. Here the logos and copyright disclaimer are more prominent then the full page images. Also, we would like to differentiate between CT-photos and normal graphs and models.

The pictures from both data sets were classified by an expert biologist (see figures 1.3 and 1.4). We estimate the overall error rate in the data set at 4%. This estimation is based on a manual random check of misclassified images and their assigned classification.

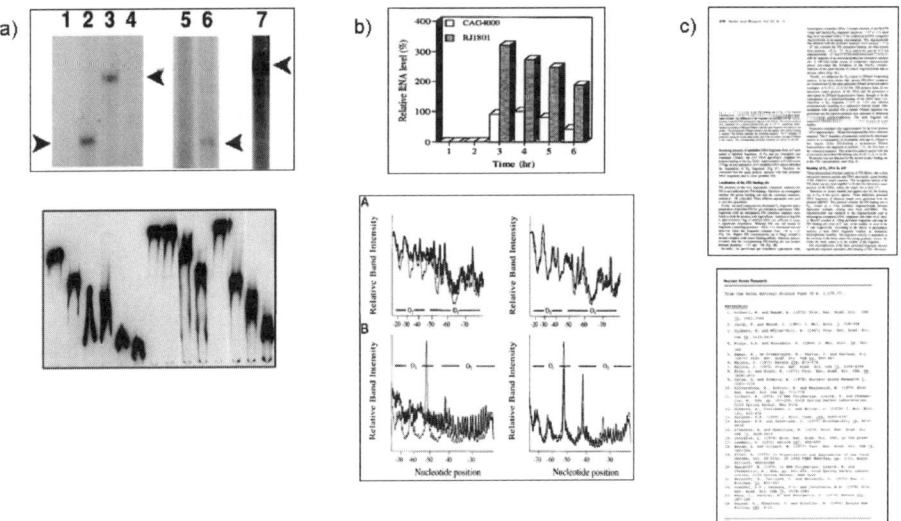

Fig. 1.3. These are sample pictures from the biological data set. Under a) are two pictures from the category raw data that show various gel pictures. b) are from the category model. c) shows full text pages.

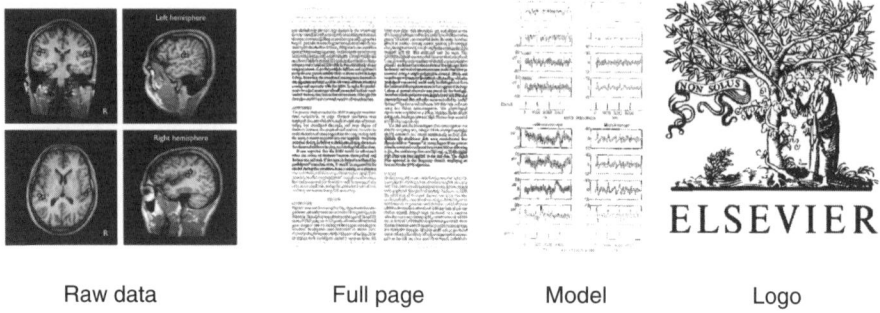

Raw data   Full page   Model   Logo

Fig. 1.4. Sample pictures from the neurological data set. The category is written below the image.

In a neurological test set, we used four categories: full pages, raw data, model and logo. In the biological test set there were no logo, hence there were only 3 categories. The distribution was quite different, instead of a large number of full pages, we had many logos. We estimate the error rate of this set in fact higher than on the biological data set, because of the increased number of categories. Unlike in the biological data set the distribution is much more biased with only 13 instances of full pages and over 1000 models.

1.5.1   Method for Classification

For image classification, a feature-based approach seems best, because we do not classify based on the object seen in the image, but on the representation of that object, e.g. gel blots vs. graph points. Other algorithms, like the random window approach, tend to repress those representation details. We base our method on [26], a method originally used to distinguish between computer-made images and real life photos, since that is a closely related problem.

In order to classify the pictures, we calculate 6 metrics or features based on the picture. The calculations for the metrics are all linear, so the calculation takes less than a second for an average picture. The small number of attributes allows fast learning and classification. An information gain estimate is given in table 1.1. The features are explained below. Also included is an interpretation of how useful these features were to our task.

1. **Number of Colours**: counts the number of occurring colours in the picture. We assume that many colours indicate slow colour changes typical for photos of experimental results, while graphs are usually black and white.

2. **Contour Sharpness**: measures the occurrence of hard changes in the colour values. First, it compares each pixel with its neighbouring pixels, to find the biggest colour difference between them. Then, all pixels with a maximum difference bigger than 0 are counted as S and those bigger then a threshold t

Table 1.1. Information gain of the metrics

biological data set	neurological data set	Metric
0.909	0.382	Number of Colors
0.854	0.473	Contour Sharpness
0.829	0.496	Smallest Dimension
0.68	0.663	Prevalent Color
0.663	0.595	Dimension Ratio
0	0.133	Saturation

are counted as C. The sharpness value is then C/S. The threshold is set halfway between black and white. While photos of experimental results tend to have fuzzy borders with slow colour changes, graphs and text have a lot of black and white changes.

3. **Smallest Dimension**: represents the actual size of the picture or in fact the minimum of width and height in centimetres. This is used to sort out very big pictures, which might indicate a full text page or very small pictures that might indicate a logo.

4. **Prevalent Colour**: is the percentage of the prevalent, usually the background, colour in the picture. The idea is that strong backgrounds indicate graphics. Unfortunately, pure gel pictures are mostly gray and mixed pictures are mostly white, so the information gain for the biological data set is limited. Still it is the most important metric for the neurological data set.

5. **Dimension Ratio**: calculates the ratio between height and width of the picture. In our test sets, it was mostly between 1 and 2, due to the standard dimension of figures. Outliers weakly indicate gel pictures, while whole pages and groups of logos have a characteristical dimension ratio.

6. **Saturation**: measures the percentage of gray-scale pixels, compared to coloured ones. While this is no help in the biological data set at all, due to complete lack of coloured pictures, it is of some help in the more modern neurological data set.

Table 1.1 also shows the decline of importance of the first three metrics over time. In fact, in the newer data set the model category are more similar to the raw data pictures. We assume that this is a result of the newer technologies, making it possible to produce models that more detailed and colourful.

1.5.2 Classification Results

We tested the method on both a neurological and a biological data set. The results (cf. table 1.2) are about the same as for the biological data set (up to 94.5%). This score can be considered almost perfect, since the estimated error rate in the data set is in the same region. Compared to the other data set, the classification task was de facto harder, as there was an additional category.

Table 1.2. Results from the classification

Data set	BayesNet	DecisionTrees	JRip	SVM
biological	92.9	93.0	92.8	89.0
neurological	91.4	93.6	94.5	84.9

Also, the neurological data set is much more biased concerning the number of instances in each category.

During the experiment we used a standard accuracy as ratio of correctly classified entities to overall entities under 10-fold cross-validation. Precision and recall were evenly distributed. The final confusion matrix in the biological data set showed some mix up between raw data and models, while the precision and recall for full pages were much higher. For the neurological data set, this trend intensified. Logos could be identified almost perfectly, while the raw data had much less accuracy (F-Score for raw data was only 0.73). Full pages was also problematic (F-Score 0.7), but that seems natural considering that there were only 13 instances to learn from.

Unlike the method proposed by [26] we decided to test several machine learning algorithms for comparison. We used Bayes net, decision trees, JRip and support vector machines with standard parameters from the Weka Tool [27]. While the rule learner JRip and decision trees generally performed similarly, Bayes Net worked significantly better on the biological data set than on the neurological data set. SVMs performance was far below average. As SVM naturally does not support multiple categories, we had to use binary classifiers between all possible combinations of categories. This lead to overlapping classifications and the complete exclusion of sparsely set categories.

1.6   Table Detection

Besides finding images and their captions, a second use for using layout information springs to mind: the analysis of tables. In order to identify the tables in a paper, we decided to take an algorithm from OCR, as already discussed in the Background section. The T-Recs algorithm was designed to find tables in scanned pages, but can be adjusted to work in a vector-based environment as well.

1.6.1   The T-Recs Algorithm

The method presented by Kieninger [28] can be split into three steps. First, possible table relationships are identified by searching for regular structures in the layout of the text. Next, some error-correcting methods are employed and finally, the actual table structure is identified and table content is separated from non-table content.

In the first step, text units are identified, by melting together words that overlap horizontally. An overlap is defined as:

$$ovl(w_1, w_2) := (w_1.x_1 < w_2.x_2) \wedge (w_1.x_2 > w_2.x_1)$$
$$\wedge (w_1.y_1 + w_1.height \geq w_2.y_2) \wedge$$
$$(w_1.y_2 - w_1.height \leq w_2.y_1))$$

Two words w_1 and w_2 overlap, if left side of w_1 is left of the right side of w_2 and vice versa. Also, they have to be in vertical proximity to each other. x_1, x_2, y_1 and y_2 represent the bounding box of the word in two points: (x_1, y_1) is the lower left corner and (x_2, y_2) the upper right corner. The zero-point is in the upper left corner of the page.

As most of the words are overlapping with each other, common text should be recognised as one unit. Table columns, in contrast, do not overlap and should thus be recognisable (also cf. Fig. 1.5). A simple distinction based on the typical number of neighbours in a unit allows a broad classification into text and table units.

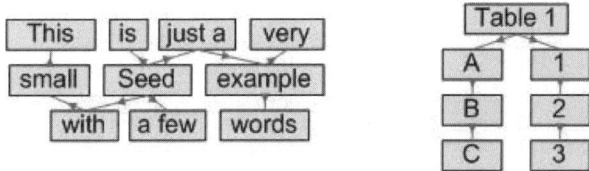

Fig. 1.5. Left: Result of the overlapping algorithm on a text unit; Right: Result of the overlapping algorithm on a table unit

The starting point for identifying the structure of the tables is the columns. These have to be aligned both vertically and horizontally. But, as you can see in Fig. 1.6 the words do not fit precisely. Instead, a margin point is rather a margin area with two border points. Since tables may come with a variety of special cases, for example, melted columns or two rows of text in a cell, the matching does not have to be precise. When a table is established, units in the proximity are tested, if they fit the pattern. That way, solitary words are reintroduced, when they fit.

We pre-evaluated our method on 86 scientific documents that included 92 tables. For the purpose of annotation it is most important to reach a high recall, as missed hits are much harder to find than sorting out wrong hits. Adjusting the parameters to that goal, we were able to reach a very high recall: 91%, although many of the tables were only found partially. Unfortunately, it lowered the precision to as much as 44%. A lot of those wrong tables were formulae with matrices or multiple lines. Another source of wrong tables was that tables were split in two, either because they were stretched out over two pages or they had a vertical gap inside, when the headline of the column was very large, while the values were quite short. We counted these as both a partial find and a wrong table. Also see table 1.3 for a subsumption of the results.

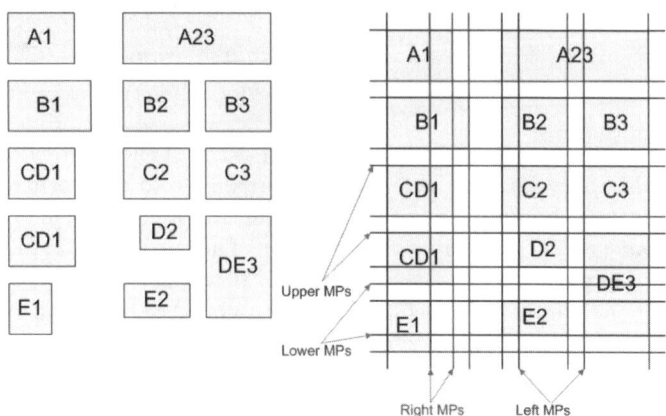

Fig. 1.6. Margin points, applied on the units 1-3

Table 1.3. Results of the table algorithm evaluation

number of documents	86
number of tables	92
number of correctly found tables	51
number of partially found tables	34
number of not found tables	8
number of found tables	236
Recall	91,73 %
Precision	43,74 %

To compare the results and further increase recall, it was also decided to store all terms in a separate index, showing them additionally to the results from the table search, together with their nearest table caption.

1.6.2 Case Study for Table Extraction

With the dawn of high-throughput analysis, biological papers have started to contain a lot more information than the actual focus of the paper would suggest. A simple experiment on diabetic rat, for example, will concentrate on a very few proteins and genes. Still, there are dozens of other proteins and genes involved in the process, but this information usually just ends up in a table in the appendix. A scientist concentrating on one of these proteins and genes may never know about the effects on diabetic rats, as this information never shows up in the abstract or via standard search methods. With our approach, we can find the tables the proteins or genes are mentioned in and show the user the table caption to have them easily assess the context.

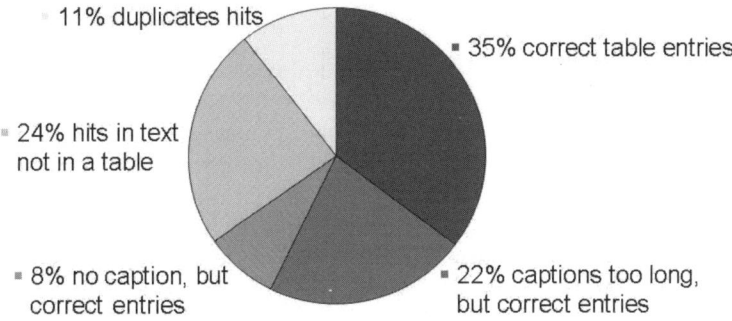

Fig. 1.7. Evaluation results

The accuracy of the workflow was tested by using randomly selected gene or protein names from tables of the PDF dataset that was downloaded by using the terms "gene expression microarray rat" on the PubMed database. The queries produced 598 hits. A sample of 16 gene or protein names was tested. The queries produced from 1 to more then 800 hits. Due to this magnitude we only tested the first 40 hits for each of the 16 queries, so that a total of 506 hits were evaluated. The hits were analysed in several aspects. The first one was to see whether it was a table entry or a text passage and if some of the analysed hits were duplications. Errors in this section usually derive from the table identification algorithm. Duplication occurred in approximately 11% of the cases. 24% of the hits were text entries (also cf. Fig. 1.7).

The next step was to see if the caption of the table entry is correct. We found different problems; the caption can be too short (this only occurred two times) or too long (about 22%). In the case of too long, usually an additional word sneaked in. This class of errors derives from problems in the PDF-text-conversion. Another problem was the lack of a genuine caption or the wrong caption for the table (the latter occurred in 8% of the cases). Both problems most likely come from the caption matching algorithm. All in all, we found that in 65% of the hits, the gene name is in a table just as intended. We believe that this is an acceptable precision, given that text entries also tend to give additional information about the gene.

All of the queries identified the table their terms were originally taken from. That places the recall at 100%. Given that the small sample size might distort that number, we tried a number of other terms, not being able to produce a failure. So, we suspect the recall to be very high, although that is hard to verify, due to the frequent high number of hits.

The participating biologist was amazed and very interested to see the diversity of contexts *her* protein was mentioned in. Most of those she would have never discovered via conventional literature review, looking only for papers already related to the problem she is working on or with the protein name in the abstract of the paper.

1.7   Conclusion

Mining the complex structure of scientific publications is a daunting task, yet achievable. Not only can the layout information be used to improve on classical tasks [9], we have also shown that it opens up new possibilities of retrieving information from the publications. Being able to specifically find images and table data, not only helps in the biological scenario of our case studies. It is also applicable to other areas, such as chemistry to find structure formulae, in engineering to find results of standardized tests in tables or in pharmacology to find which drugs did not work on a given problem.

1.8   Future Trends

The next topic of interest would be to show, whether the mistakes made by the automatic layout detection outweigh the gains when applying it to classical tasks like classification and clustering. In table recognition, it would be interesting to see, if the data from the tables can automatically extracted into databases for easier access. And these are just a few of the many possibilities that the layout analysis of scientific documents opens up.

Layout information can also be used to improve the analysis of web sites or any other textual and graphical medium, such as newspapers or magazines. As the amount of data presented in that way rises very fast and will probably continue to do so in the future, it becomes more and more important to find effective ways to handle this data and get the most out it.

References

1. Liu, Y., Navathe, S.B., Civera, J., Dasigi, V., Ram, A., Ciliax, B.J., Dingledine, R.: Text mining biomedical literature for discovering gene-to-gene relationships: a comparative study of algorithms. IEEE/ACM Trans. Comput. Biol. Bioinform. 2(1), 62–76 (2005)
2. Hu, X., Wu, D.D.: Data mining and predictive modeling of biomolecular network from biomedical literature databases. IEEE/ACM Trans. Comput. Biol. Bioinform. 4(2), 251–263 (2007)
3. Tanabe, L., Scherf, U., Smith, L.H., Lee, J.K., Hunter, L., Weinstein, J.N.: MedMiner: an Internet text-mining tool for biomedical information, with application to gene expression profiling. Biotechniques 27(6), 1210–4, 1216–7 (1999)
4. Chaussabel, D.: Biomedical literature mining: challenges and solutions in the 'omics' era. Am J. Pharmacogenomics 4(6), 383–393 (2004)
5. Natarajan, J., Berrar, D., Dubitzky, W., Hack, C., Zhang, Y., DeSesa, C., Van Brocklyn, J.R., Bremer, E.G.: Text mining of full-text journal articles combined with gene expression analysis reveals a relationship between sphingosine-1-phosphate and invasiveness of a glioblastoma cell line. BMC Bioinformatics 7, 373 (2006)
6. Faulstich, L.C., Stadler, P.F., Thurner, C., Witwer, C.: litsift: Automated text categorization in bibliographic search. In: Data Mining and Text Mining for Bioinformatics, Workshop at the ECML / PKDD 2003 (2003)

7. Schuemie, M.J., Weeber, M., Schijvenaars, B.J.A., van Mulligen, E.M., der van Eijk, C.C., Jelier, R., Mons, B., Kors, J.A.: Distribution of information in biomedical abstracts and full-text publications. Bioinformatics 20(16), 2597–2604 (2004)
8. Yeh, A., Hirschman, L., Morgan, A.: Evaluation of text data mining for database curation: lessons learned from the kdd challenge cup. Bioinformatics 19(1) (2003)
9. Shah, P.K., Perez–Iratxeta, C., Bork, P., Andrade, M.A.: Information extraction from full text scientific articles: Where are the keywords?. BMC Bioinformatics 4(20) (2003)
10. Liu, F., Jenssen, T.-K., Nygaard, V., Sack, J., Hovig, E.: FigSearch: a figure legend indexing and classification system - Textmining. Bioinformatics 20(16), 2880–2882 (2004)
11. Mathiak, B., Kupfer, A., Münch, R., Täubner, C., Eckstein, S.: Improving literature preselection by searching for images. In: Bremer, E.G., Hakenberg, J., Han, E.-H(S.), Berrar, D., Dubitzky, W. (eds.) KDLL 2006. LNCS (LNBI), vol. 3886, pp. 18–28. Springer, Heidelberg (2006)
12. Adobe Network Solutions: PDF Reference Fourth Edition(2004), http://partners.adobe.com/asn/acrobat/sdk/publicdocs/PDFReference15v6.pdf
13. GNU Ghostscript: ps2ascii (2003), http://www.cs.wisc.edu/ghost/
14. Kovacevic, M., Diligenti, M., Gori, M., Milutinovic, V.: Visual Adjacency Multigraphs - a Novel Approach to Web Page Classification. In: Proceedings of SAWM 2004 workshop, ECML 2004 (2004)
15. Cox, I., Miller, M., Minka, T., Papathornas, T., Yianilos, P.: The Bayesian Image Retrieval System, PicHunter: Theory, Implementation, and Psychophysical Experiments. In: IEEE Tran. On Image Processing, vol. 9, pp. 20–37 (2000)
16. Lehmann, T., Wein, B., Keysers, D., Bredno, J., Guld, M., Schubert, H., Kohnen, M.: Image Retrieval in Medical Applications: the IRMA-Approach. In: VISIM Workshop: Information Retrieval and Exploration in Large Collections of Medical Images, Utrecht, Netherland, p. 4 (2001)
17. Paredes, R., Keysers, D., Lehmann, T.M., Wein, B.B., Ney, H., Vidal, E.: Classification of medical images using local representations. In: Bildverarbeitung für die Medizin, pp. 171–174 (2002)
18. Green, E., Krishnamoorthy, M.: Table Image Understanding. In: Analysis, I.S., Synthesis (eds.) Proc. of the Int. Conf. on Information Systems, Analysis and Synthesis (1996)
19. Zuyev, K.: Table image segmentation. In: Proceedings of the 4th International Conference on Document Analysis and Recognition, Ulm, Germany, vol. 2, pp. 705–708. IEEE Computer Society Press, Los Alamitos (1997)
20. Kieninger, T.: Table structure recognition based on robust block segmentation. In: SPIE Proc. Document Recognition, San Jose, CA, USA, vol. 3305, pp. 22–32 (1998)
21. Hatcher, E., Gospodnetic, O.: Lucene in Action. Manning Publications (2004)
22. Apache (2005), http://jakarta.apache.org/tomcat
23. Coward, D., Yutaka: Java Servlet Specification (2003), http://jcp.org/aboutJava/communityprocess/final/jsr154/index.html
24. International Organization for Standardization: ISO/IEC 10918-1:1994: Information technology — Digital compression and coding of continuous-tone still images: Requirements and guidelines. International Organization for Standardization, Geneva, Switzerland (1994)
25. Münch, R., Hiller, K., Barg, H., Heldt, H., Linz, S., Wingender, E., Jahn, D.: Prodoric: prokaryotic database of gene regulation. Nucleic Acids Research 31(1), 266–269 (2003)

26. Athitsos, V., Swain, M., Franke, C.: Distinguishing photographs and graphics on the World Wide Web. Technical report (1997)
27. Witten, I., Frank, E., Trigg, L., Hall, M., Holmes, G., Cunningham, S.: Weka: Practical machine learning tools and techniques with java implementations (1999)
28. Kieninger, T., Dengel, A.: The T-Recs Table Recognition and Analysis System. In: Lee, S.-W., Nakano, Y. (eds.) DAS 1998. LNCS, vol. 1655, pp. 255–269. Springer, Heidelberg (1999)

2

Extracting a Fuzzy System by Using Genetic Algorithms for Imbalanced Datasets Classification: Application on Down's Syndrome Detection

Vicenç Soler and Marta Prim

Dept. Microelectrònica i Sistemes Electrònics,
Universitat Autònoma de Barcelona,
Edifici Q, Campus UAB,08193 Bellaterra, Spain
Tel.: +34 93 581 3555
{vicenc.soler, marta.prim}@uab.es

Abstract. This chapter presents a new methodology to extract a Fuzzy System by using Genetic Algorithms for the classification of imbalanced datasets when the intelligibility of the Fuzzy Rules is an issue. We propose a method for fuzzy variable construction, based on modifying the set of fuzzy variables obtained by the DDA/RecBF clustering algorithm. Afterwards, these variables are recombined to obtain Fuzzy Rules by means of a Genetic Algorithm. The method has been developed for the prenatal Down's syndrome detection during the second-trimester of pregnancy. We present empirical results showing its accuracy for this task. Furthermore, we provide more generic experimental results over UCI datasets proving that the method can have a wider applicability on imbalanced datasets.

Keywords: Imbalanced Datasets, Fuzzy Logic, RecBF, Down Syndrome, Rule Extraction.

2.1 Introduction

In this chapter, we present our research aiming at improving the nowadays existing classification of the prenatal Down's syndrome detection, during the second-trimester of pregnancy, with non invasive techniques.

Nowadays, experts (obstetricians and gynaecologists) recommend fetal chromosomal screening to all pregnant women, regardless of age, because of improvements in low-risk, non invasive screening methods. The screening methods 19 consist in evaluating data in order to achieve the maximum quantity of true positive values with the minimum loss of false positives, since a perfect evaluation cannot be obtained from the data. In particular, in Down's syndrome detection problem, a positive case (it has Down's syndrome) which is correctly classified by the screening method is a true positive, and a negative case (it has not Down's syndrome) which is incorrectly classified (the system detects that it has Down's syndrome and it is not true) is a false positive. These screening methods use statistical functions to obtain the Down's syndrome risk, from data obtained during pregnancy from the mother, like gestational

D.A. Zighed et al. (Eds.): Mining Complex Data, SCI 165, pp. 23–39.
springerlink.com   © Springer-Verlag Berlin Heidelberg 2009

age, mother age, the measure of some hormones, etc. The risk is usually indicated in odds as an expression of its certainty probability.

There are several screening methods which are applied to achieve this goal, taking into account that those methods are different depending on whether we are evaluating the first or the second-trimester of pregnancy. Nowadays, the screening methods for the first-trimester of pregnancy have an accuracy of the 80-90% and for the second-trimester just the 60-70%, with a 5% of false positives for the latter, although usually they are around and 8% 18.

The data obtained in prenatal Down's syndrome detection problem is a two-class imbalanced dataset. An imbalanced dataset is characterized by the existence of a high difference in the number of cases in one class with respect to the rest of the classes. In this case, as being a two-class dataset (it has or has not Down's syndrome), the fetus with Down's syndrome (positive class) has a much lesser number of cases than the healthy one (negative class). Regarding to the number of cases, from now on we will refer to the negative class as the *major-class* and the positive class as the *minor-class*.

The focus of our study is to improve the above results obtained by the screening methods in the second-trimester of pregnancy and, as far as possible, extract an understandable set of rules. This goal will be achieved by a new Soft Computing method based on Fuzzy Logic designed to work with imbalanced datasets.

The Soft Computing method is called *FLAGID* (Fuzzy Logic And Genetic algorithms for Imbalanced Datasets). The FLAGID method consists in using a clustering algorithm called DDA/RecBF to obtain a first set of trapezoidal Membership Functions from the dataset, recombine those functions to obtain new ones, and finally, with the recombined set of membership functions and the dataset, obtain a set of fuzzy rules by means of a Genetic Algorithm. The result is expressed as a Fuzzy System.

This chapter is structured in 6 sections. The first section is this introduction, which describes the topic which is dealt with in this chapter. In the second section we review some related work aiming at solving the imbalanced datasets problem. The third section details the development and the characteristics of our new method (FLAGID). In the following section, we present the experimental results obtained by applying the FLAGID method to the dataset corresponding to the Down's syndrome, and we also compare FLAGID accuracy with other imbalanced methods. Finally, future applications of the method and its conclusions are presented.

2.2 Related Work

The imbalanced datasets classification problem has recently received considerable attention from the machine learning community 12. Several studies have been published dealing with this classification problem, which can be divided into two important directions. The first one corresponds to the use of the traditional learning methods with some changes, in the dataset and/or the algorithm. The changes applied to the dataset, commonly deal with resampling the quantity of data of every class in order to equilibrate their number of cases. Those of the algorithm try to avoid the undesirable effects produced by an imbalanced dataset, by introducing some precise changes to the algorithm, which usually are the application of different costs (depending on the class) to the misclassified cases.

The second direction is centred on developing new learning methods that solve the imbalanced problem, which is impossible to manipulate with classic algorithms.

The resampling strategy has two approaches: oversampling and undersampling. Oversampling increases the quantity of patterns included in the minor-class (e.g. 3), and undersampling decreases the number of examples of the major-class 4. These two techniques are the most commonly used, because they allow to leverage existing classification methods that work with balanced sets of data, such as SVM 523 , neural networks 5 , and others 4. Two examples of SVM based methods are SDC and KBA. SDC 6 (SMOTE with Different Costs) combines SVM and SMOTE to solve the problem of a boundary too near to the positive instances by using different learning costs. KBA 7 (Kernel Boundary Alignment) is an algorithm based on SVM that modifies the boundary as well.

Another strategy is to add costs to misclassified patterns, depending on their class, to improve the classification accuracy. There are some reports like Domingos 8, Zadrozny and Elkan 9, Meler et al. 10, etc. which present this technique as a generic algorithm modification.

The above solutions, sometimes, do not give the wished results; it depends on the imbalanced problem. Then, there are many attempts to solve this problem using new methods developed in order to work with imbalanced datasets 11. For example, Visa and Ralescu 24 proposed a fuzzy classifier for imbalanced datasets and overlapping between classes, or Zhang et al. 12 presented a very simple and effective method, called RLSD (Rule for Learning Skewed Data) to generate rules from highly imbalanced datasets.

Regarding to the previous work with Down's syndrome detection using soft computing methods, we refer to the work done by M. Sordo 21, who used RBF networks. The obtained results were 84% of true positives and 35% of false positives, worse than our results as will be seen in the Experimental Results section.

However, there is neither method nor algorithm that has been used especifically to the Down's syndrome problem, because medical experts have openly expressed their disbelieve in such approaches, like resampling methods, which are based on the fact that they would either create new patterns which do not exist in real life or erase patterns with small details that could be relevant for the final solution.

2.3 Proposed Method

The method proposed in this chapter has the goal of achieving an accurate *Fuzzy Classification System* from an *imbalanced dataset*. The method consists of several steps, which are shown in Fig. 2.1. By using the DDA/RecBF clustering algorithm, a first set of fuzzy membership functions from the dataset can be obtained. Then, those functions are recombined by using a special method called ReRecBF (Recombined RecBF) which will be also presented in this chapter; and finally, with the recombined set of membership functions and the dataset, a set of fuzzy rules will be obtained by means of a Genetic Algorithm.

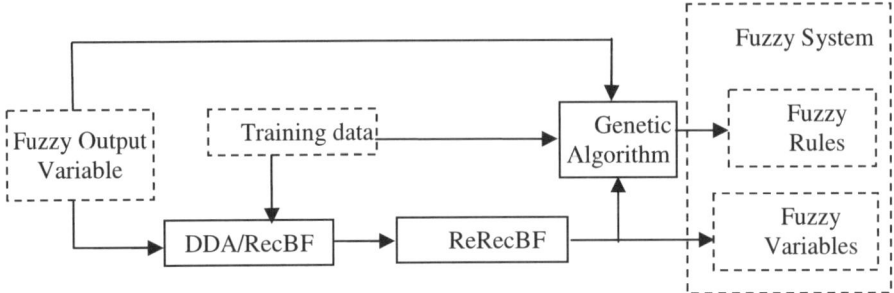

Fig. 2.1. Our method's schema. Solid line rectangles represent algorithms (DDA/RecBF, ReRecBF and Genetic Algorithm) and dashed ones represent sets (dataset, fuzzy rules set and fuzzy variables set).

The *DDA/RecBF* (Dynamic Decay Adjustment for Rectangular Basis Functions) algorithm 13 allows to obtain a RecBF Network from a dataset and a definition of the clusters (classes) of the output variable in a set of fuzzy membership functions. RecBF Networks 13 are a variation of RBF networks, in which the representation is a set of hyper-rectangles belonging to the different classes of the system. Every dimension of the hyper-rectangles represents a membership function. Finally, a network is built, representing, on every neuron, the membership function found. The algorithm which produces the RecBFN from the dataset and from the definition of the classes of the system is called DDA/RecBF. The set of membership functions defined by this method are usually well suited for classification.

These membership functions will be used to obtain a set of rules by means of a Genetic Algorithm. Although first, a previous step to transform the set of membership functions (output of the DDA/RecBF) has to be done since they need to be adapted for working with imbalanced data. This task is done through the *ReRecBF* algorithm by recombining the set of membership functions.

Genetic Algorithms 14 are used to derive fuzzy rules from the set of recombined membership functions. Genetic algorithms are methods based on principles of natural selection and evolution for global searches. Given a problem, a genetic algorithm runs repeatedly by using the three fundamental operators: reproduction, crossover and mutation. These operators, combined randomly, are based on a fitness function evolution to find a better solution in the searching space. Chromosomes represent the individuals of the genetic algorithms and a chromosome is composed of several genes. Genetic algorithms are used to find solutions to problems with a large set of possible solutions and they have the advantage of only requiring information concerning the quality of the solution. This fact makes genetic algorithms a very good method to solve complex problems.

Empirical results show that for the Down's syndrome problem the accuracy is improved with respect to the current method used. Furthermore, this method has been compared with other methods for imbalanced problems, using some UCI datasets. The results show that our method improves the accuracy in most of cases.

In summary, the FLAGID method pursues the following steps:

1. It uses the DDA/RecBF algorithm to get a first set of membership function from the dataset.
2. It recombines the membership functions found by DDA/RecBF and obtains a new set of membership functions.
3. Finally, it applies a genetic algorithm to find a small set of fuzzy rules that uses the membership functions obtained in step 2 and provide high classification accuracy. These steps will be explained deeply in the following three subsections.

2.3.1 Constructing a First Set of Membership Functions by Applying RecBFs

The *DDA/RecBF clustering algortihm*13 constructs a set of membership functions from a dataset that provides a set of linguistic labels that can be later used to concisely express relationships and dependencies in the dataset. The algorithm creates hyper-rectangles (called *Rectangular Basis Functions* or *RecBFs*) belonging to every class, from the input dataset and a defined fuzzy output variable. From these RecBFs, a set of membership functions will be extracted for every variable.

Each RecBF is defined by a *support-region* and a *core-region*. In terms of membership function, a trapezoidal membership function is composed of four points (a,b,c,d): the interval [a,d] defines the support-region and the [b,c] one, the core-region. An example is shown in Fig. 2.2.

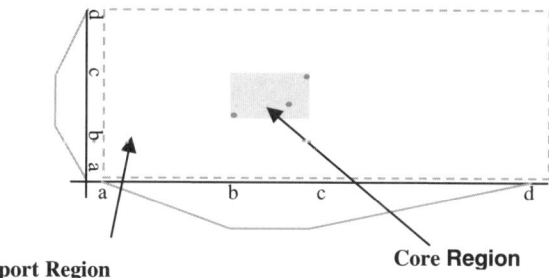

Support Region   **Core Region**

Fig. 2.2. Example of a 2-dimensional Fuzzy Point and its membership functions defined by its two regions. The a,b,c & d points are different for each dimension.

DDA/RecBF algorithm is based on three procedures, which are executed for every training pattern: *covered* is executed when the support-region of a RecBF covers a pattern, *commit* creates a new pattern if the previous condition is false and, finally, the procedure *shrink* solves possible conflicts of the pattern with respect to RecBFs of a different class. Fig. 2.3 shows the behaviour of the DDA/RecBF algorithm for one epoch.

Fig. 2.4 shows an example of the execution of the DDA/RecBF algorithm. It is shown how the patterns are passed to the DDA/RecBF algorithm and how the different RecBFs are created.

$\forall_{R^k_{i}}$ {
  $W^k_i = 0$
  $(a,b,c,d)^k_i = (a,-,-,d)^k_i$
}
$\forall_{pattern\ (x,\mu)}$ {
  $k = argmax_{1,k,c}\{\mu^k(x)\}$
  $if\ \exists_{R^k_i} : x \in [a^k_i, d^k_i]\ then$
    $W^k_i = W^k_i + 1$
    covered()
  $else$
    $m_k = m_k + 1;\ a^k_{mk} = 1$
    $(a,b,c,d)^k_{mk} = (-,x,x,-)$
  $end\ if$
  $\forall_{R^i_j}\ with\ \bullet_i(x) = 0$ {
    $if\ x \in [a^i_i, d^i_i]\ then$
      shrink()
    $end\ if$
  }
}

Reset R^k_i. R^k_i represents the ith RecBF of class k, and W^k_i represents its weight.

covered() stretch the core-region of R^k_i selected to (x,μ).

commit: a new RecBF is created, having its core-region=pattern.

If a pattern is incorrectly covered by a RecBF of another class, its support-region will be reduced until the conflict will be solved. This action is done in *shrink()*.

Fig. 2.3. One epoch of the DDA/RecBF algorithm. The algorithm iterates until stability of the RecBFs is reached.

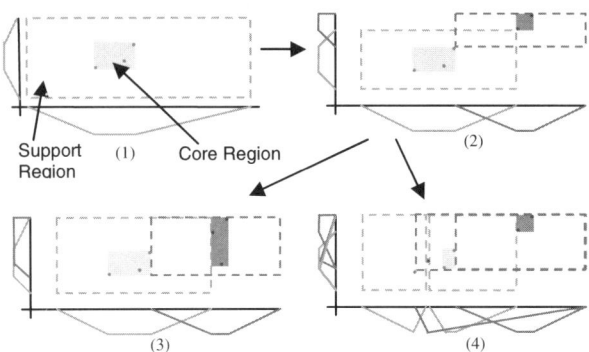

Fig. 2.4. An example of the execution of the DDA/RecBF algorithm for a 2-dimensional system. (1) shows 3 patterns from one class determining a RecBF, (2) shows 2 patterns from another class and how they cause the creation of a new RecBF and shrink the existing one, (3) and (4) show the different RecBFs created when the inclusion of new pattern is done, just varying the x coordinate: outside and inside the core-region of the other class. The x and y axis show the different membership functions created.

However, in our case we work with imbalanced or highly imbalanced datasets, and to avoid granulation of the membership functions of the minor-class, it is absolutely necessary to generalize this class, because the main problem is when the method has to classify/test patterns belonging to this class, not shown during the training process.

Therefore, the following aspects will be taken into account to train the DDA/RecBF algorithm, argued in 17:

1. The shrinking operation can be either done in all dimensions or only in one (which looses less hyper-area in its shrinking). In our case, the latter is used, as proposed by the authors of the DDA/RecBF algorithm in 13, because it causes less granulation on RecBFs.
2. The algorithm will train the dataset sorted by classes, first major-class and then minor-class. This fact will reduce the quantity of RecBFs (membership functions) of the minor-class, due to the DDA organization and RecBFs creation. So, the minor-class will have the minimum quantity of membership functions, letting the major-class organize itself in the quantity of membership functions needed.

The following subsection explains how to obtain new RecBFs by recombining the existing ones, in order to adapt them to work with imbalanced datasets.

2.3.2 ReRecBF Algorithm: Recombining Rectangular Basis Functions (RecBFs)

Since the shrinking method in RecBF algorithm is performed only in one dimension, superposed membership functions are given as result of that algorithm (Fig. 2.4(4)), when working with datasets with overlapped classes (boundary not well defined). These membership functions are not adequate to obtain an accurate set of fuzzy rules from the imbalanced dataset. The tests on different datasets demonstrated that the membership functions obtained from the DDA/RecBF algorithm were not discriminant enough and some transformations of these membership functions were needed. On these grounds we propose to (argued in 17):

1. Take only the intervals obtained by the core-regions.
2. Transform to triangles the trapezoids belonging to the minor-class.
3. If it is possible, discard the less representative RecBFs. That is, RecBFs whose core-region includes less than the 10% of the patterns of its class.

The core-regions delimit the areas where the training patterns are, and the support-regions (without taking into account the core ones) are the undefined areas between the core ones. In these undefined regions the algorithm does not know how to classify the possible patterns included there. Then, we can affirm that the only place where we are sure that a pattern belongs to a class is into the core-regions of that class.

Out of them, the RecBF represents a grading of the level in which the instance belongs to the class. In this way, the nearer the value is to the core area, the higher the possibilities to belong to the class. So, in this case, the set of overlapped membership functions belonging to a variable and a class will be split into new ones, selecting only the areas defined by the core-regions. Thanks to this operation, we will have the membership function areas divided into sectors, and this fact will improve the quantity of patterns matched by the rules found.

Finally, the recombination procedure shown in Fig. 2.4 consists in creating new membership function by splitting the existing ones by its core-region (points b and c) and eliminating the old ones. For every new membership function created, the support-area will be defined from the minimum to the maximum values of that variable. The membership functions created from examples belonging to the minor-class are transformed into triangles (having its maximum in the average point of its original trapezoid) and the major-class ones as trapezoids.

We can obtain the following conclusions about the proposed transformation:

1. More uncertainty is introduced in the definition of the membership function be-cause we do not discard that a membership function can be evaluated in the whole variable space.
2. Every value for a membership function is always greater than 0, so always the alpha-cut will be greater than 0.
3. As the value of a rule will be determined by the minimum of the alpha-cuts, and as the importance of a class is determined by the maximum of the alpha-cuts of all its rules, we can affirm that the final importance of a class in a pattern can be deter-mined by only one rule, and that rule has to have all membership function as near as possible the values of the example treated.
4. The minor-class membership functions introduce more uncertainty because they are treated as triangles. If we do not have much information about how that class is (the case for highly imbalanced datasets), we can consider an area centred in the average of the membership function's core-region and its belonging value decreas-ing as far away it is.

Thanks to these characteristics, the membership functions are not so rigid and they have many more possibilities to participate in every rule. Thus, we give to the system more feasible tools to adapt itself to the given dataset.

The following subsection explains how the rule set will be extracted by a genetic algorithm, once the membership functions have been calculated by the previously specified procedure.

2.3.3 Obtaining the Rule Set: The Genetic Algorithm

The codification of one chromosome of our genetic algorithm is expressed in the following line: $(x_{1,1},...,x_{1,n},x_{2,1},....,x_{2,n},x_{m,1},...,x_{m,n})$ where n is the number of variables (input variables plus output variables) and m is the number of rules. $x_{i,j}$ is the value that a gene can take, which is an integer value compressed in the interval $[0,n\_fuzzysets_j]$ and $n\_fuzzysets_j$ is the number of membership functions of the jth variable. If a $x_{i,j}$ has value 0 it expresses that this variable is not present in the rule. If the 0 value is assigned to the output fuzzy set, the rule is not taken into account to evaluate the rules. So, the system is able to find a set of rules less than m, just putting 0 in the output fuzzy set of the rule. Every $x_{i,1},..., x_{i,n}$ corresponds to a rule of the system.

The initial population is either taken randomly or by an initial set of rules. Every gene of a chromosome is generated randomly in the interval $[0,n\_fuzzysets_j]$, but some rules can be fixed for the entire simulation or just given as an initial set of rules, if needed. If a fuzzy set is 0, it means that the variable is not taken into account in the rule.

In the case of this chapter, the initial rules have been generated randomly and no restriction has applied on them.

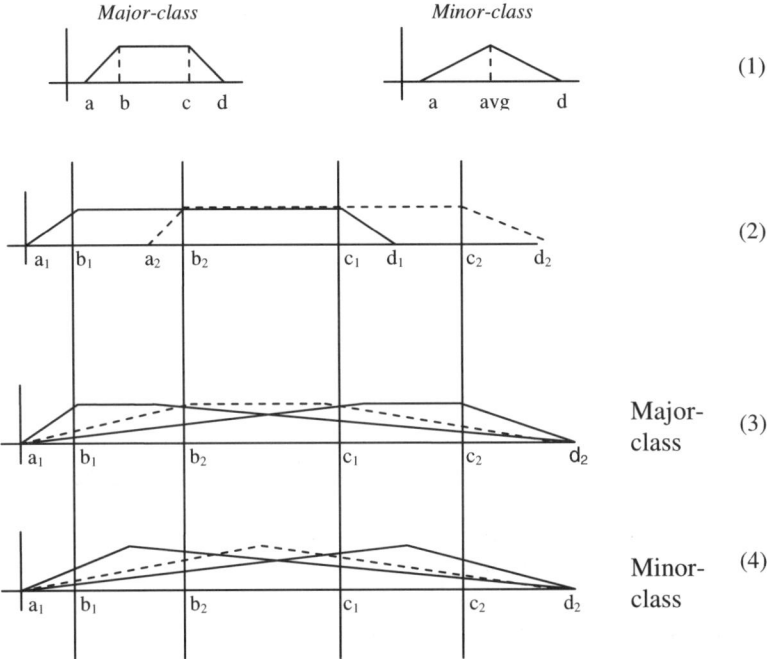

Fig. 2.5. (1) Trapezoids for major-class membership functions and triangles for minor-class membership functions. (2) Example of 2 overlapped trapezoids. (3) Three trapezoids will be obtained from above if the membership functions would belong to the major-class and (4) Three triangles if would belong to the minor-class. Both (3) and (4) have membership functions with the minimum and the maximum in the limits of the variable.

The genetic algorithm uses the g-means metric (1), suggested by Kubat et al. 11, as fitness value for imbalanced datasets.

$$g = \sqrt{acc^+ \cdot acc^-} \qquad (1)$$

The g-means is the most common measure used to evaluate results in imbalanced datasets, where acc^+ is the accuracy classification on the positive instances, and acc^- the accuracy on the negative ones.

The classification of the data, in a fuzzy logic system depends on the shape of the membership functions of the input and output variables and the rules. The membership functions of the input variables are obtained by the ReRecBF method, the rules by means of this genetic algorithm and the output variable has to be defined from the beginning. So, another important fact in the matching of the examples by the rule set is the shape of the output trapezoids. We can choose between symmetric (they have the same area and shape) or non-symmetric (they have different shapes and areas) (Fig. 2.5), that is, both classes will have either the same importance or not.

In this chapter, 4 different types of asymmetry were tested. They have been chosen in order to represent different degrees of asymmetry to provide one degree more of freedom in the learning process. Fig. 2.6 (1) represents a high degree of asymmetry, and Fig. 2.6 (2) is symmetric.

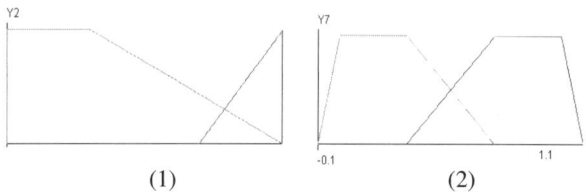

(1)   (2)

Fig. 2.6. Example of (1) non-symmetric and (2) symmetric 2-class output

The symmetric one is much better, because the fuzzy rules will be much easier intelligible.

Regarding the stopping criteria, it is difficult to apply an early-stopping method based on the test of a small quantity of data, since usually the minor-class is provided with a reduce number of patterns. In this case, the genetic algorithm will stop when a determinate threshold value in its fitness function is reached.

2.4 Experimental Results

The goal of the tests made, mainly, is to find a Fuzzy System that tries to improve the results of the method used nowadays for the detection of the Down syndrome for the second-trimester of pregnancy (called age/LR 18). In this section, we will apply the FLAGID method to two types of problems. In the first subsection, the results of the FLAGID method on the Down's syndrome dataset will be compared with those obtained with the current method (age/LR). In the second subsection, the FLAGID method will be compared with two of the best methods that work with imbalanced datasets, SDC and KBA, in order to demonstrate that the new proposed method works well with imbalanced datasets.

2.4.1 Experimental Results for Down's Syndrome Detection

The dataset has been obtained from the University Hospital Dr. Josep Trueta of Girona (Spain). The data contain many variables, and only a few are used: the race of the mother, the number of fetuses, the age of the mother, its weight, the gestational age of the fetus, the existence of diabetes, the consumption degree of tobacco and alcohol, and the measure of the AFP and hCG hormones. Current methods also use a reduction of variables used in medicine called MoM (Multiple of Median) 18. This reduction is commonly used in this problem and consists in a regression based on the median of the 2 hormones. The median values are previously modified by other variables, which produces 2 new variables that include the value of the rest of variables except the age of the mother. This variable is not included because it will be an error to reduce the quantity of information of its contribution to solve the problem. Thus, in case of MoM just 3 variables will be taken into account. With respect to the output variable, it only indicates if the fetus has or has not the Down's syndrome. From now on, the non-MoM variables will be called physical variables.

From these variables, the medical team made us to consider take into account: white race women[1] and with a single fetus, in order to simplify the problem, since the changes in race or number of fetuses produces significant changes in the value of the hormonal markers.

Therefore, the input variables are the age of the mother, its weight, the gestational age of the fetus, the existence of diabetes, the consumption degree of tobacco and alcohol, and the hormonal markers AFP and hCG, and their respective MoMs. As the MoMs include almost all the input variables, the variables have been divided into two groups:

1. One formed by the age of the mother and both MoM (MoM-AFP and MoM-hCG).
2. Other formed by the 8 variables that are not expressed in MoMs: the age of the mother, its weight, the gestational age of the fetus, the existence of diabetes, the hormonal consumption degree of tobacco and alcohol, and AFP and hCG markers.

The data is divided into two groups: one with 3109 cases (3096 negatives and 13 positives) and another with 4995 (4980 negatives and 15 positives), ordered chronologically, so the cases of the second group are later than the cases of the first group. The data is numerical, has 2 output classes (has or has not Down's syndrome) and its imbalance ratio is approximately 1:300, i.e., highly imbalanced dataset. In Table 2.1 it is shown the characteristics of the two datasets and the total. The last set, the Down_total, is the sum of the two previous ones.

Table 2.1. Characteristics of the 2 datasets and the total

Name	#patterns	#neg	#pos	%neg	%pos
Down_3109	3109	3096	13	99.60%	0.40%
Down_4995	4995	4980	15	99.70%	0.30%
Down_total	8922	8892	30	99.66%	0.34%

In the current method age/LR, the hormones markers and its MoM are truncated by their upper and lower limits, due to an adjustment of the function to achieve a Gaussian shape, since it is fulfilled in the central part but not in the ends. In order to fulfill the Gaussian shape, the approximate limits are settled down in 3 times the variance. In our case, we considered that just the upper limits have to be truncated, and the truncation is 5 times the MoM (e.g. in case of MoM the upper limit is 5), due to the upper values are really spread, producing a big scope of the variables which have the information concentrated in a smaller area (the 95% of the cases concentrated in the 10% of its space). This fact may cause problems in the extraction of a learning model and for that reason the data is truncated. Table 2.2 shows the new upper limits.

---

[1] In Spain, the white race is majority.

Table 2.2. Upper limits of both hormonal markers

	MoM	No MoM
AFP	5	115
hCG	5	85

Both datasets Down_3109 and Down_4995 have differences. We pointed out above that they are chronologically different, appearing a degradation in the acquisition of the data. Down_4995 has %TP=40% and %FP=7.71%, whilst Down_3109 has normal rates, specified in the Introduction section.

The learning/test process has been done by selecting the stratified (the same proportion of data of both classes) 50% of the Down_3109 dataset for learning and the other 50% for testing. With the best solutions, a second test with the Down_4195 has been done. Several generations of Fuzzy Systems have been extracted by selecting different variables. In case of the physical variables the best solutions have been obtained from using a set of 5 variables (physical: age of the mother, weight of the mother, gestational age of the fetus, and the measures of two hormones), and in case of MoM just have been considered 3 variables (age of the mother and the two MoM hormones). In this case, the fuzzy system has been obtained from a dataset that matches the results of the age/LR method and that it is also tested with a noisy dataset, preventing this situation in the future. Thus, we can affirm that the fuzzy system extracted will be robust under noisy conditions.

The variables of tobacco and alcohol consumption did not produce good results, as well as the diabetes one. The first two variables may produce noisy due to their value depends on the truth of the women telling whether the consumption of alcohol is null, 1-5 times a week or 1-5 times a day; and whether they smoke 5 cigarettes or more a day or they do not smoke. With respect to the diabetes, we have rejected its use due to noisy problems as well.

Several classification methods have been tested with this dataset in order to discard the possibility of finding a good solution with them: Neural Networks (BayesNN, Backpropagation, etc.), classical methods of Fuzzy Rule Extraction and other methods, like decision trees or SVM 9,10,15,16. The results always were negative, because of the treatment of the minor-class patterns: either because they did specialize neurons/rules/etc. in the cases belonging to the minor-class or because the minor-class patterns were ignored (they tried always to match the major-class patterns without taking into account the minor-class patterns).

In Table 2.3, the best results for the Down's syndrome problem using the FLAGID method and ordered by %TP (True Positives) are shown. The first two columns refer to a test done with the Down_4815 dataset not included in the Down_3071 used for learning. For every % of correct positives found, it is shown the % of false positives, the type of set with better results (5 or 3 variables), if discarded membership functions needed, the number of rules found and the %FP (False Positives) and %TP from the first dataset. The last two columns show the accuracy for the Down_3071 dataset after testing for whole training and test patterns.

Table 2.3 shows that the results are very similar to those obtained by the age/LR method (60%-70% TP and 10% FP). Screening methods, as commented at the

Table 2.3. Results for the Down's syndrome problem using the FLAGID method. The first 2 columns express the accuracy of the test of 4815 patterns not included in the set of 3071 training patterns. The type of the set refers to the type of dataset: using MoM or physical. The type of output can be symmetric or non-symmetric. Discarded RecBFs indicates if the solution was found discarding the less representative RecBFs. The last 2 columns refer to the accuracy of the training/test dataset, training always with the stratified half of patterns.

%TP (4815)	%FP (4815)	type of dataset	type of output	discarded RecBFs	#rules	%TP (3071)	%FP (3071)
60%	9.69%	physical	Symmetric	0%	4	81.82%	8.39%
66.66%	10.21%	MoM	Symmetric	0%	6	81.82%	7.25%
73.33%	13.56%	MoM	Symmetric	0%	6	90.91%	10.49%
80%	14.48%	physical	Symmetric	0%	4	100%	12.87%

beginning of this chapter, want to find a good solution dealing with the %TP and the %TN. In case of Down's syndrome problem, rather than %TN, the %FP will be taking into account. Finding the best solution, a threshold in one of both indexes has to be placed.

The different rows in Table 2.3 show the best %FP for different thresholds of %TP. The best results are in the first three rows, which minimizes the %FP. A FP, in Down's syndrome problem, is the case that the method classifies a fetus as positive but in reality is negative, and in this case the mother would try to do an invasive test, which has 1% of probability of loosing the child, to be 100% sure of the results.

In all cases shown in Table 2.3, no RecBF obtained was discarded and the output variable has a symmetric distribution of its membership function. However, the results which improve the current methods are focused in the very small quantity of rules found: between 4 and 6. This fact makes the system very understandable and hence very adequate for the task of extracting intelligible fuzzy rules.

2.4.2 Comparison with Other Methods

In order to know if the FLAGID method can be applied to the classification of any imbalanced dataset, it is needed a comparison with other methods specialized in dealing with imbalanced datasets.

Table 2.4 shows this comparison with other two methods for imbalanced datasets: KBA and SDC. These two methods are two of the best methods for imbalanced datasets, with very good results with datasets of the UCI repository 25. These datasets will be used to do this comparison.

The **SDC** method (Smote with Different Costs) 6 combines SVM and SMOTE to solve the problem that appears in SVM when the dataset is imbalanced: the border is located always too near to the minor-class. This algorithm applies the modified SVM function proposed by Veropoulos, Campbell and Cristianini 26, shown in Equation (2). This SVM function uses different costs to the errors in the positive class and in the negative class. The SDC method uses this function in combination with an oversampling method called SMOTE 3.

$$L_p(w,b,\alpha) = \frac{|w|^2}{2} + C^+ \sum_{i/y_{ii}=+1}^{n^+} \xi_i + C^- \sum_{j/y_j=-1}^{n^-} \xi_j - \sum_{i=1}^{n} \alpha_i [y_i(w \cdot x_i + b) - 1 + \xi_i] - \sum_{i=1}^{n} \beta_i \xi_i \qquad (2)$$

The authors of the *KBA* (Kernel Boundary Alignment) aim at the potential problems in SVM like the border is slanted towards the minor-class, due to its lack of representation in the dataset. In such case, they propose to modify the function of nucleus adapting it to the distribution of classes and thus reshaping the boundary.

For KBA, the authors divided the dataset into 7 parts: 6 for training and 1 for testing, and 7:3 in SDC case. The columns with the SVM and SMOTE methods are included because they can be compared with the rest of the results (both authors included them in their results). Both KBA and SDC are kernel methods, based on modifications of a SVM.

The UCI datasets selected to do this comparison are: Segmentation, Glass, Car, Yeast, Abalone and Sick. All of the selected UCI datasets do not have binary output, and they were transformed into binary datasets by choosing one class as the negative one and the rest of classes were grouped all as one. We chose the same class than the authors of both methods just to compare the results.

In Table 4 the number that appears with the name of the dataset corresponds to the chosen class. The first four columns of the table explains the type of the dataset, the following three columns are the g-means results from the KBA and the next 3 columns the same results from the SDC. The last two columns reflect the g-means results from the method proposed in this paper (FLAGID) and the different number of rules of the systems found.

These results show that for the Glass, Segmentation and Yeast datasets our method (FLAGID) is better, in terms of the g-means metric whilst for the Car and Abalone it is worst. The results are equal for the Sick dataset. In case of Abalone FLAGID improves KBA but it is worst than SDC. Always the system found has a small number of rules, i.e., so it has a high probability of being understandable by a human expert.

To prove that the method scales good when the imbalancing level increases, we imbalanced even more some datasets. For instance, we imbalanced much more the

Table 2.4. Comparison of our method (FLAGID) with the SVM, SMOTE, KBA & SDC methods for some UCI datasets, by means of the g-means measurement. In first column, the number of the class made as minor-class is included with the name of the dataset. The columns 2, 3 & 4 express the characteristics of every dataset. The last column indicates the number of rules of the fuzzy system found, that shows that it is reduced.

Dataset	#attrib	#pos	#neg	SVM (6:1)	SMOTE (6:1)	KBA (6:1)	SVM (7:3)	SMOTE (7:3)	SDC (7:3)	FLAGID	#rules
Segmentation(1)	19	30	180	0.98	0.98	0.98	0.99	0.99	0.97	1	3,4,5,9
Glass(7)	10	29	185	0.89	0.91	0.93	0.86	0.87	0.94	0.97	3,6
Car(3)	6	69	1659	0.99	0.99	0.99	0	0.98	0.984	0.94	3,4
Yeast(5)	8	51	1433	0.59	0.69	0.82				0.83	3,6
Abalone(19)	8	32	4145	0.0	0.0	0.57	0	0	0.74	0.72	11,13
Sick(2)	27	231	3541				0	0.40	0.86	0.86	3,5

Car dataset, letting just take from the minor-class the half of the patterns taken in the method expressed in Table 2.4. If the ratio for training/test was 7:3, the positive patterns chosen for training were 24, rather than 48 (it is the 3/10 of 69). The results after applying this reduction in the number of patterns belonging to the minor-class during the training phase, were practically the same than for 48 (g=0.93). Similar results were found for the Hepatitis dataset (taking only 30 and 5 patterns of both classes).

2.5 Future Trends

From the point of view of the accomplishment of this work, some research lines have been opened. The first one is to try to solve the overfitting problem that appears when training, by evaluating the existing overlapping in the classes. To do it, it is considered the use of the measurement proposed by S. Visa and A. Ralescu in 24, which measures the degree of overlapping between datasets. Batista et al. 22 concluded in their paper that the main problem in learning from imbalanced datasets is their degree of overlapping between classes (the case that a boundary cannot be well defined). Therefore, on the basis of this measurement a research line can be opened, in order to anticipate whether there will or will not exist possibilities of finding overfitting, and raising solutions for it. Within the same researching line, a new measurement also could be proposed.

Another opened line, that has relation with the previous one, is to determine the ideal number of rules (approximated) for a dataset. Now it is determined being based on the number of membership function generated. In addition to this number of rules, also this research line would be focused on determining the best values of the parameters that are used in the FLAGID method: order of the patterns, reshrink operation, discarding Fuzzy Points, etc. Now this process is made by means of carried out tests with the training set, but it could be productive that some of these parameters were calculated automatically from the dataset.

One of the problems in SVM with respect to its variation applied to the imbalanced datasets, is the one of knowing which is the ideal C^-/C^+ ratio. All the publications pointed out that choosing the ratio between the number of patterns of each class as the C^-/C^+ ratio, already gave good results to them. In this case, a researching line is opened in this field, since the Down's syndrome dataset does not fulfil this empirical rule.

A proposal to try to improve the probability of success in the evaluation of knowing if a fetus is affected of Down's syndrome would be to try to combine both methods, the method age/LR and the result of the FLAGID method.

Another line of research lies in trying to modify and to improve the FLAGID method to work better for imbalanced datasets, improving its results. A way to apply an improvement to it could be the combination of the results of different solutions using bagging or the combination of this method with others by using boosting.

2.6 Conclusions

We have presented a method to work with imbalanced or highly imbalanced datasets, called FLAGID. The method has been shown to give the same results than the used methods on the Down's syndrome detection problem. Furthermore, it has been proven to be competitive with the state of the art of classifiers for imbalanced datasets, on UCI repository.

This method extracts information in the dataset, and expresses it in a fuzzy system. This information is expressed in a small number of rules, as can be seen in tables 2.3 and 2.4. This fact means that the rules are not specialized in cases of the minor-class, but they are distributed among both classes. Normally, we can find more or many more rules belonging to the major-class than to the minor-class. This was one of the goals to be achieved by the method.

Finally, we can conclude that the new method presented in this chapter and called FLAGID (Fuzzy Logic And Genetic algorithms for Imbalanced Datasets) is a very good method to deal with imbalanced datasets.

References

1. Japkowicz, N., Stephen, S.: The Class Imbalance Problem: A Systematic Study. Intelligent Data Analysis 6(5), 429–450 (2002)
2. Chawla, N., Japkowicz, N., Kolcz, A. (eds.): Learning from Imbalanced Data Sets, ACM SIGKDD Explorations 6(1) (June 2004) (special issue)
3. Chawla, N., Bowyer, K., Hall, L., Kegelmeyer, W.: SMOTE: Synthetic Minority Over-sampling Technique. Journal of Artificial Intelligence Research 16, 321–357 (2002)
4. Japkowicz, N.: The Class Imbalance Problem: Significance and Strategies. In: Proceedings of the 2000 International Conference on Artificial Intelligence: Special Track on Inductive Learning, Las Vegas, Nevada (2000)
5. Kecman, V.: Learning & Soft Computing, Support Vector Machines, Neural Networks and Fuzzy Logic Systems. MIT Press, Cambridge (2001)
6. Akbani, R., Kwek, S., Japkowicz, N.: Applying Support Vector Machines to Imbalanced Datasets. In: Boulicaut, J.-F., Esposito, F., Giannotti, F., Pedreschi, D. (eds.) ECML 2004. LNCS (LNAI), vol. 3201. Springer, Heidelberg (2004)
7. Wu, G., Chang, E.Y.: KBA: Kernel Boundary Alignment Considering Imbalanced Data Distribution. IEEE Transactions on knowledge and data engineering (2005)
8. Domingos, P.: MetaCost: a general method for making classifiers cost-sensitive. In: Vth ACM SIGKDD International Conference on Knowledge Discovery and Data Mining (KDD 1999), San Diego, USA, pp. 155–164 (1999)
9. Zadrozny, B., Elkan, C.: Learning and making decisions when costs and probabilities are both unknown. In: VII ACM SIGKDD International Conference on Knowledge Discovery and Data Mining (KDD 2001), San Francisco, USA, pp. 204–213 (2001)
10. Merler, S., Furlanello, C., Larcher, B., Sboner, A.: Automatic model selection in costsensitive boosting. Information Fusion 4(1), 3–10 (2003)
11. Kubat, M., Holte, R., Matwin, S.: Learning when negative examples abound. In: van Someren, M., Widmer, G. (eds.) ECML 1997. LNCS, vol. 1224, pp. 146–153. Springer, Heidelberg (1997)
12. Zhang, J., Bloedorn, E., Rosen, L., Venese, D.: Learning rules from highly unbalanced data sets. In: IVth IEEE International Conference on Data Mining (ICDM 2004), Brighton, UK, pp. 571–574 (2004)
13. Berthold, M., Huber, K.P.: Constructing Fuzzy Graphs from Examples. Intelligent Data Analysis 3, 37–53 (1999)

14. Goldberg, D.E.: Genetic Algorithms in Search, Optimization, and Machine Learning. Addison-Wesley, Reading (1989)
15. Wu, T.P., Chen, S.M.: A New Method for Constructing membership functions and Fuzzy Rules from Training Examples. IEEE Transactions on Systems, Man and Cybernetics-Part B: Cybernetics 29(1), 25–40 (1999)
16. Wang, L.X., Mendel, J.M.: Generating Fuzzy Rules by Learning from Examples. IEEE Transactions on Systems, Man and Cybernetics 22(6), 1414–1427 (1992)
17. Soler, V., Roig, J., Prim, M.: Adapting Fuzzy Points for Very-Imbalanced Datasets. In: NAFIPS 2006 Conference (2006)
18. Sabrià, J.: Screening bioquímico del segundo trimestre. Nuestra experiencia, Progresos en Diagnóstico Prenatal 10(4), 147–153 (1998)
19. Norgaard-Pedersen, L., et al.: Maternal serum markers in screening for Down syndrome (1990)
20. Soler, V., Roig, J., Prim, M.: Finding Exceptions to Rules in Fuzzy Rule Extraction. In: KES 2002, Knowledge-based Intelligent Information Engineering Systems, Part 2, pp. 1115–1119 (2002)
21. Sordo, M.: Neural Nets for Detection of Down's Syndrome. MSc Thesis, Department of Artificial Intelligence, University of Edinburgh, UK (1995)
22. Batista, G.E.A.P.A., Prati, R.C., Monard, M.C.: A study of the behavior of several methods for balancing machine learning training data. SIGKDD Explorations 6(1), 20–29 (2004)
23. Vapnik, V.: Statistical learning theory. Wiley, New York (1998)
24. Visa, S., Ralescu, A.: Learning imbalanced and overlapping classes using fuzzy sets. In: Workshop on Learning from Imbalanced Datasets (ICML 2003), Washington, DC, USA (2003)
25. Blake, C.L., Merz, C.J.: UCI repository of machine learning databases, University of California Irvine, USA (1998),
    http://www.ics.uci.edu/~mlearn/MLRepository.html
26. Veropoulos, K., Campbell, C., Cristianini, N.: Controlling the sensitivity of support vector machines. In: Proceedings of the International Joint Conference on AI, pp. 55–60 (1999)

A Hybrid Approach of Boosting Against Noisy Data

Emna Bahri, Stephane Lallich, Nicolas Nicoloyannis, and Maddouri Mondher

University of Lyon 2
ERIC Laboratory- 5, avenue Pierre Mendès-France
69676 Bron cedex - France
INSAT
zone urbaine la charguia II Tunis
1002 Tunisie
{ebahri, slallich}@eric.univ-lyon2.fr,
Mondher.Maddouri@fst.rnu.tn

Abstract. To reduce error in generalization, a great number of work is carried out on the classifiers aggregation methods in order to improve generally, by voting techniques, the performance of a single classifier. Among these methods of aggregation, we find the Boosting which is most practical thanks to the adaptive update of the distribution of the examples aiming at increasing in an exponential way the weight of the badly classified examples. However, this method is blamed because of overfitting, and the convergence speed especially with noise. In this study, we propose a new approach and modifications carried out on the algorithm of AdaBoost. We will demonstrate that it is possible to improve the performance of the Boosting, by exploiting assumptions generated with the former iterations to correct the weights of the examples. An experimental study shows the interest of this new approach, called hybrid approach.

3.1 Introduction

The great emergence of the modern databases and their evolution in an exponential way as well as the evolution of transmission systems result in a huge mass of data which exceeds the human processing and understanding capabilities. Certainly, these data are sources of relevant information and require means of synthesis and interpretation. As a result, researches were based on powerful systems of artificial intelligence allowing the extraction of useful information helping us in decisions making. Responding to this need, data mining was born. It drew its tools from the statistics and databases. The methodology of data mining gives the possibility to build a model of prediction. This model is a phenomenon starting from other phenomena more easily accessible, based on the process of the knowledge discovery from data which is a process of intelligent data classification. However, the built model can sometimes generate errors of classification that even a random classification does not make. To reduce these errors, a great amount of research in data mining and specifically in machine learning has been carried out on classifiers aggregation methods having as goal

D.A. Zighed et al. (Eds.): Mining Complex Data, SCI 165, pp. 41–54.
springerlink.com   © Springer-Verlag Berlin Heidelberg 2009

to improve by voting techniques the performance of a single classifier. These aggregation methods are good for compromised Skew-variance, thanks to the three fundamental reasons explained in [6]. These methods of aggregation are divided into two categories. The first category refers to those which merge preset classifiers, such as simple voting [2], the weighted voting [2], and the weighted majority voting [12]. The second category consists of those which merge classifiers according to data during the training, such as adaptive strategies (Boosting) and the basic algorithm AdaBoost [21] or random strategies (Bagging) [3].

We are interested in the method of Boosting, because of the comparative study [7] that shows, in little noise, AdaBoost is seemed to be working against the overfitting. In fact, AdaBoost tries to optimize directly the weighted votes. This observation has been proved not only by the fact that the empirical error on the training set decreases exponentially with iterations, but also by the fact that the error in generalization also decreases, even when the empirical error reached its minimum. However, this method is blamed because of overfitting, and the speed of convergence especially with noise. In the last decade, many studies focused on the weaknesses of AdaBoost and proposed its improvement. The important improvements were carried on the modification of the weight of examples [19], [18], [1], [20], [14], [8], the modification of the margin [9], [20], [17], the modification of the classifiers' weight [15], the choice of weak learning [5], [24] and the speed of convergence [22], [13], [18]. In this paper, we propose a new improvement to the basic Boosting algorithm AdaBoost. This approach aims exploiting assumptions generated with the former iterations of AdaBoost to act both on the modification of the weight of examples and the modification of the classifiers' weight. By exploiting these former assumptions, we think that we will avoid the re-generation of a same classifier within different iterations of AdaBoost. Thus, consequently, we expect a positive effect on the improvement of the speed of convergence. The paper is organized in three sections. In the following section, we describe the studies whose purpose is to improve the Boosting against its weaknesses. In the third section, we describe our improvement of boosting by exploiting former assumptions. In the fourth section, we present an experimental study of the proposed improvement by comparing its error in generalization, its recall and its speed of convergence with AdaBoost, on many real databases. We study also the behavior of the proposed improvement on noisy data. We present also comparative experiments of our proposed method with BrownBoost (a new method known that it improves AdaBoost with noisy data). Lastly, we give our conclusions and perspectives.

3.2   State of Art

Due to the finding of some weaknesses, such as the overfitting and the speed of convergence, met by the basic algorithm of boosting AdaBoost, several researchers have tried to improve it.

Therefore, we make a study of main methods having as purpose to improve boosting relatively to these weaknesses. With this intention, the researchers try to use the strong points of Boosting such as the update of the badly classified examples, the maximization of the margin, the significance of the weights that AdaBoost associates the hypothesis and finally the choice of weak learning.

3.2.1 Modification of the Examples' Weight

The distributional adaptive update of the examples, aiming at increasing the weight of those badly learned by the preceding classifier, makes it possible to improve the performance of any training algorithm . Indeed, with each iteration, the current distribution supports the examples having been badly classified by the preceding hypothesis, which characterizes the adaptivity of AdaBoost. As a result, several researchers proposed strategies related to a modification of weight update of the examples, to avoid the overfitting.

Indeed, we can quote for example MadaBoost [8] whose aim is to limit the weight of each example by its initial probability. It acts thus on the uncontrolled growth of the weight of certain examples (noise) which is the problem of overfitting.

Another approach which make the algorithm of boosting resistant to the noise is Brownboost [14], an algorithm based on Boost-by-Majority by incorporating a time parameter. Thus for an appropriate value of this parameter, BrownBoost is able to avoid the overfitting. Another approach, which adapts to AdaBoost a logistic regression model, is Logitboost [18].

An approach, which produces less errors of generalization compared with the traditional approach but with the cost of an error of training slightly more raised , is the Modest boost [1]. In fact, its update is based on the reduction in the contribution of classifier, if that functions "too well" on the data correctly classified. This is why the method is called Modest AdaBoost - it forces the classifiers to be "modest" and it works only in the field defined by a distribution.

An approach, which tries to reduce the effect of overfitting by imposing limitations on the distribution produced during the process of boosting is used in SmoothBoost [20]. In particular, a limited weight is assigned to each example individually during each iteration. Thus, the noisy data can be excessively underlined during the iterations since they are assigned to the extremely large weights.

A last approach, Iadaboost [19], is based on the idea of building around each example a local information measurement, making it possible to evaluate the overfitting risks, by using neighboring graph to measure information around each example. Thanks to these measurements, we have a function which translates the need for updating the example. This function makes it possible to manage the outliers and the centers of clusters at the same time.

3.2.2 Modification of the Margin

Certain studies, analyzing the behavior of Boosting, showed that the error in generalization still decreases even when the errors in training are stable. The explanation is that even if all the examples of training are already well classified, Boosting tends to maximize the margins [20].

Following this, some studies try to modify the margin either by maximizing it or by minimizing it with the objective of improving the performance of Boosting against overfitting.

Several approaches followed such as AdaBoostReg [17] which tries to identify and remove badly labeled examples, or to apply the constraint of the maximum margin to examples supposed to be badly labeled, by using the Soft Margin.

In the algorithm, proposed by [9], the authors use a weighting diagram which exploits a margin function that grows less quickly than the exponential function.

3.2.3 Modification of the Classifiers' Weight

During the performance evaluation of Boosting, researchers wondered about the significance of the weights $\alpha(t)$ that AdaBoost associates with the produced hypotheses.

However, they noted at the time of experiments on very simple data that the error in generalization decreased further whereas the weak learning had already provided all the possible hypotheses. In other words, when a hypothesis appears several times, it votes finally with a weight, office sum of all $\alpha(t)$, which is perhaps absolute. So several researchers hoped to approach these values by a nonadaptive process , such as locboost [15] an alternative to the construction of the whole representations of experts which allows the coefficients $\alpha(t)$ to depend on the data.

3.2.4 Choice of Weak Learner

A question that several researchers posed against the problems of boosting is that of weak learner and how to make a good choice of this classifier?

A lot of research moves towards the study of choosing the basic classifier of boosting, such as GloBoost [24]. This approach use a weak learner which produces only correct hypotheses. RankBoost [5] is also an approach which is based on weak learner which accepts as data attributes functions of rank.

3.2.5 The Speed of Convergence

In addition to the problem of overfitting met by boosting in the modern databases mentioned above, we find another problem : the speed of convergence of Boosting especially AdaBoost.

Indeed, in the presence of noisy data, the optimal error of the training algorithm used is reached after a long time. In other words, AdaBoost "loses" iterations, and thus time, with reweighing examples which do not deserve in theory any attention, since it is a noise.

Thus research was made to detect these examples and improve the performance of Boosting in terms of convergence such as: iBoost [22] which aims at specializing weak hypotheses on the examples supposed to be correctly classified.

The IAdaBoost approach also contributes to improve AdaBoost against its speed of convergence. In fact, the basic idea of the improvement is the modification of the theorem [18]. This modification is carried out in order to integrate the risk of Bayes. The effects of this modification are a faster convergence towards the optimal risk and a reduction of the number of weak hypotheses to build. Finally, RegionBoost [13] is a new weighting strategy of each classifier. This weighting is evaluated at the voting time by a technique based on K Nearest Neighbors of the example to label. This approach makes it possible to specialize each classifier on areas of the training data.

3.3 Boosting by Exploiting Former Assumptions

To improve the performance of AdaBoost and to avoid forcing it to learn either from the examples that contain noise, or from the examples which would become too difficult to learn during the process of Boosting, we propose a new approach.

This approach is based on the fact that for each iteration, Adaboost, builds hypotheses on a defined sample, it makes its updates and it calculates the error of training according to the results given only by these hypotheses. In addition, it does not exploit the results provided by the hypotheses already built on other samples to the former iterations. This approach is called AdaBoostHyb

Program Code

- Input X_0 to classify
- $S = (x_1, y_1), \ldots, (x_n, y_n)$ Sample
- For i=1,n Do
- $p_0(x_i) - 1/n$,
- End FOR
- $t \leftarrow 0$
- While $t \leq T$ Do
- Learning sample S_t from S with probabilities p_t.
- Build a hypotheses h_t on S_t with weak learning A.
- ϵ_t apparent error of h_t on S with $\epsilon_t = \sum weight \ of \ examples$ **such that** $argmax(\sum_{i=1}^{t} \alpha_i h_i(x_i) \neq y_i)$. $\alpha_t = 1/2ln((1 - \epsilon_t)/\epsilon_t)$.
- For i=1, m Do
- $P_{t+1}(x_i) \leftarrow (p_t(x_i)/Z_t)e^{-\alpha_t}$ **if** $argmax(\sum_{i=1}^{t} \alpha_i h_i(x_i)) = y_i$ **(correctly classified)**
  $P_{t+1}(xi) \leftarrow (p_t(x_i)/Z_t)e^{+\alpha_t}$ **if** $argmax(\sum_{i=1}^{t} \alpha_i h_i(x_i)) \neq y_i$ **(badly classified)**
  (Z_t normalized to $\sum_{i=1}^{n} p_t(x_i) = 1$)
- End For
- $t \leftarrow t + 1$
- End While
- Final hypotheses :
  $H(x) = argmax \ y \in Y \sum_{t=1}^{T} \alpha_t$

The modification within the algorithm is made through two ways

The first way is during the modification of the weights of the examples: Indeed, this strategy, with each iteration, is based on the opinion of the experts already used (hypotheses of the former iterations) for the update of the weight of the examples.

In fact, we do not compare only the class predicted by the hypothesis of the current iteration with the real class but also the sum of the hypotheses balanced from the first iteration to the current iteration. If this sum votes for a class different from the real class, an exponential update such as in the case of AdaBoost is applied to the badly classified example. Thus, this modification lets the algorithm be interested only in the examples which are either badly classified or not classified yet. So, results related to the improvement the speed of convergence are awaited, similarly for the reduction of the error of generalization, because of the richness of the space of hypotheses to each iteration.

The second way is during the error analysis $\epsilon(t)$ of the hypothesis to the iteration T: Indeed, this other strategy is rather interested in the classifiers' coefficient (hypothesis) to each iteration $\alpha(t)$.

In fact, this coefficient depends on the apparent error analysis $\epsilon(t)$. This method, with each iteration, takes into account hypotheses preceding the current iteration during the calculation of $\epsilon(t)$. So the apparent error with each iteration is the weight of the examples voted badly classified by the hypotheses weighted of the former iterations by comparison to the real class.

Results in improving the error of generalization are expected since the vote of each hypothesis (coefficient $\alpha(t)$) is calculated from the other hypotheses.

3.4 Experiments

The objective of this part is to compare our new approach and especially its contribution with the original approach of Adaboost and to look further into this comparison by the choice of a version improved of Adaboost (Brown Boost [14]).

Our Choice of BrownBoost was based on its robustness against the problems of noisy data. In fact, BrownBoost is an adaptive algorithm which use a function that depends on the iteration number K (execution time), the Current iteration i, the number of times that the example has already been correctly predicted r, and the probability of success $1 - \gamma$

$\alpha_r^i = \binom{k-i-1}{k/2-r}(1/2 + \gamma)^{(k/2)-r}(1/2 - \gamma)^{(k/2)-i-1+r}$ instead on the exponential function.

So by a good estimation of K parameter BrownBoost is capable of avoiding overfitting. The advantage of this approach is that the noised data would be detected at some point, and their weights stop rising.

The comparison criterions chosen in this article are the error rate, the recall, the p-value, the average gain compared to AdaBoost, the speed of convergence and the sensitivity to noise.

Table 3.1. Databases Description

Databases	Nb. Inst	Attrib	Cl. Pred	Miss.VaL
IRIS	150	4 numeric	3	no
NHL	137	8 numeric and symbolic	2	yes
VOTE	435	16 boolean valued	2	yes
WEATHER	14	4 numeric and symbolic	2	no
CREDIT-A	690	16numeric and symbolic	2	yes
TITANIC	750	3 symbolic	2	no
DIABETES	768	8 numeric	2	no
HYPOTHYROID	3772	30 numeric and symbolic	4	yes
HEPATITIS	155	19 numeric and symbolic	2	yes
CONTACT-LENSES	24	4 nominal	3	no
ZOO	101	18 numeric and boolean	7	no
STRAIGHT	320	2 numeric	2	no
IDS	4950	35 numeric and symbolic	12	no
LYMPH	148	18 numeric	4	no
BREAST-CANCER	286	9 numeric and symbolic	2	yes

To do this experimental comparison, we used the C4.5 algorithm as a weak learner (according to the study of Dietterich [6]). To estimate without skew the theoretical success rate, we used a procedure of cross-validation in 10 folds (according to the study [11]). In order to choose the databases for our experiments, we considered the principle of diversity. We have considered 15 databases of the UCI. Some databases are characterized by theirs missing values (NHL, Vote, Hepatitis, Hypothyroid). Some others concern the problem of multi-class prediction (Iris: 3 classes, Diabetes: 4 classes, Zoo: 7 classes, IDS: 12 classes). We choose the IDS database [23] especially because it has 35 attributes. Table 1 describes the 15 databases used in the experimental comparison.

3.4.1 Comparison of Generalization Error

Graphic 1 indicated the error rates in 10-fold cross-validation corresponding to the algorithm AdaBoost M1,BrownBoost and the proposed one. We used the same samples for the tree algorithms in cross-validation for comparison purposes. The results are obtained while having chosen for each algorithm to carry out 20 iterations. The study of the effect of the number of iterations on the error rates of the tree algorithms will be presented in the section 4.3, where we will consider about 1000 iterations.

The results in graphic 1 show already that the proposed modifications improve the error rates of AdaBoost. Indeed, for 14 databases out of 15, the proposed algorithm shows an error rate lower or equal to AdaBoost M1. We remark, also, a significant improvement of the error rates corresponding to the three databases NHL, CONTACT-LENS and BREAST-CANCER. For example, the error rate corresponding to the BREAST-CANCER database goes from 45.81% to 30.41%.

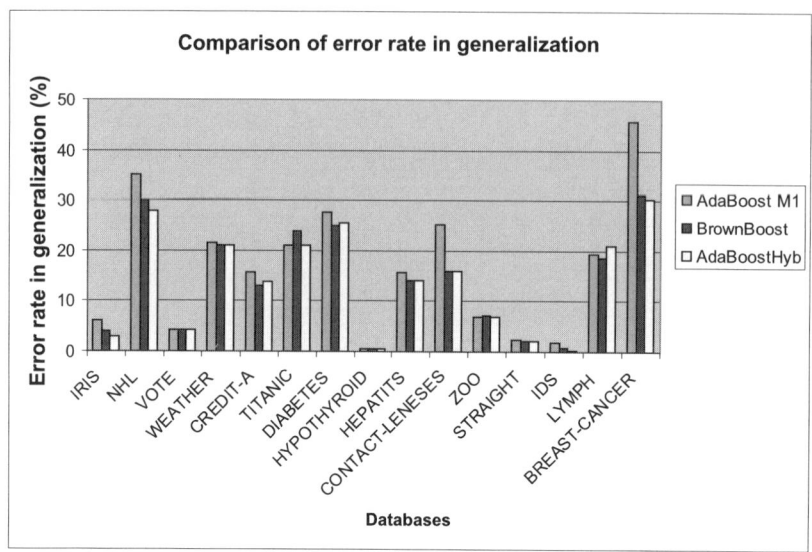

Fig. 3.1. Rate of error of generalization

According to this results, we use test-student and we find a significant p-value 0.0034. We have also a average gain of 2.8 compared to AdaBoost.

Even, if we compare the proposed algorithm with BrownBoost, we remark that for 11 databases out of 15 the proposed algorithm shows an error rate lower or equal to BrownBoost. Using test of student, we find a p-value not significant, But a average gain of 2.5 compared to AdaBoost.

This gain shows that by exploiting hypotheses generated with the former iterations to correct the weights of the examples, it is possible to improve the performance of the Boosting. This can be explained by the calculation of the precision of the error analysis $\epsilon(t)$ and consequently the calculation of the coefficient of the classifier $\alpha(t)$ as well as the richness of the space of the hypotheses to each iteration since it acts on the whole of the hypotheses generated by the preceding iterations and the current iteration.

3.4.2   Comparison of Recall

The encouraging results, found previously, enable us to proceed further within the study of this new approach. Indeed, in this part we try to find out the impact of the approach on the recall, since our approach does not really improve Boosting if it acts negatively on the recall.

Graphic 2 indicates the recall for the algorithms AdaBoost M1, Brownboost and the proposed one. We remark that the proposed algorithm has the best recall overall the 14 for 15 studied databases. This result confirms the preceding ones. We remark also that it increases the recall of the databases having less

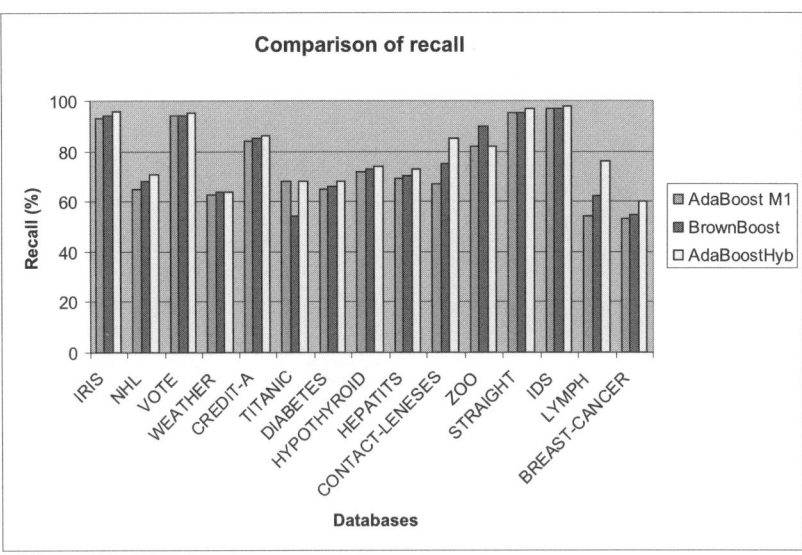

Fig. 3.2. Rate of recall

important error rates. Based on this results, we use test-student and we find a significant p-value 0.0010. We have a significant average gain of 4.8 compared to AdaBoost.

Considering Brownboost, we remark that it improves the recall of AdaBoostM1, overall the data sets (except the TITANIC one). However, the recall rates given by our proposed algorithm are better than those of BrownBoost. Except, with the zoo dataset. In this case, we have a significant p-value 0.0002 and not a significant average gain (1.4)compared to AdaBoost according to the results given by AdaBoosthyb.

It is also noted that our approach improves the recall in the case of the Lymph base where the error was more important. It is noted though that the new approach does not act negatively on the recall but it improves it even when it can not improve the error rates.

3.4.3   Comparison with Noisy Data

In this part, we are based on the study already made by Dietterich [6] by adding random noise to the data. This addition of noise of 20% is carried out, for each one of these databases, by changing randomly the value of the predicted class by another possible value of this class.

Graphic 3 shows us the behavior of the algorithms with noise. We notice that the hybrid approach is also sensitive to the noise since the error rate in generalization is increased for all the databases.

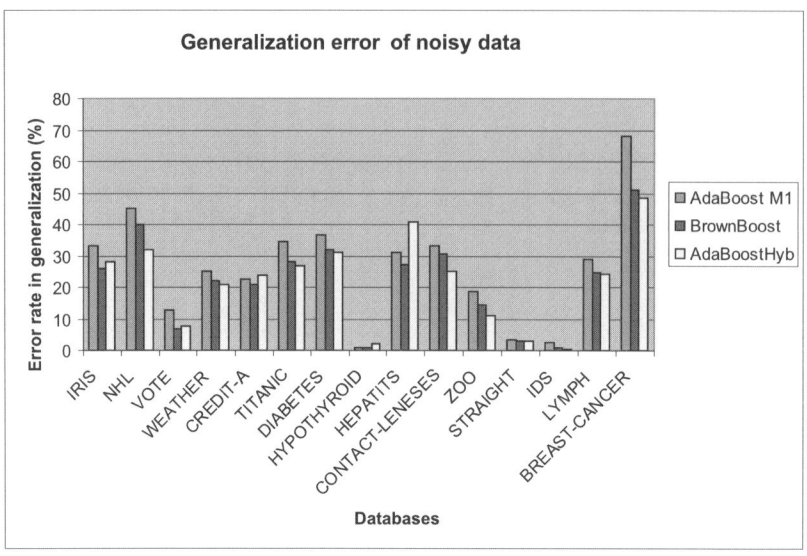

Fig. 3.3. Rate of error on Noisy data

However this increase remains always inferior with that of the traditional approach except for the databases such as Credit-A, Hepatitis and Hypotyroid.

So, we studied these databases and we observed that all these databases have missing values. In fact, Credited, Hepatitis and Hypothyroid have respectively 5%, 6% and 5,4% of missing values. It seems that our improvement loses its effect with accumulation of two types of noise: missing values and artificial noise, although the algorithm AdaBoost*Hyb* improves the performance of AdaBoost against the noise. Using test-student, we find a significant p-value 0.0352 and a average gain of 4.6 compared to AdaBoost. Considering Brownboost, we remark that it gives better error rates that AdaboostM1 on all the noisy data sets. However, It gives better error rates than our proposed method, only with 6 data sets. Our proposed method gives better error rates with the other 9 data sets. We haven't a significant p-value but a average gain of 4.4 compared to AdaBoost. This encourages us to study in details the behavior of our proposed method on noisy data.

3.4.4 Comparison of Convergence Speed

In this part, we are interested in the number of iterations that allow the algorithms to converge, i.e. where the error rate is stabilized. Table 2 and graphic 4 shows us that the hybrid approach allows AdaBoost to converge more quickly. Indeed, the error rate of AdaBoost M1 is not stabilized even after 100 iterations, whereas Adaboost Hyb converges after 20 iterations or even before.

Table 3.2. comparison of speed convergence

-	AdaBoost M1				BrownBoost				AdaBoost hyb			
Nb. iterations	10	20	100	1000	10	20	100	1000	10	20	100	1000
Iris	7,00	6,00	5,90	5,85	3.96	3.89	3,80	3,77	3,50	3,00	3,00	3,00
Nhl	37,00	35,00	34,87	34,55	30,67	30,01	29,89	29,76	31,00	28,00	28,00	28,00
Weather	21,50	21,42	21,40	14,40	21,10	21,00	20,98	21,95	21,03	21,00	21,00	21,00
Credit-A	15,85	15,79	15,75	14,71	13,06	13,00	12,99	12,97	14,00	13,91	13,91	13,91
Titanic	21,00	21,00	21,00	21,00	24,08	24,00	23,89	23,79	21,00	21,00	21,00	21,00
Diabetes	27,70	27,61	27,55	27,54	25,09	25,05	25,03	25,00	25,56	25,56	25,56	25,56
Hypothyroid	0,60	0,51	0,51	0,50	0,62	0,60	0,59	0,55	0,43	0,42	0,42	0,42
Hepatitis	16,12	15,60	14,83	14,19	14,15	14,10	14,08	14,04	14,03	14,00	14,00	14,00
Contact-Lenses	26,30	24,80	24,50	16,33	15,90	15,86	15,83	15,80	16,00	16,00	16,00	16,00
Zoo	7,06	7,00	7,00	7,00	7,25	7,23	7,19	7,15	7,00	6,98	7,00	7,00
Straight	2,50	2,46	2,45	2,42	2,12	2,00	1,98	1,96	0,42	0,42	0,42	0,42
IDS	2,00	1,90	1,88	1,85	0,7	0,67	0,65	0,63	0,7	0,67	0,65	0,63
Lymph	19,53	19,51	19,51	19,50	18,76	18,54	18,50	18,45	18,76	18,54	18,50	18,45
Breast-Cancer	45,89	45,81	45,81	45,79	31,10	31,06	31,04	31,00	31,10	31,06	31,04	31,00

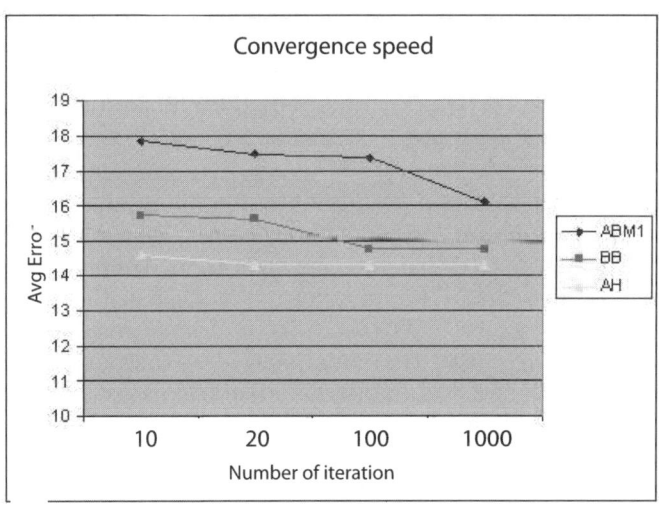

Fig. 3.4. Average Error by iteration

For this reason we choose for the first part 20 iterations to carry out the comparison in terms of error and recall. These results are also valid for the database Hepatitis. In fact, This database has a lot of missing values (Rate 6%). These missing values always present a problem of convergence. Moreover, the same results appear on databases of various types (several attributes, the class to be predicted with K modalities, important sizes).

This makes us think that due to the way of calculating the apparent error, the algorithm reaches stability more quickly. Finally, we remark that BrownBoost does'nt converge even after 1000 iterations. This remark prove the fact that the BrownBoost problem is the speed of convergence.

These results are confirmed by graphic 4. In fact, this graphic show us that firstly, the average error in generalization of AdaBoosthyb is less than the average error of other algorithms and secondly adaboosthyb converge more quickly than BrownBoost and specially AdaBoost M1 that don't converge even after 1000 iterations.

3.5 Conclusion

In this paper, we proposed an improvement of AdaBoost which is based on the exploitation of the hypotheses already built with the preceding iterations. The experiments carried out and the results show that this approach improves the performance of AdaBoost in error rate, in recall, in speed of convergence and in sensibility to the noise. However, it proved that this same approach remains sensitive to the noise.

We did an experimental comparison of the proposed method with BrownBoost (a new method known that it improves AdaBoost M1 with noisy data). The results show that our proposed method improves the recall rates and the speed of convergence of BrownBoost overall the 15 data sets. The results show also that BrownBoost gives better error rates with some datasets, and our method gives better error rates with other data sets. The same conclusion is reached with noisy data.

To confirm the experimental results obtained, more experimentations are planned. We are working on further databases that were considered by other researchers in theirs studies of the boosting algorithms. We plan to choose weak learning methods other than C4.5, in order to see whether the obtained results are specific to C4.5 or general. We plan to compare the proposed algorithm to new variants of boosting, other than AdaBoost M1. We can consider especially those that improve the speed of convergence like IAdaBoost and RegionBoost. In the case of encouraging comparisons, a theoretical study on convergence will be done to confirm the results of the experiments.

Another objective which seems important to us consists in improving this approach against the noisy data. In fact, the emergence and the evolution of the modern databases force the researchers to study and improve the boosting's capacities of tolerance to the noise . Indeed, these modern databases contain a lot of noise, due to new technologies of data acquisition such as the Web. In parallel, studies such as [5], [16] and [18] , show that AdaBoost tends to overfit the data and especially the noise. So, a certain number of recent work tried to limit these risks of overfitting. These improvements are based primarily on the concept that AdaBoost tends to increase the weight of the noise in an exponential way. Thus two solutions were proposed to reduce the sensibility to noise. One is by detecting these data and removing them based on the heuristic and selection

of prototypes such as research presented in [4]and [25]. The other solution is
by detecting these data through the process of boosting, in which case we speak
about a good management of noise. According to the latest approach, we plan
to improve the proposed algorithm against the noisy data, by using neighboring
graphs or using update parameters.

Finally, a third perspective work aims at studying the Boosting with a weak
learner that generates several rules (Rule learning [10]). Indeed, the problem of
this type of learners is the production of conflicting rules within the same iteration of boosting. These conflicting rules will have the same weights (attributed
by the boosting algorithm). In the voting procedure, we are thinking about a
combination of the global weights (those attributed by the boosting algorithm)
and the local weights (those attributed by the learning algorithm).

References

1. Vezhnevets, A., Vezhnevets, V.: Modest adaboost: Teaching adaboost to generalize
   better, Moscow State University (2002)
2. Bauer, E., Kohavi, R.: An empirical comparison of voting classification algorithms:
   Bagging, boosting, and variants. Machine Learning 24, 173–202 (1999)
3. Breiman, L.: Bagging predictors. Machine Learning 26, 123–140 (1996)
4. Brodley, C.E., Friedl, M.A.: Identifying and eliminating mislabeled training instances. In: AAAI/IAAI, vol. 1, pp. 799–805 (1996)
5. Dharmarajan, R.: An effecient boosting algorithm for combining preferences. Technical report, MIT (September 1999)
6. Dietterich, T.G.: An experimental comparison of three methods for constructing
   ensembles of decision trees: bagging, boosting, and randomization. Machine Learning, 1–22 (1999)
7. Dietterich, T.G.: Ensemble methodes in machine learning. In: First International
   Workshop on Multiple ClassifierSystems, pp. 1–15 (2000)
8. Domingo, C., Watanabe, O.: Madaboost: A modification of adaboost. In: Proc.
   13th Annu. Conference on Comput. Learning Theory, pp. 180–189. Morgan Kaufmann, San Francisco (2000)
9. Friedman, J., Hastie, T., Tibshirani, R.: Additive logistic regression: a statistical
   view of boosting. Dept. of Statistics, Stanford University Technical Report (1998)
10. Friedman, J.H., Popescu, B.E.: Predictive learning via rule ensembles (technical
    report). Stanford University (7) (2005)
11. Kohavi, R.: A study of cross-validation and bootstrap for accuracy estimation
    and model selection. In: International Joint Conference on Artificial Intelligence
    (IJCAI) (1995)
12. Littlestone, N., Warmuth, M.K.: The weighted majority algorithm. In: Information
    and computation, vol. 24, pp. 212–261 (1994)
13. Maclin, R.: Boosting classifiers regionally. In: AAAI/IAAI, pp. 700–705 (1998)
14. McDonald, R., Hand, D., Eckley, I.: An empirical comparison of three boosting
    algorithms on real data sets with artificial class noise. In: Fourth International
    Workshop on Multiple Classifier Systems, pp. 35–44 (2003)
15. Meir, R., El-Yaniv, R., Ben-David, S.: Localized boosting. In: Proc. 13th Annu.
    Conference on Comput. Learning Theory, pp. 190–199. Morgan Kaufmann, San
    Francisco (2000)

16. Rtsch, G.: Ensemble learning methods for classification. Master's thesis, Dep. of computer science, University of Potsdam (April 1998)
17. Rätsch, G., Onoda, T., Müller, K.-R.: Soft margins for adaboost. Mach. Learn. 42(3), 287–320 (2001)
18. Schapire, R.E., Singer, Y.: Improved boosting algorithms using confedence rated predictions. Machine Learning 37(3), 297–336 (1999)
19. Sebban, M., Suchier, H.M.: Tude sur amlioration du boosting: rduction de l'erreur et acclration de la convergence. Journal lectronique d'intelligence artificielle, 200–214 (2003)
20. Servedio, R.A.: Smooth boosting and learning with malicious noise. In: Helmbold, D.P., Williamson, B. (eds.) COLT 2001 and EuroCOLT 2001. LNCS (LNAI), vol. 2111, pp. 473–489. Springer, Heidelberg (2001)
21. Shapire, R.: The strength of weak learnability. Machine Learning 5, 197–227 (1990)
22. Kwek, S., Nguyen, C.: iboost: Boosting using an instance-based exponential weighting scheme. In: Thirteenth European Conference on Machine Learning, pp. 245–257 (2002)
23. Stolfo, S.J., Fan, W., Lee, W., Prodromidis, A., Chan, P.K.: Cost-based modeling and evaluation for data mining with application to fraud and intrusion detection (1999)
24. Torre, F.: Globoost: Boosting de moindres gnraliss. Technical report, GRAppA - Universit Charles de Gaulle - Lille 3 (September 2004)
25. Wilson, D.R., Martinez, T.R.: Reduction techniques for instance-based learning algorithms. Machine Learning 38(3), 257–286 (2000)

4

Dealing with Missing Values in a Probabilistic Decision Tree during Classification

Lamis Hawarah, Ana Simonet, and Michel Simonet

Institut d'Ingénierie et de l'Information de Santé (TIMC)
Faculté de Médecine
38700 La Tronche, France
{lamis.hawarah,ana.simonet,michel.simonet@imag.fr}

This chapter deals with the problem of missing values in decision trees during classification. Our approach is derived from the ordered attribute trees method, proposed by Lobo and Numao in 2000, which builds a decision tree for each attribute and uses these trees to fill the missing attribute values. Our method takes into account the dependence between attributes by using Mutual Information. The result of the classification process is a probability distribution instead of a single class. In this chapter, we explain our approach, we then present tests performed of our approach on several real databases and we compare them with those given by Lobo's method and Quinlan's method. We also measure the quality of our classification results. Finally, we calculate the complexity of our approach and we discuss some perspectives.

4.1 Introduction

In classification, the goal of a learning algorithm is to build a classifier from a training set. Each example in such a training set is assigned to a class. Classification is the task of assigning objects to their respective categories. Decision Trees are one of the most popular classification algorithms currently in use in Data Mining and Machine Learning. Decision Trees belong to supervised classification methods. Once built, decision trees are used to classify new cases. A case is classified by starting at the root node of the tree, testing the attribute specified by this node, then moving through the tree until a leaf is encountered; the case is classified by the class associated with the leaf. It may happen that some objects have no value for some attributes. We can encounter this problem, known as the problem of missing values, both during the construction phase and the classification phase of a decision tree. In the latter situation, when classifying an object, if the value of a particular attribute which was branched on in the tree is missing in the object, it is not possible to decide which branch to take in order to classify this object, and the classification process cannot be completed.

Our objective is to classify an object with missing values. Our work is situated in the framework of probabilistic decision trees [1, 5, 23]. We aim at using

D.A. Zighed et al. (Eds.): Mining Complex Data, SCI 165, pp. 55–74.
springerlink.com

the dependencies between attributes to predict missing attribute values. Therefore, we are interested in the type of approaches which use decision trees to fill in missing values [16, 21], because if a decision tree is able to determine the class of an instance thanks to the values of its attributes, then it can be used to determine the value of an unknown attribute from its dependent attributes. We have extended the *Ordered Attribute Trees* method *(OAT)*, proposed by Lobo and Numao [16]. They use decision trees to deal with missing values and they fill missing values both in training data and test data [16]. Our approach to deal with missing values uses decision trees during the classification phase. The result of classification is a probabilistic distribution according to the class values.

In our experimentation, we have tested our approach on several databases [20], aiming at measuring and evaluating the quality of our classification results. For this purpose, we have compared each instance in the test data with all the instances in the training data by calculating the *distance* between them. Our approach is based on an algorithm called *Relief* [12] and its extension *ReliefF* [14, 24], to calculate the distance between two instances. For each test instance, we calculate the frequency of its nearest instances from each class. This frequency is compared with the classification results obtained by our approach and the C4.5 method for the same test instance. The *Relief* algorithm, developed by [12], and its extension *ReliefF* [14, 24] are measures for classification. They take into account the context of other attributes when estimating the quality of an attribute with respect to the class.

In this chapter, we first present the work in the domain, and particularly Lobo's approach *(OAT)* and C4.5's method. We then describe our method to estimate missing values that uses the dependencies between attributes and gives a probabilistic result; we then present the tests performed on several databases, using our approach, *OAT's* method and Quinlan's method. We also measure the quality of our classification results. Finally, we calculate the complexity of our method and we present some perspectives.

4.1.1 Related Work

We present in this section the methods to deal with missing values using decision trees[1]. The general idea in filling missing values is to infer them from other known data. We can distinguish several approaches to deal with missing values. The simplest one is to ignore instances containing missing values [15]. The second type of technique consists in replacing a missing value with a value considered as adequate in the situation. For example, [13] proposes a method that uses class information to estimate missing attribute values during the training phase; the idea is to assign the most probable value of the attribute to the missing value, given the class membership of the case concerned. [22] fills in the missing values of an attribute with its most common known value in the training set during the

---

[1] The methods which deal with missing values in the statistical domain are not presented in this chapter [2].

classification phase. The third type of technique replaces missing values with a distribution of probability. For example, Quinlan's method [5] assigns probability distributions to each node of the decision tree when learning from training instances. The fourth type of technique focuses on the classification phase and uses another attribute instead of the one that is unknown, in order to keep on classifying the current case; the selected attribute is then correlated with the unknown attribute. For example, the CART method [1], which constructs binary decision trees, consists in using a surrogate split when an unknown value is found in the attribute originally selected. A surrogate split is a split that is similar to the best split in the sense that it makes a similar partition of the cases in the current node. Algorithms for constructing decision trees, such as [1, 5], create a single best decision tree during the training phase, and this tree is then used to classify new instances. The fifth type of technique constructs the best classification rule instead of constructing the whole decision tree. For example, the dynamic path generation method [15]. produces only the path (i.e., the rule) needed to classify the case currently under consideration, instead of generating the whole decision tree beforehand. This method can deal with missing values in a very flexible way. Once a missing value is found to be present in an attribute of a new instance, such an attribute is never branched on when classifying the instance. Similarly, the lazy decision tree method [7] conceptually constructs the best decision tree for each test instance. In practice, only a classification path needs to be constructed. Missing attribute values are naturally handled by considering only splits on attribute values that are known in the test instance. Training instances with unknowns filter down and are excluded only when their value is unknown for a given test in a path. The last type of approach uses decision trees to fill in missing values. For example, Shapiro's method [21] constructs a decision tree for an unknown attribute by using the subset of the original training set consisting of those instances whose value of the unknown attribute is defined. The class is regarded as another attribute and it participates in the construction of the decision tree for this attribute. This method is used only in the building phase. We now present the *Ordered Attribute Trees (OAT)* method [16],which also deals with missing values and which we have studied in more detail.

4.1.2   Ordered Attribute Trees Method

Ordered Attribute Trees (OAT) is a supervised learning method to fill missing values in categorical data. It uses decision trees as models for estimating unknown values. This method constructs a decision tree for each attribute, using a training subset that contains instances with known values for the attribute. These cases in the training subset, for a target attribute, are described only by the attributes whose relation with the class has lower strength than the strength of the relation between the target attribute and the class. The resulting decision tree is called an *attribute tree*. This method uses *Mutual Information* [3] as a measure of the strength of relations between the attributes and the class[2]. There is an order for the construction of the attribute trees. This order is guided by the *Mutual Information* between the attributes and the class. The method orders the attributes from those with low mutual information to those with high mutual information. It constructs attribute trees according to this order. These trees are used to determine unknown values for each attribute. The first attribute tree constructed is a one-node tree with the most frequent value among the values of the attribute. An attribute tree is constructed for an attribute A_i using a training subset, which contains instances with known values for the attribute A_i, and the attributes whose missing values have already been filled before. Consequently, the attributes A_k for which $MI(A_i, C) < MI(A_k, C)$ are excluded [16]. During the calculation of $MI(A_i, C)$, instances which have missing values for the attribute A_i are ignored [18]. This method is not general enough to be applicable to every domain [18]. The domains in which there are strong relations between the attributes appear to be the most suitable to apply the *OAT* method. In this method, the idea to start by dealing with the attribute which is the less dependent on the class [16, 17, 18] is interesting, because it is the attribute which has the least influence on the class.

4.1.3   C4.5's Method

Quinlan's method [5] assigns probability distributions to each node of the decision tree when learning from training instances. The probability distribution, for the values of the attribute involved in the test in a node, is estimated from the relative frequencies of the attribute values among the training instances collected at that node. The result of the classification is a class distribution instead of a single class. This approach works well when most of the attributes are independent, because it depends only on the prior distribution of the attribute values for each attribute being tested in a node of the tree [5, 23].

4.1.4   Conclusion

We observe that the methods above have some drawbacks. For example, [5, 13, 22] determine the missing attribute values only once for each object with this unknown attribute. The Dynamic path generation method and the lazy

---

[2] Mutual Information (MI) between two categorical random variables X and Y is the average reduction in uncertainty about X that results from learning the value of Y:

$$MI(X,Y) = - \sum_{x \in D_x} P(x)log_2 P(x) + \sum_{y \in D_y} P(y) \sum_{x \in D_x} P(x|y)log_2 P(x|y)$$

D_x and D_y are the domains of the categorical random variables X and Y. $P(x)$ and $P(y)$ are the probability of occurrence of $x \in D_x$ and $y \in D_y$, respectively. $P(x|y)$ is the conditional probability of X having the value x once Y is known to have the value y.

decision tree method [7, 15] do not resolve the missing values problem during the construction of the tree, but they classify the object using only its known attribute. Shapiro's method [21] makes good use of all the information available from the class and all the other attributes, but there is a difficulty that arises if the same case has missing values on more than one attribute [15]: during the construction of a tree to predict an unknown attribute, if a missing value is tested for another attribute, another tree must be constructed to predict this attribute, and so on. This method cannot be used practically, because this recursion process of constructing a decision tree once we find missing values for an attribute, leads to eliminating too many training cases when there are many unknown attributes. By constructing the attribute trees according to an order relying only on mutual information between the attributes and the class, Lobo and Numao provide a solution which can work in every situation [18]. However, they do not take into account all the dependencies between attributes, because they are built in an ordered manner. Therefore, It seems to make sense to build an attribute tree from the attributes which are dependent on it.

4.2 Probabilistic Approach

Our approach to estimate missing values during classification uses a decision tree to predict the value of an unknown attribute from its dependent attributes [8]. This value is represented by a probability distribution. We made two proposals. The first one, called *Probabilistic Ordered Attribute Trees (POATs)*, simply extends Lobo's *OATs* [16] with probabilistic data. In this proposal, we construct a probabilistic attribute tree for each attribute in the training data. These trees are constructed according to an order guided by the Mutual Information between the attributes and the class. The attributes used to build a *POAT* for an attribute A_i are those whose *attributes trees* have already been built before and are dependent on A_i. The result of classifying an object with missing values using *POAT* is a class distribution instead of a single class. These trees give a probabilistic result which is more refined than Lobo's initial *OATs*. However, they do not take into account all the dependencies between attributes, because they are built in the same ordered manner that is used by Lobo's *OAT*. Therefore, we suggested another approach, called *Probabilistic Attribute Trees (PATs)*, which uses the dependence between attributes and also gives a probabilistic result [8]. In the *PATs* approach, we calculate the Mutual Information between each pair of attributes in order to determine for each attribute its dependent attributes. A *Probabilistic Attribute Tree (PAT)* is constructed for each attribute, using all the attributes depending on it.

4.3 Classification Algorithm

To classify an instance with missing values using the final probabilistic decision tree[3], we start tracing the decision tree from its root until we reach a leaf by

---

[3] A final decision tree is the tree which corresponds to all the training set.

following the branches according to attributes values of the instance. When we encounter a missing value for a test-attribute (test-node), we must trace all the paths corresponding to the values of this attribute. In this case, we reach several leaves in the tree, and not only one leaf as in classical classification. For this purpose, it is necessary to calculate the class probability on each one of these leaves.

Let us assume that the class has two values A, D, and for a path from the root of the tree to a leaf F, we go through the branches B_1, B_2,.., B_n.

P(class A at leaf F) = P(A | path from the root to F) = $P(A|B_1, B_2, .., B_n)$

P(class D at leaf F) = P(D | path from the root to F) = $P(D|B1, B2, .., Bn)$

$$P(\texttt{A in the tree}) = \sum_i P(A|F_i) * P(F_i)$$

$$P(\texttt{D in the tree}) = \sum_i P(D|F_i) * P(F_i)$$

where i = 1,..,m (m is the number of leaves in the tree).

The probability $P(A|F_i)$ is the conditional probability of class A at this leaf; the probability $P(F_i)$ is the joint probability of the attributes in the path which starts from the root until the leaf F_i.

To simplify, let us consider that the path from the root of the tree until F_i goes through only the branches B_1 and B_2:

$P(F_i) = P(B_1, B_2) = P(B_1) * P(B_2|B_1)$; B_1 is less dependent on the class than B_2[4].

4.3.1   Calculating the Joint Probability $P(B_1, B_2)$ Using Our Approach

To calculate this joint probability, we distinguish the following cases:

- B_1 and B_2 are independent:
  $P(B_2|B_1) = P(B_2)$ and $P(B_1, B_2) = P(B_1) * P(B_2)$
  Consequently, the *PAT* of B_1 is constructed without B_2 and the *PAT* of B_2 is constructed without B_1. We calculate the probability of the attribute B_1 from its *PAT*. The probability of B_2 is also calculated from its *PAT*.
- B_1 and B_2 are dependent and the *POAT* of B_1 is constructed without B_2 because B_1 is less dependent on the class than B_2: $P(B_1|B_2) \neq P(B_1)$. The probability of B_1 is calculated from its *POAT*. Note that the *PAT* of B_2 is constructed using B_1. Therefore, we calculate the conditional probability of B_2 given B_1 $P(B_2|B_1)$ from the *PAT* of B_2.

---

[4] In our work, when two attributes are dependent and unknown at the same time *(Cycle problem)*, we deal first with the attribute which is less dependent on the class by using its *POAT*. Then, for the other attribute, we use its *PAT*.

- B_1 et B_2 are dependent, B_1 is the less dependent on the class. There is another missing attribute G which is dependent on B_1 and B_2, G is less dependent on the class than B_1 and B_2[5].

$$P(B_1) = \sum_i P(B_1|G_i) * P(G_i)$$

$$P(B_2|B_1) = \sum_i P(B_2|B_1, G_i) * P(G_i|B_1)$$

$$P(B_1, B_2) = \sum_i P(B_1, B_2, G_i)$$

$$= \sum_i P(G_i) * P(B_1|G_i) * P(B_2|B_1, G_i) \qquad (4.1)$$

- B_1 and B_2 are independent but they are dependent on another missing attribute G. G is less dependent on the class than B_1 and B_2:

$$P(B_1, B_2) = \sum_i P(B_1, B_2, G_i)$$

$$= \sum_i P(G_i) * P(B_1|G_i) * P(B_2|G_i)$$

B_2 and B_1 are conditionally independent, given G.

4.4 Experiment

In our experiment, we tested our approach on several databases from the UCI repository [20]. Each database was tested on several thresholds. To choose a threshold, we calculate the average Normalized Mutual Information[6] calculated between each attribute and the class. We then choose some thresholds that are closest to this average value [9]. We compared our classification results with those generated by Quinlan's method [5]; we found that our results are equal or better than those given by C4.5. We present the tests performed on the *vote* database [20]. A training set, which has 232 instances with 16 discrete attributes (all are Boolean and take the values y or n), is used to construct our trees (*POATs* and *PATs*). The class in this database can take two values: (*Democrat* and *Republican*). This training data does not have any missing values, but the test data we used contains 240 objects with missing values. The average value of Normalized Mutual Information is 0.26. Therefore, we have tested our approach

---

[5] We also can calculate the joint probability given in the equation 4.1 as following: $P(B_1, B_2) = P(B_1) * P(B_2|B_1) = P(B_1) * \sum_i P(B_2|B_1, G_i) * P(G_i|B_1) = \sum_i P(B_2|B_1, G_i) * P(G_i|B_1) * P(B_1) = \sum_i P(B_2|B_1, G_i) P(B_1|G_i) P(G_i)$

[6] We use *Normalized Mutual Information* as proposed by Lobo and Numao [18] instead of *Mutual Information*. *Normalized Mutual Information* is defined as:

$$MI_N(X, Y) \equiv \frac{2MI(X, Y)}{\log ||D_x|| + \log ||D_y||}$$

Table 4.1. Tests performed on *vote* database

	Threshold	well classif.	¬ well classif.	50%
	0.2	**91.25%**	**08.33%**	**0.41%**
	0.3	90%	09.16%	0.83%
PAT	0.4	88.33%	11.66%	
	0.5	87.08%	12.91%	
C4.5		**83.75%**	**16.25%**	
OAT		**91.66%**	**08.33%**	

Table 4.2. The confusion matrix of *Vote* database using *PAT, C4.5, OAT*

a	b	¡- classified as
154	18	a=democrat
2	66	b=republican

a	b	¡- classified as
166	6	a=democrat
34	34	b=republican

a	b	¡- classified as
153	19	a=democrat
1	67	b=republican

Table 4.3. Tests performed on *Nursery* database

	Threshold	well classif.	¬ well classif.	50%
	0.001	**68.25%**	**23.80%**	**07.93%**
PAT	0.02	**67.46%**	**32.53%**	
	0.03	**67.46%**	**32.53%**	
C4.5		**68.25%**	**23.80%**	**06.34%**
OAT		**72.22%**	**27.77%**	

on several thresholds: 0.2, 0.3, 0.4 and 0.5. The result of the tests is shown in Table 4.1. The column 50% in Table 4.1 contains the percentage of objects having a probability of 0.5 for each class value. Our results are better than the results given by C4.5, which are presented in the same table. But our results are equal to those given by *OAT* when the threshold is 0.2. Generally, when we decrease the threshold, we increase the degree of dependence between attributes, and consequently we use more attributes to construct our trees, which decreases the number of instances on each leaf in each tree. In Table 4.1, we note that when we decrease the threshold, our results improve and the best results are obtained by *PAT* with a threshold of 0.2.

In Table 4.2, we show the complete confusion matrix of the Vote database using *PAT*, C4.5 and *OAT* from the left to the right, respectively. We find that 34 objects are misclassified with C4.5 when the class is *Republican*.

In Table 4.3, we present the tests performed on the *Nursery* database [20], which has 8 discrete attributes. The class takes 5 values: *not-recom, recommend, very-recom, priority, spec-prior*. The training set has 12960 instances, without missing values. The test set has 126 instances where the missing values rates are

Table 4.4. The confusion matrix of the *Nursery* database using *PAT, C4.5 and OAT*

a	b	c	d	e	¡- classified as
26	0	0	8	8	a = not-recom
0	0	0	0	0	b = recommend
2	0	1	3	0	c = very-recom
2	0	0	34	3	d = priority
4	0	0	8	27	e = spec-prior

a	b	c	d	e	¡- classified as
26	0	0	8	8	a = not-recom
0	0	0	0	0	b = recommend
2	0	1	3	0	c = very-recom
2	0	0	34	3	d = priority
2	0	0	10	27	e = spec-prior

a	b	c	d	e	¡- classified as
24	0	1	12	5	a = not-recom
0	0	0	0	0	b = recommend
0	0	6	0	0	c = very-recom
0	0	3	36	0	d = priority
0	0	0	14	25	e = spec-prior

Table 4.5. Tests performed on *lymphography* database

	Threshold	well classif.	¬ well classif.	50%
	0.06	91.93%	8.06%	
PAT	**0.07**	**95.16%**	**4.83%**	
	0.08	83.87%	12.90%	
C4.5		**79.03%**	**19.35%**	**01.61%**
OAT		**79.03%**	**19.35%**	**01.61%**

35% for the attribute *parents*, 37% for *has-nur*, 39% for *health* and 13% for *form*. We remark that our results are closer to those given by C4.5 when the Threshold is 0.02 or 0.03. Our results do not improve when decreasing the threshold because all the attributes in this database are independent. The confusion matrix are shown in Table 4.4.

We find that *PAT* and *C4.5* give nearly the same classification error when the attributes are independent. *OAT* is better when the class is *very-recom*. We have also tested our approach on the *lymphography* database [20]. The training set has 148 objects and 18 discrete attributes. The class takes 4 values: *normal, metastases, malign-lymph, fibrosis*. We also use a test data which has 62 objects. The missing values rates in the test data are: 56% for the attribute *block-of-affere*, 30% for *lym-nodes-dimin*, 40% for *changes-in-node*, 12% for *early-uptake-in*, 13% for *special-forms*, 11% for *changes-in-stru*, 17% for *defect-in-node* and 11% for the attribute *lym-nodes-enlar*. Table 4.5 contains the tests performed on the *lymphography* database [20] using the *PAT* approach, C4.5 and *OAT*.

From Table 4.6, we find that generally the performance of our *PAT* approach is closer to those given by *C4.5* and *OAT*, but it is better than themwhen the class is *malign-lymph*. We have also tested our approach on the *Mushroom* database [20]. The training data has 5644 instances and 22 discrete attributes. The class takes

Table 4.6. The confusion matrix of the *Lymphography* database using *PAT*, *C4.5* and *OAT*

a	b	c	d	¡- classified as
0	0	0	0	a = normal
0	35	4	0	b = metastases
0	1	20	0	c = malign-lymph
0	0	0	2	d = fibrosis

a	b	c	d	¡- classified as
0	0	0	0	a = normal
0	35	4	0	b = metastases
0	9	12	0	c = malign-lymph
0	0	0	2	d = fibrosis

a	b	c	d	¡-classified as
0	0	0	0	a = normal
0	33	6	0	b = metastases
0	7	14	0	c = malign-lymph
0	0	0	2	d = fibrosis

Table 4.7. Tests performed on *Mushroom* database

	Threshold	well classif.	¬ well classif. 50%
PAT %	**0.1**	**80.82%**	**19.17 %**
	0.2	75.34%	24.65 %
C4.5		**58.90%**	**41.09%**
OAT		**67.12%**	**32.87%**

Table 4.8. The confusion matrix of the *Mushroom* database using *PAT*, *C4.5* and *OAT*

a	b	¡-classified as
33	5	a = e
13	22	b = p

a	b	¡-classified as
33	5	a = e
25	10	b = p

a	b	¡-classified as
35	3	a = e
21	14	b = p

2 values *e*, *p*. The test data has 73 objects. The missing values rates are: 75% for the attribute *odor*, 26% for *stalk-shape*, 52% for *stalk-root*, 23% for *veil-type*, 27% for *spore-print-color* and 19% for *ring-type*. We remark that in Tables 4.7, 4.8 our results are better than those given by C4.5 and *OAT*.

Finally, we present the tests performed on the *Zoo* database [20]. The training data has 101 instances and 17 attributes. The class takes 7 values: *mammal, bird, reptile, fish, amphibian, insect, invertebrate*. The test data has 65 instances. The missing values rates in the test data are: 21% for the attribute *feathers*, 29% for *milk*, 16% for *airborne*, 16% for *aquatic*, 13% for *predator* and 13% for the attribute *legs*. The result of testing is given in Table 4.9. From Tables 4.10, 4.11 we remark that when the class is *bird*, we have 17 objects in the test data, 16 of which are misclassified with C4.5, 4 of which are misclassified with *OAT* and only 1 object is misclassified with *PAT*.

Table 4.9. Tests performed on *Zoo* database

	Threshold	well classif.	¬ well classif. 50%
	0.1	98.46%	1.53%
PAT	0.2	100%	0%
	0.3	96%	4%
C4.5		78.46%	21.53%
OAT		88.73%	11.26%

Table 4.10. The confusion matrix of the *Zoo* database using *PAT and C4.5*

a	b	c	d	e	f	g	¡- classified as		a	b	c	d	e	f	g	¡- classified as
27	0	0	0	0	0	0	a = mammal		27	0	0	0	0	0	0	a = mammal
0	16	0	0	0	0	1	b = bird		6	1	10	0	0	0	0	b = bird
1	0	2	0	1	0	0	c = reptile		1	0	2	0	0	0	1	c = reptile
0	0	0	10	0	0	0	d = fish		1	0	0	9	0	0	0	d = fish
0	0	0	0	3	0	0	e = amphibian		1	0	0	1	0	0	1	e = amphibian
0	1	0	0	0	3	0	f = insect		1	0	0	0	0	3	0	f = insect
0	0	0	0	0	0	6	g = invertebrate		2	0	0	0	1	0	3	g = invertebrate

Table 4.11. The confusion matrix of the *Zoo* database using *OAT*

a	b	c	d	e	f	g	¡– classified as
25	0	2	0	0	0	0	— a = mammal
0	13	3	0	1	0	0	— b = bird
0	0	2	0	2	0	0	— c = reptile
0	0	0	10	0	0	0	— d = fish
0	0	0	0	3	0	0	— e = amphibian
0	0	0	0	0	4	0	— f = ínsect
0	0	0	0	0	0	6	— g = invertebrate

In our experiment, we have also calculated the *Root Mean Squared Error*[7] which is a metric for comparing the accuracy of probability estimates [4]. Table 4.12 shows RMSE for each method. Since RMSE is a measure of error, smaller is better. RMSE for Both C4.5 and *OAT* are bigger than the RMSE for *PAT*.

---

[7] The root mean squared error for an instance x is given by the following equation:

$$RMSE = \sqrt{\frac{1}{n}\sum_{j=1}^{j=n}(t(j|x) - P(j|x))^2} \qquad (4.2)$$

where x is the instance, j is the class value, $t(j|x)$ is the true probability of class j for x and $P(j|x)$ is the probability estimated by the method for instance x and class j. For test data where the true classes are known, but not probabilities, $t(j|x)$ is defined to be 1 if the class of x is j and 0 otherwise.

Table 4.12. The Root Mean Squared Error

DataBase	PAT	C4.5	OAT
Vote	**0.310443**	0.52039	0.315079
Nursery	0.4456728	0.44999	**0.436149**
Lymphography	**0.260477**	0.477835	0.420603
Mushroom	0412543	0.643147	0.535865
Zoo	**0.133817**	0.44812	0.245

4.5   Measure of the Quality of the Classification Results

Our approach is based on the dependence between attributes. The results of classification given by our approach are probabilistic. We measured the quality of our classification results in order to improve the performance of our approach. For this purpose, we considered an algorithm [12] called *Relief*, which has been shown to be very efficient in estimating attributes. We were interested in *Relief* because it relies entirely on statistical analysis and employs few heuristics. On the other hand, the classical measures for classification[8] evaluate the quality of an attribute with respect to the class independently of the context of other attributes [25]. However, *Relief* takes into account the context of other attributes when estimating the quality of an attribute with respect to the class. The basic idea of *Relief*, when analysing training instances, is to take into account not only the difference in attribute values and the difference in classes, but also the *distance* between instances. In this section, we first present the algorithm *Relief*, its extension *ReilefF* and the *Distance* function used to calculate the distance between two instances. We then propose an algorithm which calculates for each test instance in the test data the frequency of its nearest instances from each class. Finally, we give some examples.

4.5.1   Relief

The key idea of Relief is to estimate attributes according to how well their values distinguish among instances that are close to each other. For that purpose, given a randomly selected instance R from m instances, *Relief* [14] searches for its two nearest neighbors: one H from the same class and the other M from a different class. It uses a function *diff* that calculates the difference between the values of Attribute for two instances. For a discrete attribute this difference is either 1 when the values are different or 0 when the values are equal. Estimating the quality $W[A]$ of attribute A is defined as shown below:

$$W[A] = W[A] - diff(A, R, H)/m + diff(A, R, M)/m \qquad (4.3)$$

Relief updates the quality estimation $W[A]$ for all the attributes A depending on their values for R, M and H. This is repeated m times according to the m

---

[8] As information gain, gain ration, distance measure and Gini-index, etc.

instances. The original *Relief* can deal with discrete and continuous attributes. However, it cannot deal with incomplete data and is limited to two-class problems. Its extension, which solves these and some other problems, is called *ReliefF* [14, 24]. *ReliefF* is able to deal with incomplete and noisy data and can be used for evaluating the attribute quality in multi-class problems. *ReliefF* also generalizes the function *diff(A, Instance$_1$, Instance$_2$)* to deal with missing values. This function becomes for a discrete attribute A:

$$diff(A, I_1, I_2) = \begin{cases} 0 \text{ if } V^{(A,I_1)} = V^{(A,I_2)} \\ 1 \text{ if } V^{(A,I_1)} \neq V^{(A,I_2)} \\ 1 - P(V^{(A,I_2)}|Class_{I_1}) \\ \text{if A is unknown in } I_1 \end{cases} \tag{4.4}$$

Where:

- $V^{(A,I_j)}$ is the value of A in the instance I_j
- $Class_{I_1}$ is the value of the class in the instance I_1
- $1 - P(V^{(A,I_2)}/Class_{I_1})$ is the probability that two instances I_1 and I_2 have different values for the given attribute A when one of the instances (I_1 here) has unknown value for A.

We notice that this function also calculates the probability that two instances I_1 and I_2 have different values for the given attribute A when both instances have unknown attribute values, but we do not explain it in equation 4.4. However, we can find it in [14, 24].

4.5.2 Calculating the Distance between Instances

Relief and *ReliefF* were the basis of our approach to calculate the distance between two instances using the function given in equation 4.4. The first instance is from the test data with missing values, the other one is from the training data without unknown attributes. The total distance is simply the sum of differences over all the attributes [14]. The Distance function is shown in equation 4.5 below:

$$Distance(I_1, I_2) = \sum_{j=1}^{j=n} diff(A_j, I_1, I_2)^9 \tag{4.5}$$

For example, if the distance between two instances is 5, it means that there are 5 attributes whose values are different in the two instances.

In our experiment[11], to measure the quality of our classification results, we compare each instance in the test data with all the instances in the training data by calculating the distance between them by using the function given in equation 4.5. Then, for each test instance, we calculate the frequency of its nearest instances from each class. This frequency, which is a statistical result,

---

[9] n is the number of attributes.

is to be compared with the classification result obtained by the *PAT* approach for the same test instance. For this purpose, we propose an algorithm which is presented below[10].

4.5.3   Instance Analysis Algorithm

```
Input: Inst test instance ,
       n training instances I;
Output: for Inst:
frequency of nearest instances from the same class and frequency of
nearest instances from the different class;

Function Instance-Analysis(Inst:test instance,
                 I:array[1..n] of instances): Pc:array[1..2] of real;
Const    near=5;
Var      nbSCL, nbDCL, k, near: integer;
                       dis: real;
  begin
  nbSCL=0, nbDCL=0;
   For k:=1  to n  do
    begin
    dis= Distance(Inst,I[k])
    If dis < near    {the two instances are nearest neighbor}
    then
    If(both Inst and I[k] are from same Class)
       then nbSCL++
       else nbDCL++;
   end; (*for k*)
  Pc1= P(nearest instances from same class)     = nbSCL/(nbDCL+nbSCL)
  Pc2= P(nearest instances from different class)= nbDCL/(nbDCL+nbSCL)
  end;
  return(Pc);
```

In the above algorithm, we present only the treatment of two-class problems. However, in our experiment, we also deal with the mutli-class problem. The constant *near* is fixed by the user. We consider that two instances are *nearest* if the distance between them is lower than *near*. For a test instance, this algorithm tells us statistically about the proportion of its nearest instances from each class. We then compare this frequency with the classification result obtained by the *PAT* approach for the same test instance.

Results: To illustrate our experience using the *Instance Analysis Algorithm* proposed above, we present only the result of testing this algorithm on three examples as presented in Table 4.13 from the *vote* database. The *vote* database

---

[10] This algorithm is a K-Nearest Neighbour method from instance-based learning. We use the distance function in equation 4.5 because it is the most appropriate to our problem.

Table 4.13. Result of testing *Instance Analysis Algorithm* on *Vote* database

attributes	instance 1	instance 2	instance 3
physician-fee	?	?	?
el-salvador	y	?	y
education	y	n	n
crime	y	n	n
near=8	**(16%, 83%)**	**(91%, 08%)**	**(92%, 07%)**
near=10	(29%, 70%)	(84%, 15%)	(75%, 24%)
near=12	(38%, 61%)	(70%, 29%)	(57%, 42%)
PAT	**(11%,89%)**	**(99%,01%)**	**(85%,15%)**
C4.5	(53%,47%)	(53%,47%)	(53%,47%)

has 16 attributes, so the constant *near* may be 8, 9, 10, 11, or 12. Table 4.13 contains the results of testing this algorithm only when *near* is 8, 10 and 12. It also contains the results of testing the *PAT* approach and C4.5 on the same examples. By comparing the results of *PAT* and C4.5 with the statistical results given by the *Analysis-Instance* algorithm for the same examples, we remark that *PAT's* results in Table 4.13 are closer to the statistical results in Table 4.13 when *near* is 8. Therefore, they are better than the C4.5 results when *near* is 8, 10 or 12.

We note that the only attribute used to construct the decision tree using the C4.5 method is *physician-fee-freeze*, which has the greatest influence on the decision. However, if this attribute is unknown, as in Table 4.13, C4.5 calculates its frequency in all the training data without taking into account the other attributes which depend on it. Consequently, in the test data each object which has a missing value for *physician-fee-freeze* is classified *Democrat* with probability 0.53 and *Republican* with probability 0.47 (Table 4.13). For the threshold 0.4, our *probabilistic decision tree* corresponding to the training data is constructed using the attributes *physician-fee-freeze*, *el-salvador-aid* and *education-spending*. The *PAT* for *physician-fee-freeze* is constructed using *el-salvador-aid*, *education-spending* and *crime*. Consequently, when *physician-fee-freeze* is unknown, we calculate its probability according to its dependent attributes, and so on. For example, in Table 4.13, we notice that in the *PAT* approach, the probability distribution of each object depends on the other attributes values. However, with C4.5, this distribution depends only on *physician-fee-freeze*'s frequency.

4.6 The Complexity of Constructing a Decision Tree

The complexity of a decision tree depend on the size of the training data and the number of attributes in this data. For example, if we have training data that contains n objects and m attributes without the class, the computational complexity of the decision tree constructed using this training data is [19]:

$$O(n \times (m + 1) \times log(n)) \tag{4.6}$$

In the *POAT* approach, we construct a *POAT* for each attribute in the training data. Therefore, for a training set which has m attributes without the class attribute and n instances, we construct m decision trees. The first tree is a one-leaf tree constructed using one attribute. The second tree uses two attributes, and so on; the last tree is constructed using m attributes[11]. Since we obtain m decision trees with an increasing number of attributes, the complexity of these trees is:

$O(1\ n\ log(n) + 2\ n\ log(n) + ..+ m\ n\ log(n)) = O((\sum_{i=1}^{m} i)\ n\ log(n)) = O(\frac{m(m+1)}{2}\ n\ log(n))$

For the same training set, we construct another m decision trees using the *PAT* approach. Each tree is constructed for an attribute by using all the attributes dependent on it. In the worst-case, all the attributes are mutually dependent. In this case, each decision tree is constructed using m attributes. Therefore, the complexity of these m trees is:

$O(m\ n\ log(n) + m\ n\ log(n) + ... + m\ n\ log(n)) = O(\ m\ m\ n\ log(n)) = O(m^2\ n\ log(n))$.

4.6.1   The Complexity of Classifying a New Instance

Once a decision tree has been built, classifying a new instance is extremely fast, with a worst-case complexity of $O(h)$, where h is the maximum depth of the tree [6]. This is true when we classify a new instance without missing values, because we trace only one path in the decision tree from its root-node to a leaf-node according to the outcome of each attribute node in this path. Let us assume that L is the number of leaves in the tree. Therefore, the height of this tree is superior or equal to $log_v(L)$ [10].

The complexity of classifying a new case is $O(log_v(L))$. This complexity changes when the new instance has missing values for some attributes, because we trace several paths in the tree instead of only one path. In the worst-case, we trace all the possible paths. The complexity of tracing all the possible paths in the tree becomes $O(L\ log_v(L))$, which is the classification complexity of C4.5.

The complexity of our classification algorithm

In our method, during the classification process, when we encounter a missing attribute-test we trace all the paths corresponding to its values. When we reach a leaf in the final tree, we calculate the probability of each missing attribute encountered in such a path by calling its tree[12]. The way in which we calculate the probability of each missing attribute is not arbitrary. We always start by dealing with the attribute which is the less dependent on the class. Therefore, in

---

[11] In the worst-case, when all the *m-1* attributes are dependent on the attribute *m*.

[12] *POAT* or *PAT* according to the situation. We take into account the other missing attributes encountered in such a tree.

each tree which is called, we trace only one path. At the end of our classification process, we have called several *attribute trees*. We can calculate the complexity of our classification algorithm as follows: Let us assume that at the end of our classification, we have called $nbAT$ *attribute trees*. Therefore, the complexity of these $nbAT$ trees is:

$$O(\sum_{i=1}^{i=nbAT} ComplexityTree(Tree_i)) \tag{4.7}$$

$$ComplexityTree(Tree) = \begin{cases} O(0) \text{ if one leaf}/L_{Tree_i} = 1/ \\ O(log(L_{Tree_i})) \text{ else} \end{cases} \tag{4.8}$$

where L_{Tree_i} is the number of leaves in this tree.

To calculate $nbAT$ which is the number of *attribute trees* called during the classification process, let us assume that in our test instance, we have m missing attributes; attribute i takes v_i values. If all of these attributes are dependent, we can calculate the number of *attribute trees* called during our classification process as follows:

$$nbAT = m \times (v_1 \times v_2 \times ... \times v_m)$$

Assuming that $\bar{v} = \frac{v_1+v_2+...+v_m}{m}$ is the average number of possible values for each attribute. The number of called trees becomes:

$$nbAT = m \times \bar{v}^m \tag{4.9}$$

From equations 4.7, 4.8 and 4.9, we calculate the complexity of all the attribute trees called during the classification as:

$$O(\sum_{i=1}^{i=nbAT} ComplexityTree(Tree_i)) - O(\sum_{i=1}^{i=nbAT} log(L_{Tree_i}))$$

$$= O(log(\prod_{i=1}^{i=nbAT} L_{Tree_i})) = O(log(\prod_{i=1}^{i=nbAT} \bar{L}_T))$$

$$= O(log(\bar{L}_T^{nbAT})) = O(nbAT \times log(\bar{L}_T))$$

$$= O(m \times \bar{v}^m log(\bar{L}_T)) \tag{4.10}$$

Where :

- $\bar{L}_T$ is the average number of leaves in an *attribute tree*.
- m is the number of missing attributes in the instance.

We remark that the complexity in equation 4.10 increases when the number of missing values in the instance increases.

Finally, the complexity of classifying an instance in the final decision tree using our approach is equal to the complexity of tracing this decision tree + the complexities of all the *attribute trees* called during the classification process:

$$Complexity(Trees) = O(nbPath_{Tree} \ Log(L_{Tree}))$$
$$+ O(\sum_{i=1}^{i=nbAT} ComplexityTree(Tree_i))$$

$$(4.11)$$

$$nbPath_{Tree} = \begin{cases} 1 \text{ if no missing values} \\ L_{Tree} \text{ if all are missing} \\ nbPT \text{ else} \end{cases} \qquad (4.12)$$

From equation 4.11, we remark that $nbPath_{Tree}$ may be L_{Tree} in the worst-case when all the attributes are missing. Therefore, we can note that the complexity $O(nbPath_{Tree} \ Log_v(L_{Tree}))$ in this equation is quasilinear according to the number of leaves in the final decision tree.

Consequently, the total complexity of classifying a new instance with missing values becomes:

$$Complexity(Trees) = O(nbPath_{Tree} \ Log(L_{Tree}))$$
$$+ O(m \times \bar{v}^m log(\bar{L}_T))$$

This complexity is exponential in the number of missing attributes in the instance to classify and quasilinear in the number of leaves in the final decision tree.

4.7   Conclusion and Perspectives

In this chapter, we have introduced a probabilistic approach to fill missing values in decision trees during classification. We proposed replacing an unknown attribute with a probability distribution and taking into account the dependence between attributes. We presented the results of tests performed on several databases and we compared our results with those given by C4.5 and Lobo's approach for the same databases. We have observed that the results of our approach are better than those obtained by OAT and C4.5. To measure the quality of our classification results, we used an approach derived from Relief and its extensions to calculate the distance between two instances with missing values. For each instance in the test data, our *Analysis-Instance* algorithm gives the frequency of its nearest instances from each class. We then compared the results obtained by the *Analysis-Instance* algorithm with *PAT* and C4.5 results. We observed that our classification results are closer to the *Analysis-Instance* algorithm results and better than those given by C4.5. The complexity of constructing our attributes trees in the two approaches *(PAOTs, PATs)* and the complexity of classifying a new instance with missing values using our approach were briefly presented in this chapter. We found that the complexity of our classification algorithm is exponential according to the number of missing attributes in the test instance [10].

In the future, we aim at using a statistical test to calculate the dependence between attributes instead of Mutual Information. We are also going to test our approach on a higher dimensional space and compare our results with those of C4.5 and *OAT*. Finally, we aim at comparing our results with those given by statistical methods [2].

References

1. Breiman, L., Friedman, J.H., Olshen, R.A., Stone, C.J.: Classification and regression trees. Wadsworth International Group, CA (1984)
2. Roderick, J.A.L., Donald, B.R.: Statistical Analysis with Missing Data, 2nd edn. Wiley-Interscience, Chichester (2002)
3. Shannon, C., Weaver, W.: Théorie mathématique de la communication. Les classiques des sciences humaines (1949)
4. Witten Ian, H., Frank, E.: Data Mining: Practical Machine Learning Tools and Techniques, 2nd edn. Morgan Kaufmann, San Francisco (2005)
5. Quinlan, J.R.: C4.5: Programs for Machine Learning. Morgan Kaufmann, San Diego (1993)
6. Tan, M.S.P.N., Kumar, V.: Introduction to Data Mining. Addison-Wesley, Reading (2006)
7. Friedman, J.H., Kohavi, R., Yun, Y.: Lazy Decision Trees. In: Proc. 13th National Conference on Artificial Intelligence and the Eighth Innovative Applications of Artificial Intelligence Conference, pp. 717–724. AAAI press, Menlo Park (1996)
8. Hawarah, L., Simonet, A., Simonet, M.: A probabilistic approach to classify incomplete objects using decision trees. In: Galindo, F., Takizawa, M., Traunmüller, R. (eds.) DEXA 2004. LNCS, vol. 3180, pp. 549–558. Springer, Heidelberg (2004)
9. Hawarah, L., Simonet, A., Simonet, M.: Evaluation of a probabilistic approach to classify incomplete objects using decision trees. In: Bressan, S., Küng, J., Wagner, R. (eds.) DEXA 2006. LNCS, vol. 4080, pp. 193–202. Springer, Heidelberg (2006a)
10. Hawarah, L., Simonet, A., Simonet, M.: The complexity of a probabilistic approach to deal with missing values in a decision tree. In: 8th International Symposium on Symbolic and Numeric Algorithms for Scientific Computing (SYNASC 2006), Romania, pp. 26–29 (September 2006b)
11. Hawarah, L., Simonet, A., Simonet, M.: Dealing with Missing Values in a Probabilistic Decision Tree during Classification. In: The Sixth IEEE International Conference on Data Mining-Workshops (ICDM Workshops 2006), Hong Kong, China, December 18-22 (2006c)
12. Kira, K., Rendell, L.A.: A practical approach to feature selection. In: ML 1992: Proceedings of the ninth international workshop on Machine learning, San Francisco, CA, USA, pp. 249–256 (1992)
13. Kononenko, I., Bratko, I., Roskar, E.: Experiments in Automatic Learning of Medical Diagnostic Rules. Technical Report, Jozef Stefan Institute, Ljubljana, Yugoslavia (1984)
14. Kononenko, I.: Estimating attributes: Analysis and extensions of relief. In: ECML: European Conference on Machine Learning, pp. 171–182. Springer, Heidelberg (1994)
15. Liu, W.Z., White, A.P., Thompson, S.G., Bramer, M.A.: Techniques for Dealing with Missing Values in Classification. In: Liu, X., Cohen, P., Berthold, M. (eds.) Advances on Intelligent Data Analysis. Springer, Heidelberg (1997)

16. Lobo, O., Numao, M.: Ordered estimation of missing values. In: PAKDD 1999: Proceedings of the Third Pacific Asia Conference on Methodologies for Knowledge Discovery and Data Mining, pp. 499–503. Springer, London (1999)
17. Lobo, O., Numao, M.: Ordered estimation of missing values for propositional learning. The Japanese Society for Artificial Intelligence 1, 162–168 (2000)
18. Lobo, O., Numao, M.: Suitable domains for using ordered attribute trees to impute missing values. IEICE TRANS INF. and SYST, E84-D, no. 2 (February 2001)
19. Martin, J.K., Hirschberg, D.S.: The time complexity of decision tree induction. Technical Report. ICS-TR-95-27 (1995)
20. Newman, D., Hettich, S., Blake, C., Merz, C.: UCI Repository of machine learning databases (1998), http://www.ics.uci.edu/~mlearn/MLRepository.html
21. Quinlan, J.R.: Induction of decision trees. Machine Learning 1, 81–106 (1986)
22. Quinlan, J.R.: Unknown attribute values in induction. In: Proc. Sixth International Machine Learning Workshop. Morgan Kaufmann, San Francisco (1989)
23. Quinlan, J.R.: Probabilistic decision trees. Machine Learning: an Artificial Intelligence Approach 3, 140–152 (1990)
24. Robnik-Sikonja, M., Kononenko, I.: Attribute dependencies, understandability and split selection in tree based models. In: Machine Learning: Proceedings of the Sixteenth International Conference. ICML 1999, pp. 344–353 (1999)
25. Robnik-Sikonja, M., Kononenko, I.: Theoretical and empirical analysis of relieff and rrelieff. Mach. Learn. 53(1-2), 23–69 (2003)

5

Kernel-Based Algorithms and Visualization for Interval Data Mining

Thanh-Nghi Do[1,2] and François Poulet[3]

[1] CIT, CanTho University, VietNam
    dtnghi@cit.ctu.edu.vn
[2] INRIA Futurs/LRI, Université de Paris-Sud, Orsay, France
    Thanh-Nghi.Do@lri.fr
[3] IRISA, Rennes, France
    Francois.Poulet@irisa.fr

Abstract. Our investigation aims at extending kernel methods to interval data mining and using graphical methods to explain the obtained results. Interval data type can be an interesting way to aggregate large datasets into smaller ones or to represent data with uncertainty. No algorithmic changes are required from the usual case of continuous data other than the modification of the Radial Basis Kernel Function evaluation. Thus, kernel-based algorithms can deal easily with interval data. The numerical test results with real and artificial datasets show that the proposed methods have given promising performance. We also use interactive graphical decision tree algorithms and visualization techniques to give an insight into support vector machines results. The user has a better understanding of the models' behavior.

5.1 Introduction

In recent years, real-world databases have increased rapidly [1], so that the need to extract knowledge from very large databases is increasing. Data mining [2] can be defined as the particular pattern recognition task in the knowledge discovery in databases process. It uses different algorithms for classification, regression, clustering or association rules. The support vector machines algorithms (SVM) proposed by Vapnik [3] are a well-known class of algorithms using the idea of kernel substitution. They have shown practical relevance for classification, regression and novelty detection tasks. The successful applications of SVM and other kernel-based methods have been reported for various fields like facial recognition, text categorization, bioinformatics, etc. [4].

  While SVM and kernel-based methods are a powerful paradigm, they have some difficulty to deal with the challenge of large datasets. The learning task is accomplished through the resolution of quadratic problem. Therefore, the computational cost of a SVM approach is at least equal to the squared number of training data points and the memory requirement makes them intractable with very large datasets. We propose to scale up their training tasks based on the interval data concept [5]. We summarize the massive datasets into interval data. Then, we must adapt the kernel-based algorithms, e.g. SVM to deal with

D.A. Zighed et al. (Eds.): Mining Complex Data, SCI 165, pp. 75–91.
springerlink.com   © Springer-Verlag Berlin Heidelberg 2009

this new data type. We construct a new radial basis kernel function (RBF) for interval data used for classification, regression and novelty detection tasks. The numerical test results are obtained on real and artificial datasets.

Furthermore, for many applications, sampling data corrupted with noise makes the input data uncertain. The interval data concept can also represent this uncertainty. So the kernel-based methods and SVM can also deal with uncertain data.

Although SVM gives high quality results, the interpretation of these results is not so easy. The support vectors found by the algorithms provide limited information. Most of the time, the user only obtains information regarding support vectors and accuracy. He cannot explain or understand why a model constructed by SVM makes a good prediction. Understanding the model obtained by the algorithm is as important as the accuracy because the user has a good comprehension of the knowledge discovered and more confidence in this knowledge [6], [7]. Our investigation aims at using visualization methods to try to explain the SVM results. We use interactive graphical decision tree algorithms and visualization methods [6], [8] to give an insight into classification, regression and novelty detection tasks with SVM. We illustrate how to combine some strengths of different visualization methods to help the user to improve the comprehensibility of SVM results.

This paper is organized as follows. In section 2, we present a new Gaussian RBF kernel construction to deal with interval data. In section 3, we briefly introduce classification, regression and novelty detection for interval data with SVM algorithms and other kernel-based methods. Section 4 presents a way to explain SVM results by using interactive decision tree algorithms. We propose to use an approach based on different visualization methods to try to interpret SVM results in section 5 before the conclusion and future work.

5.2 Non Linear Kernel Function for Interval Data

SVM and kernel-based methods are a powerful paradigm and have shown practical relevance for classification and regression, but the learning task is not easy to perform with the challenge of large datasets. We propose to scale up their training tasks based on the interval data concept. Large datasets are aggregated into smaller data sizes, we need to use more complex data type, e.g. interval type instead of standard ones.

The simplest way depicted in figure 5.1 is to summarize large datasets into high-level data type, e.g. clusters using clustering algorithm (e.g. k-means [9]). We can use the interval data concept to represent the clusters where an interval vector corresponds to a cluster, the low and high values of an interval are computed by low and high bound of data points inside this cluster. Then, we need to construct non-linear kernel function for dealing with interval datasets.

We are interested in RBF kernel function because it is general and efficient [10]. Assume we have two data points x and $y \in R^n$. The RBF kernel formula

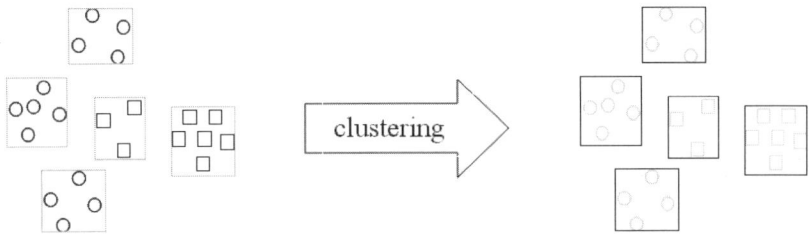

Fig. 5.1. Aggregation of a large dataset

(kernel width σ) in (1) of two data vectors x and y of continuous type, $K(x,y)$ is based on the Euclidean distance between these vectors, $d_E(x, y) = \parallel x - y \parallel$.

$$K\langle x, y \rangle = \exp\left(-\frac{\parallel x - y \parallel^2}{2\sigma^2}\right) \tag{5.1}$$

For dealing with interval data, we only need to measure the distance between two vectors of interval type and then we substitute this distance measure for the Euclidean distance into the RBF kernel formula (1). Thus the new RBF kernel can deal with interval data. We propose to use the Hausdorff (1868-1942) distance to measure the dissimilarity between two data vectors of interval type.

Suppose we have two intervals represented by low and high values: $I_1 = [low_1, high_1]$ and $I_2 = [low_2, high_2]$, the Hausdorff distance between two intervals I_1 and I_2 is defined by (2):

$$d_H(I_1, I_2) = \max\left(|low_1 - low_2|, |high_1 - high_2|\right) \tag{5.2}$$

Let us consider two data vectors u, $v \in \Omega$ having n dimensions of interval type:

$u = ([u_{1,low}, u_{1,high}], [u_{2,low}, u_{2,high}], \ldots, [u_{n,low}, u_{n,high}])$
$v = ([v_{1,low}, v_{1,high}], [v_{2,low}, v_{2,high}], \ldots, [v_{n,low}, v_{n,high}])$

The Hausdorff distance between two vectors u and v is defined by (3):

$$d_H(u, v) = \sqrt{\sum_{i=1}^{n} \max\left(|u_{i,low} - v_{i,low}|^2, |u_{i,high} - v_{i,high}|^2\right)} \tag{5.3}$$

By substituting the Hausdorff distance measure d_H into RBF kernel formula, we obtain a new RBF kernel for dealing with interval data. This modification tremendously changes kernel algorithms for mining interval data. No algorithmic changes are required from the usual case of continuous data other than the modification of the RBF kernel evaluation. All the benefits of the original kernel methods are kept. Kernel-based learning algorithms including Support Vector

Machines [3], Kernel Fisher's Discriminant Analysis (KFDA) [11], Kernel Principal Component Analysis (KPCA) [12] and Kernel Partial Least Squares (KPLS) [13] can use the RBF function to build interval data models in classification, regression and novelty detection.

5.3 Interval Data Mining with Kernel-Based Methods

5.3.1 Support Vector Classification (SVC)

Let us consider a binary linear classification task depicted in figure 5.2 with m data points in a *n-dimensional* input space $x_1, x_2, \ldots, x_m$ having corresponding labels $y_i = \pm 1$. SVM classification algorithm aims to find the best separating surface as being furthest from both classes. It is simultaneously to maximize the margin between the support planes for each class and minimize the errors. This can be accomplished through the quadratic program (4).

$$\min (1/2) \sum_{i=1}^{m} \sum_{j=1}^{m} y_i y_j \alpha_i \alpha_j K \langle x_i, x_j \rangle - \sum_{i=1}^{m} \alpha_i$$

$$s.t. \sum_{i=1}^{m} y_i \alpha_i = 0 \tag{5.4}$$

$$C \geq \alpha_i \geq 0 \; (i = 1, \ldots, m)$$

where C is a positive constant used to tune the margin and the errors.

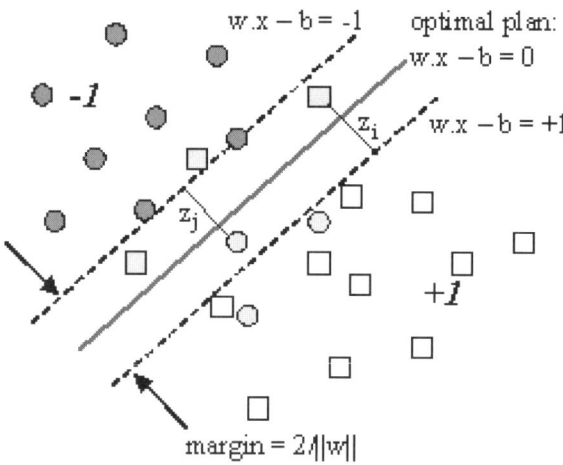

Fig. 5.2. Linear support vector classification

From the α_i obtained by the solution of (4), we can recover the separating surface and the scalar b determined by the support vectors denoted by #SV (for which $\alpha_i > 0$). By changing the kernel function K as a linear inner product, a polynomial, a radial basis function or a sigmoid neural network, we can get different classification model. The classification of a new data point x is based on:

$$f(x) = sign(\sum_{i=1}^{\sharp SV} y_i \alpha_i K \langle x, x_i \rangle - b)$$

5.3.2 Novelty Detection with Support Vector Machines (One-Class SVM)

For one-class (novelty detection), the SVM algorithm has to find an hypersphere with a minimal radius R and center c which contains most of the data and then we test if novel points lie outside the boundary of the hypersphere. One-class SVMs pursue these goals with the quadratic program (5):

$$\min \sum_{i=1}^{m} \sum_{j=1}^{m} \alpha_i \alpha_j K \langle x_i, x_j \rangle - \sum_{i=1}^{m} \alpha_i K \langle x_i, x_i \rangle$$

$$s.t. \sum_{i=1}^{m} \alpha_i = 1 \tag{5.5}$$

$$1/m\nu \geq \alpha_i \geq 0 \ (i = 1, \dots, m)$$

where a constant $\nu \in (0, 1)$ is used to tune the number of support vectors and outliers.

A new datapoint x is novel if:

$$f(x) = K(x, x) - 2 \sum_{i=1}^{\sharp SV} \alpha_i K \langle x, x_i \rangle + \sum_{i=1}^{\sharp SV} \sum_{j=1}^{\sharp SV} \alpha_i \alpha_j K \langle x_i, x_j \rangle - R^2 \geq 0$$

where R^2 is computed by a training datapoint which is non bound and setting decision function $f(x)$ to zero.

5.3.3 Support Vector Regression (SVR)

SVM can also be applied to regression problem by the introduction of an alternative loss function. By using an $\epsilon - insensitive$ loss function proposed by Vapnik, Support vector regression (SVR) aims to find a predictive function $f(x)$ that has at most ϵ deviation from the actual value y_i. This task is also accomplished through the quadratic program (6):

$$\min (1/2) \sum_{i=1}^{m} \sum_{j=1}^{m} (\alpha_i - \alpha_i^*)(\alpha_j - \alpha_j^*) K \langle x_i, x_j \rangle - \sum_{i=1}^{m} (\alpha_i - \alpha_i^*) y_i + \epsilon \sum_{i=1}^{m} (\alpha_i + \alpha_i^*)$$

$$s.t. \sum_{i=1}^{m}(\alpha_i - \alpha_i^*) = 0 \qquad (5.6)$$

$$C \geq \alpha_i, \alpha_i^* \geq 0 \ (i = 1, \ldots, m)$$

The solution of (6) gives α_i, α_i^*. Thus the regression function is given by:

$$f(x) = \sum_{i=1}^{\sharp SV}(\alpha_i - \alpha_i^*)K\langle x, x_i \rangle - b$$

where the scalar b is determined by the support vectors.

A survey [14] and the book [15] provide more details about SVM and others kernel-based learning methods.

These SVMs only deal with continuous data. To deal with interval data no algorithmic changes are required from the usual case of continuous data other than the substitution of the RBF kernel function for interval data described in section 2 into the classical SVM algorithms including SVC, One-class SVM, SVR. All the benefits of the classical SVMs are kept. Thus they can be used to deal with interval data.

For the evaluation of our proposed approach, we have added the new non-linear kernel for interval data to the publicly available toolkit, LibSVM [16]. The software program is able to deal with interval data in classification, regression and novelty detection tasks. To apply the SVM algorithms to the multi-class classification problem (more than 2 classes), LibSVM uses one-against-one strategy. Assume that we have k classes, LibSVM construct $k*(k-1)/2$ models: a model separates i^{th} class against j^{th} class. Then to predict the class for a new data point, LibSVM just predicts with each model and finds out which one separates the furthest into the positive region. We have used datasets from Statlog [17], the UCI machine learning repository [18], regression datasets [19] and Delve [20]. By using k-means algorithm [9], the large datasets are aggregated into smaller ones. A data point in interval datasets corresponds to a cluster, the low and high values of an interval are computed by the cluster data points. Some other methods for creating interval data can be found in [5]. Furthermore, we generated uncertain data set for evaluating our algorithm. This dataset called Ringnoise is $4 - dimensional$ with 2 classes where class 1 is multivariate normal with mean 0 and covariance 4 times the identity matrix and class 2 has unit covariance and mean $(0.5, 0.5, 0.5, 0.5)$. Then Gaussian noise is added with mean $(0, 0, 0, 0)$ and covariance matrix $\sigma_i I$ where σ_i is randomly chosen from $[0.1, 0.8]$, the matrix I denotes the 4x4 identity matrix. The interval data concept can also represent this dataset uncertainty. Table 5.1 presents the dataset description and aggregations (interval data). We report the cross validation accuracy of classification results and mean squared error of regression results in table 5.2. The results of novelty detection task are presented in table 5.3 with the number of outliers (furthest from other data points in the dataset). According to our knowledge, there is no other available algorithm being able to deal with interval data in non-linear

Table 5.1. Dataset description

Dataset	Original size	Dims	Classes	Aggregation size	Evaluation
Wave	300	21	3	30	leave-1-out
Pima	768	8	2	77	leave-1-out
Shuttle	58000	9	7	594	10-fold
Segment	2310	19	7	319	10-fold
Ringnoise	50000	4	2	500	10-fold
Bank8FM	4499	8	continuous	450	10-fold

Table 5.2. SVM classification and regression results

Dataset	Accuracy %	Mean squared error
Wave	80.00	0.46
Pima	79.22	0.21
Shuttle	94.78	1.10
Segment	91.22	1.70
Ringnoise	84.20	0.19

Table 5.3. One-class SVM results

Dataset	Outliers
Shuttle	9
Bank8FM	6

classification, regression and novelty detection tasks. There is no experimental result with interval data mining provided by other algorithms. Therefore, we only report results obtained by our approach, it is difficult to compare with the other ones.

5.3.4   Other Kernel-Based Methods

Many multivariate statistics algorithms based on generalized eigenproblems can also be kernelized [21], e.g. Kernel Fisher's Discriminant Analysis (KFDA), Kernel Principal Component Analysis (KPCA) or Kernel Partial Least Squares (KPLS). These kernel-based methods can also use the RBF kernel function described in section 2 to deal with interval data. We use KPCA and KFDA to visualize datasets in the embedding space where the user can intuitively see the separating boundary between the classes based on human pattern recognition capabilities.

The eigenvectors of the data can be used to detect directions of maximum variance, and thus, linear PCA is to project data onto principal components by solving an eigenproblem. By using a kernel function instead of the linear inner

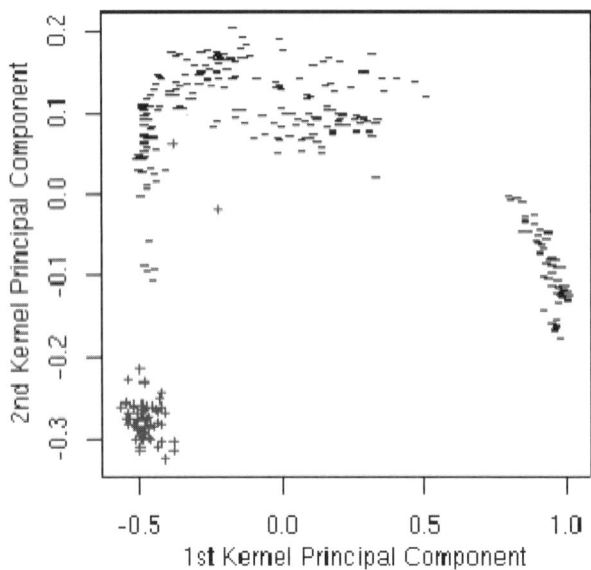

Fig. 5.3. Visualization of Kernel PCA with the Segment dataset

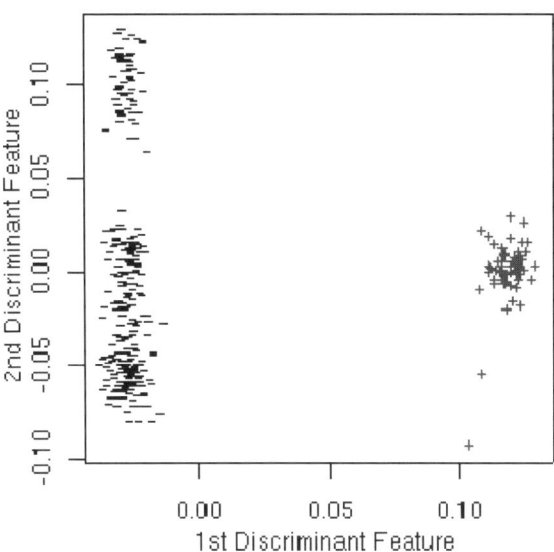

Fig. 5.4. Visualization of Kernel FDA with the Segment dataset

product in the formula, we obtain non-linear PCA (KPCA). An example of the visualization of the Segment interval dataset (class 7 against all) with KPCA using the RBF kernel function is shown in figure 5.3.

In linear FDA, we consider projecting all the multidimensional data onto a generic direction w, and then separately observing the mean and the variance of the projections of the two classes. By substituting the kernel function for a linear inner product into the linear FDA formula, we have non-linear FDA (KFDA). An example of the visualization of the Segment interval dataset (class 7 against all) with KFDA using the RBF kernel function is shown in figure 5.4.

These kernel-based methods are also extended for learning model in which the input data is corrupted with noise, e.g. sampling data corrupted with noise makes the input data uncertain. The interval data concept can also represent them. So the kernel-based methods and SVM using the RBF kernel function in section 2 can also deal with uncertain data.

5.4 Inductive Rules Extraction for Explaining SVM Results

Although SVM algorithms have shown to build accurate models, their results may be very difficult to understand. Most of the time, the user only obtains information regarding the support vectors being used as "black box" to classify the data with a good accuracy. The user does not know how SVM models can work. For many data mining applications, understanding the model obtained by the algorithm is as important as the accuracy even if up to now very few methods have been proposed [6], [8] and [22].

We propose here to use interactive decision tree algorithms [23], [24] for trying to explain the SVM results. The SVM performance in classification task is deeply understood by the way of IF-THEN rules extracted intuitively from the graphical representation of the decision trees that can be easily interpreted by humans.

Figure 5.5 is an example of the inductive rule extraction explaining support vector classification results with the Segment interval dataset. The SVM algorithm using the RBF kernel function classifies the class 7 (considered as +1 class) against all other classes (considered as -1 class) with 100.00 % accuracy. CIAD uses 2D scatter-plot matrices [25] for visualizing interval data: the data points are displayed in all possible pair-wise combinations of dimensions in 2D scatter-plot matrices. For $n - dimensional$ data, this method visualizes $n(n - 1)/2$ matrices. A data point in two interval dimensions is represented by a cross and color corresponds to the class.

The user interactively chooses the best separating split (parallel to an axis) to interactively construct the decision tree (based on human pattern recognition capabilities) or with the help of automatic algorithms. The obtained decision tree having 4 leaves (corresponding to 4 rules) can explain the SVM model. One rule is created for each path from the root to a leaf, each dimension value along a path is added in a conjunction and the leaf node holds the class prediction.

And thus, the non-linear SVM is interpreted in the way of the following four inductive rules (IF-THEN) that will be easy to understand:

IF (A19 > 0.617975) THEN class=7

IF (A19 ≤ 0.617975) AND (A2 ≤ 0.634350) THEN class≠7

IF (A19 ≤ 0.617975) AND (A2 > 0.634350) AND (A10 ≤ 0.163654) THEN class=7

IF (A19 ≤ 0.617975) AND (A2 > 0.634350) AND (A10 > 0.163654) THEN class≠7

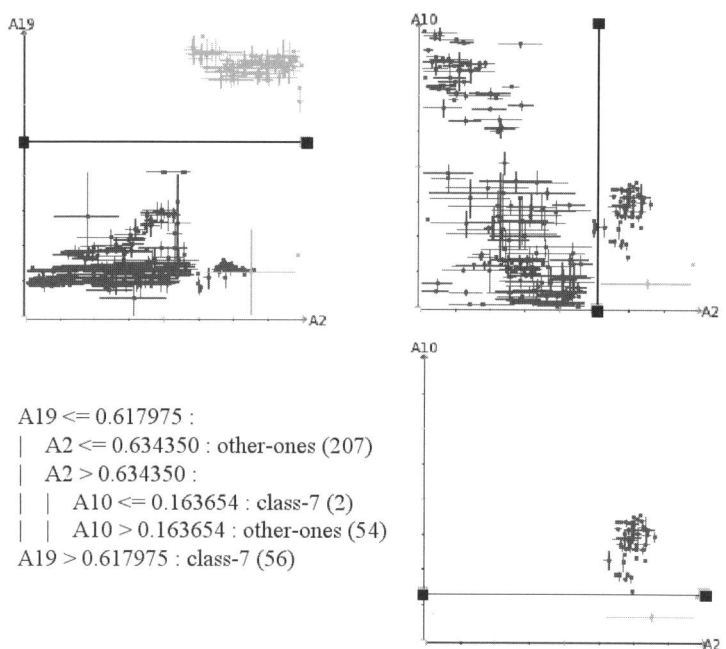

A19 <= 0.617975 :
| A2 <= 0.634350 : other-ones (207)
| A2 > 0.634350 :
| | A10 <= 0.163654 : class-7 (2)
| | A10 > 0.163654 : other-ones (54)
A19 > 0.617975 : class-7 (56)

Fig. 5.5. Visualization of the decision tree explaining the SVM result with the Segment dataset

5.5   Visualization Tools for Explaining SVM Results

We have studied some ways to try to explain SVM results by using graphical representation of high dimensional data. The information visualization methods guide the user towards the most appropriate visualizations for viewing mining results (post-processing step). There are many possibilities to visualize data by using different visualization methods, but all of them have strengths and weaknesses. We use the linking technique to combine different visualization methods to overcome the single one. The same information is displayed in different views with different visualization techniques providing useful information to the user.

The interactive brushing technique allows the user to focus on an area (brush) in the displayed data to highlight groups of data points. And thus, the linked multiple views provide more information than the single one. We use the interactive brushing and linking techniques and different visualization methods to try to explain SVM results.

5.5.1 Support Vector Classification Results

For classification tasks with SVM algorithms, understanding the margin (furthest distance between +1 class and -1 class) is one of the most important keys of the support vector classification. For this purpose, we need to display the points near the separating boundary between the two classes. To achieve this goal, we propose to use the data distribution according to the distance from the separating surface. While the classification task is processed (based on the support vectors), we also compute the data distribution according to the distance from the separating surface. For each class, the positive distribution is the set of correctly classified data points and the negative distribution is the set of misclassified data points. The data points being near the frontier correspond to the bar charts near the origin. When the bar charts corresponding to the points near the frontier are selected, the data points are also selected in the other views (visualization methods) by using the brushing and linking technique. We use 2D scatter-plot matrices for visualizing interval data. The user can see approximately the boundary between classes and the margin width. This helps the user to evaluate the robustness of the model obtained by support vector classification. He can also know the interesting dimensions (corresponding to the projections providing a clear boundary between the two classes) in the obtained model.

Figure 5.6 is an example of visualizing support vector classification results with the Segment interval dataset (class 7 against all). From data distribution according to the distance from the separating surface, the four bar charts near the origin are brushed, and then the corresponding points are linked and displayed in 2D scatter-plot matrices. From the upper part of figure 5.6, we can conclude there is a clear boundary between the two classes (there is no misclassified data point), and from the lower part, we can see that dimensions 2 and 16 showing a clear boundary between the two classes are interesting in the obtained model.

5.5.2 Support Vector Regression Results

We have extended this idea for visualizing support vector regression results. We have also computed the data distribution according to the distance from the regression function. Then we combine the histogram with 2D scatter-plot matrices for visualization. When the user selects the data points far from the regression function, he can know how the function fits data. If the function well predicts the data points in high-density region then the obtained model is interesting.

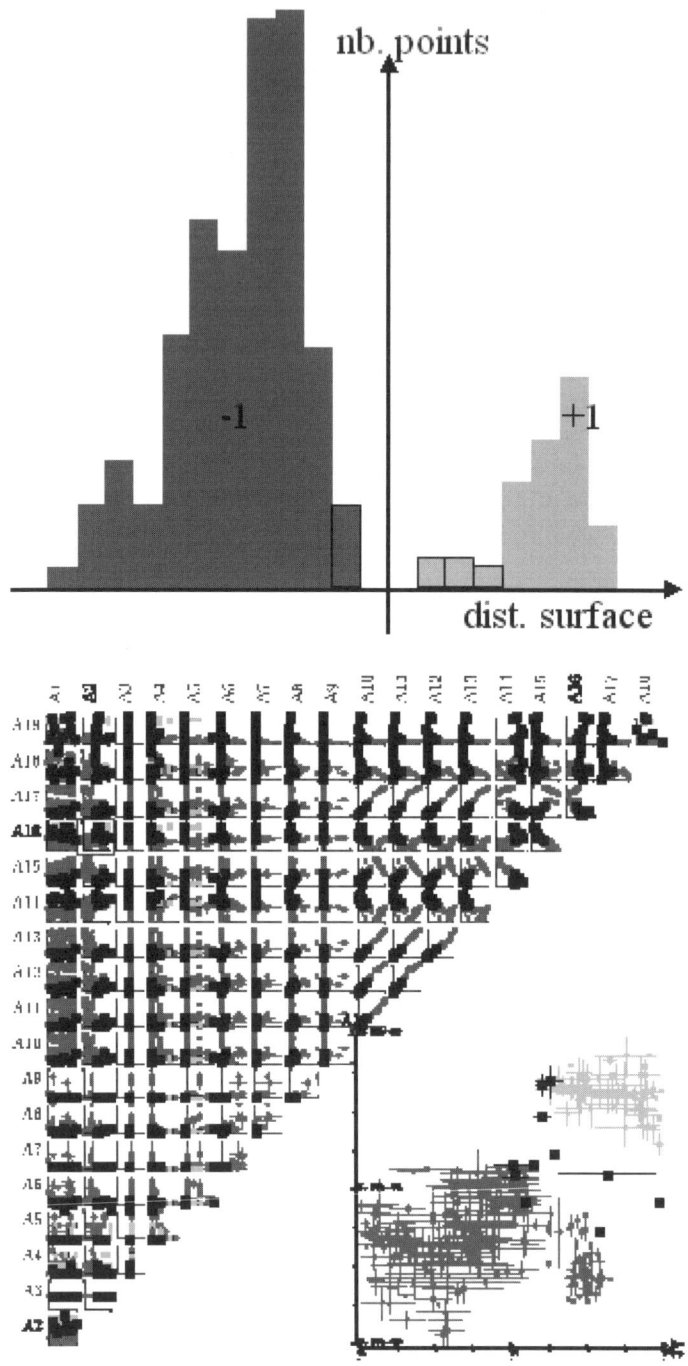

Fig. 5.6. Visualization of the classification result with the Segment dataset

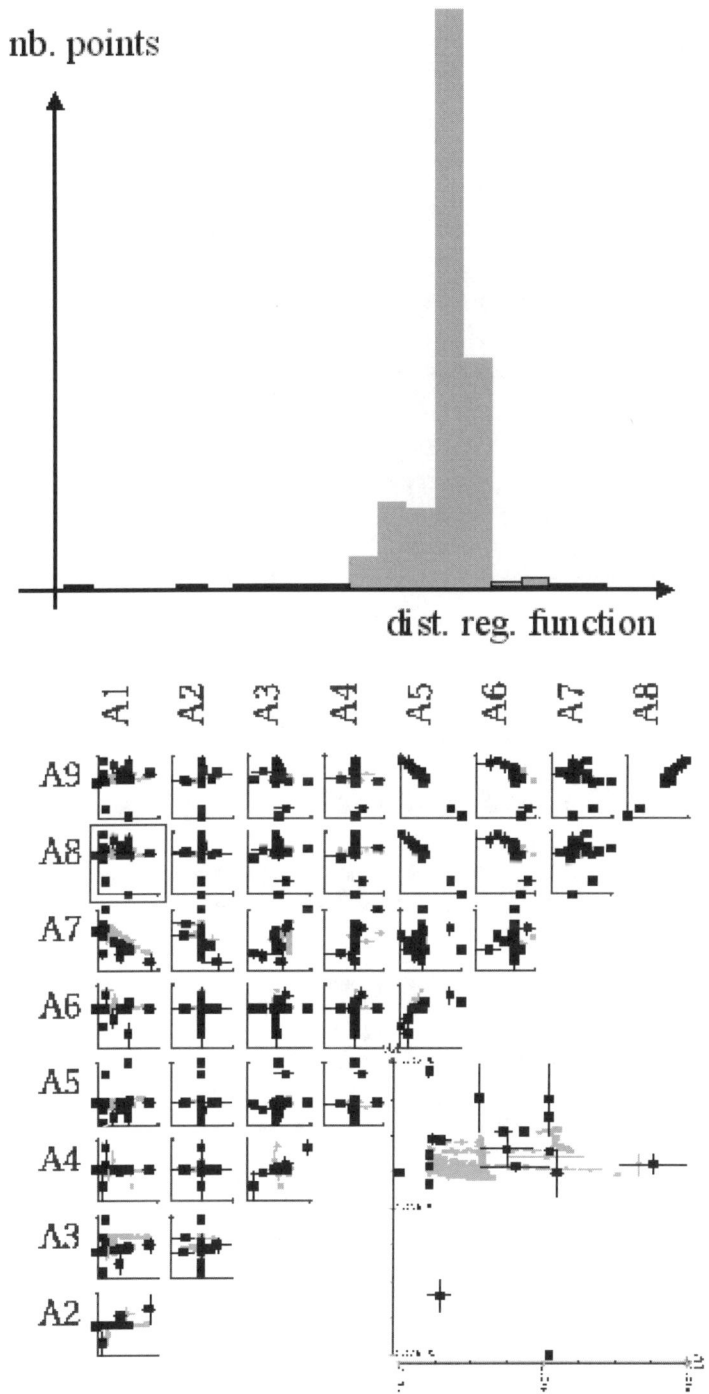

Fig. 5.7. Visualization of the regression result with the Shuttle dataset

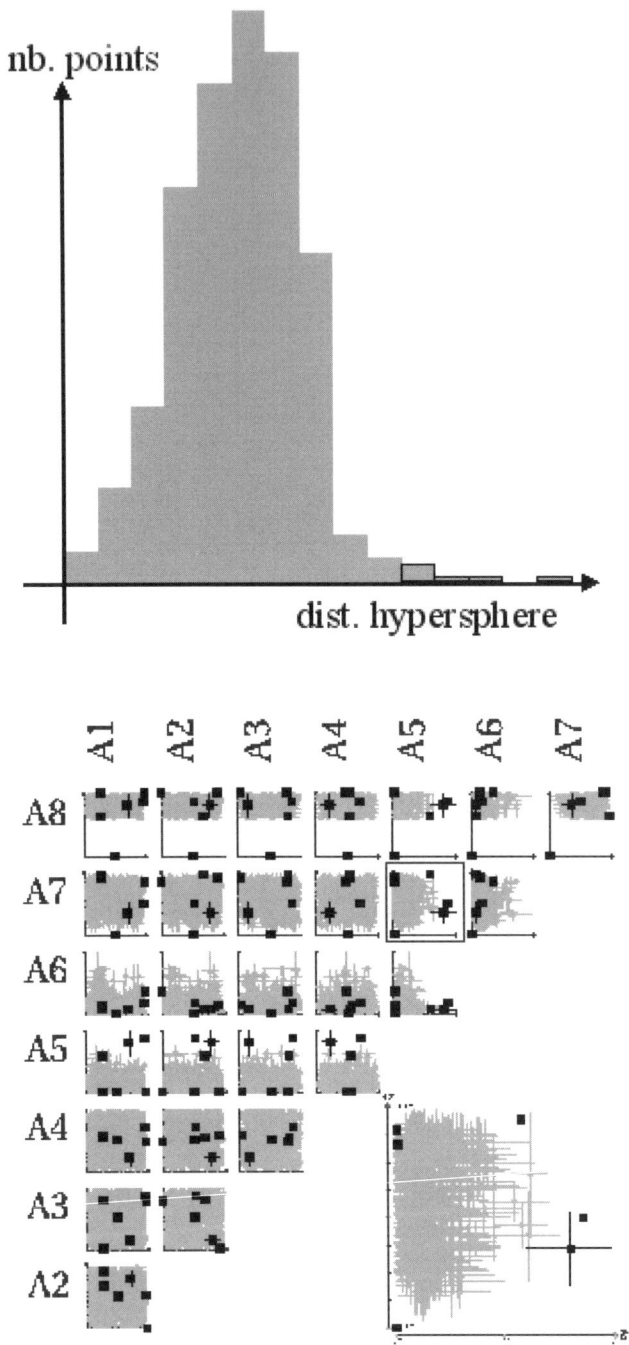

Fig. 5.8. Visualization of the one-class SVM result with the Bank8FM dataset

For example, we visualize regression result with the Shuttle interval dataset in figure 5.7. The data points far from the regression function are outliers. Thus, the model has high quality. The dimensions 1 and 8 are interesting in the obtained model.

5.5.3 Novelty Detection Results

With a novelty detection task, we visualize the outliers allowing the user to validate them. The approach is based on the interactive linking and brushing of the histogram and 2D scatter-plot views. The histogram displays the data distribution according to the distance from the hyper-sphere obtained by one class SVM. The data points far from the hyper-sphere are brushed in the histogram view, thus they are automatically selected in 2D scatter-plot view. The user can interpret and validate the outliers. And then, the dimensions corresponding to the projection present clearly the outliers and are interesting in the obtained model.

Figure 5.8 is a visualization of one class SVM result on Bank8FM. The user can verify the outliers obtained by the novelty detection with SVM. And for example, the dimensions 5 and 7 corresponding to the projection present clearly some outliers and are interesting in the obtained model.

5.6 Conclusion and Future Works

We have presented an interval data mining approach using kernel-based and visualization methods. Our investigation aims at scaling up kernel-based algorithms to mine very large datasets and data corrupted with noise. The approaches are based on the interval data concept. The massive datasets or the uncertain data is represented by the interval data concept. Thus, we have proposed to construct a new RBF kernel for interval data. This modification tremendously changes kernel-based algorithms. No algorithmic changes are required from the usual case of continuous data other than the modification of the RBF kernel evaluation. The kernel-based algorithms can deal with interval data in classification, regression and novelty detection. It is extremely rare to find algorithms being able to construct non-linear models on interval data for the three problems: classification, regression and novelty detection.

We have also proposed two ways to try to explain SVM results that are well-known "black boxes". The first one is to use interactive decision tree algorithms to explain SVM results. The user can interpret the SVM performance in the way of IF-THEN rules extracted intuitively from the graphical representation of the decision trees that can be easily interpreted by the user. The second one is based on a set of different visualization techniques combined with linking and brushing techniques giving an insight into classification, regression and novelty detection tasks with SVM. The graphical representation shows the interesting dimensions in the obtained model.

For dealing with histogram data type, [26] proposed to represent each histogram individual with k bins by a succession of k interval individuals (the first one included in the second one, the second one included in the third one and so on). Once the histogram data have been represented by interval data, our proposed approaches can naturally deal with histogram data. A forthcoming improvement will be to extend our approaches to taxonomical or mixed data types. Another one will be to use high-level representative data for tuning the SVM parameters. This approach drastically reduces the cost compared with the research in initial large datasets.

References

1. Fayyad, U., Piatetsky-Shapiro, G., Uthurusamy, R.: Summary from the kdd-03 panel - data mining: The next 10 years. SIGKDD Explorations 5(2), 191–196 (2004)
2. Fayyad, U., Piatetsky-Shapiro, G., Smyth, P.: From data mining to knowledge discovery in databases. AI Magazine 17(3), 37–54 (1996)
3. Vapnik, V.: The Nature of Statistical Learning Theory. Springer, Heidelberg (1995)
4. Guyon, I.: Svm application list (1999),
   http://www.clopinet.com/isabelle/Projects/SVM/applist.html
5. Bock, H., Diday, E.: Analysis of Symbolic Data. Springer, Heidelberg (2000)
6. Do, T., Poulet, F.: Enhancing svm with visualization. In: Int. Conf. on Discovery Science, pp. 183–194 (2004)
7. Do, T., Poulet, F.: Very large datasets with svm and visualization. In: Int. Conf. on Entreprise Information Systems, pp. 127–134 (2005)
8. Poulet, F.: Svm and graphical algorithms: a cooperative approach. In: Proceedings of IEEE International Conference on Data Mining, pp. 499–502 (2004)
9. MacQueen, J.: Some methods for classification and analysis of multivariate observations. In: Berkeley Symposium on Mathematical Statistics and Probability, vol. (1), pp. 281–297. University of California Press (1967)
10. Lin, C.: A practical guide to support vector classification (2003)
11. Mika, S., Rätsch, G., Weston, J., Schölkopf, B., Müller, K.R.: Fisher discriminant analysis with kernels. Neural Networks for Signal Processing IX, 41–48 (1999)
12. Schölkopf, B., Smola, A., Müller, K.R.: Nonlinear component analysis as a kernel eigenvalue problem. Neural Computation 10, 1299–1319 (1998)
13. Rosipal, R., Trejo, L.J.: Kernel partial least squares regression in reproducing kernel hilbert space. Journal of Machine Learning Research 2, 97–123 (2001)
14. Bennett, K., Campbell, C.: Support vector machines: Hype or hallelujah? SIGKDD Explorations 2(2), 1–13 (2000)
15. Cristianini, N., Shawe-Taylor, J.: An Introduction to Support Vector Machines and Other Kernel-based Learning Methods. Cambridge University Press, Cambridge (2000)
16. Chang, C., Lin, C.: Libsvm – a library for support vector machines (2001),
    http://www.csie.ntu.edu.tw/~cjlin/libsvm/
17. Michie, D., Spiegelhalter, D.J., Taylor, C.: Machine Learning, Neural and Statistical Classification. Ellis Horwood (1994)
18. Blake, C.L., Merz, C.J.: UCI repository of machine learning databases (1998),
    http://www.ics.uci.edu/~mlearn/MLRepository.html
19. Torgo, L.: Regression data sets (2003),
    http://www.liacc.up.pt/~ltorgo/Regression/DataSets.html
20. Delve: Data for evaluating learning in valid experiments (1996),
    http://www.cs.toronto.edu/~delve

21. Shawe-Taylor, J., Cristianini, N.: Kernel Methods for Pattern Analysis. Cambridge University Press, Cambridge (2004)
22. Caragea, D., Cook, D., Wickham, H., Honavar, V.: Visual methods for examining svm classifiers. Visual Data Mining: Theory, Techniques, and Tools for Visual Analytics (2008)
23. Poulet, F.: Cooperation between automatic algorithms, interactive algorithms and visualization tools for visual data mining. In: Proceedings of VDM@ECLM/PKDD 2002, 2nd International Workshop on Visual Data Mining, pp. 67–79 (2002)
24. Poulet, F.: Interactive decision tree construction for interval and taxonomical data. In: Proceedings of VDM@ICDM, 3nd Workshop on Visual Data Mining, pp. 183–194 (2003)
25. Carr, D.B., Littlefield, R.J., Nicholson, W.L.: Scatterplot matrix techniques for large n. Journal of the American Statistical Association 82(398), 424–436 (1987)
26. Rodriguez, O., Diday, E., Winsberg, S.: Generalization of the principal components analysis to histogram data. In: Proceeding of PKDD Workshop on Symbolic Data Analysis: Theory, Software and Applications for Knowledge Mining (2000)

Part II

Rules Extraction

6

Evaluating Learning Algorithms Composed by a Constructive Meta-learning Scheme for a Rule Evaluation Support Method

Hidenao Abe[1], Shusaku Tsumoto[1], Miho Ohsaki[2], and Takahira Yamaguchi[3]

[1] Department of Medical Informatics, Shimane University, School of Medicine
   abe@med.shimane-u.ac.jp, tsumoto@computer.org
[2] Faculty of Engineering, Doshisha University
   mohsaki@mail.doshisha.ac.jp
[3] Faculty of Science and Technology, Keio University
   yamaguti@ae.keio.ac.jp

Abstract. The Post-processing of mined patterns such as rules, trees and so forth is one of the key operations in a data mining process. However, it is difficult for human experts to completely evaluate several thousand rules from a large dataset with noise. To reduce the cost of this kind of rule evaluation task, we have developed a rule evaluation support method with rule evaluation models that learn from objective indices for mined classification rules and evaluations by a human expert for each rule.

In this paper, we present an evaluation of the learning algorithms of our rule evaluation support method for the post-processing of mined patterns with rule evaluation models based on objective indices. To enhance the adaptability of these rule evaluation models, we introduced a constructive meta-learning system for the construction of appropriate learning algorithms. We then have performed case studies using the meningitis as an actual problem. Furthermore, we evaluated our method with the eight rule sets obtained from eight UCI datasets. With regard to these results, we show the applicability of the constrictive meta-learning scheme as a learning algorithm selection method for our rule evaluation support method.

6.1 Introduction

In recent years, enormous amounts of data have been stored on information systems in natural science, social science, and business domains. People have been able to obtain valuable knowledge due to the development of information technology. In addition, data mining techniques combine different kinds of technologies such as database technologies, statistical methods, and machine learning methods. Data mining has become well known as a method for utilizing the data stored on database systems. In particular, if-then rules, which are produced by rule induction algorithms, are considered to be a highly usable and readable output of data mining. However, for large datasets with hundreds of attributes including noise, the process often obtains many thousands of rules. From such a large rule set, it is difficult for human experts to find out valuable knowledge, which is rarely included in the rule set.

D.A. Zighed et al. (Eds.): Mining Complex Data, SCI 165, pp. 95–111.
springerlink.com   © Springer-Verlag Berlin Heidelberg 2009

To support such a rule selection, much effort has gone into the use of objective rule evaluation indices such as recall, precision, and other interestingness measurements (hereafter, we refer to these indices as "objective indices"). Further, it is difficult to estimate the criterion of a human expert using a single objective rule evaluation index; this is because his/her subjective criterion, such as the interestingness or importance for his/her purpose, is influenced by the amount of his/her knowledge and/or the passage of time. In addition, rule selection methods have never explicitly re-used the history of each rule evaluation, such as focused items and the relationships between items, which are only stored in the mind of the human expert.

With regard to the above-mentioned issues, we have developed an adaptive rule evaluation support method for human experts that use rule evaluation models. This method predicts the experts' criteria based on objective indices by re-using the results of the evaluations by human experts. In Section 6.3, we describe the rule evaluation model construction method based on objective indices. Then, in Section 6.4, we present a performance comparison of learning algorithms to obtain rule evaluation models. With the results of the comparisons, we discuss the applicability of the constructive meta-learning scheme as a learning algorithm selection method for our rule evaluation model construction approach.

6.2   Interestingness Measures and Related Work

Much effort has been expended to develop a method to select valuable rules from large mined rule set based on objective rule evaluation indices. Some of these works suggest indices for discovering interesting rules from such a large number of rules.

To avoid confusing real human interest, objective index, and subjective index, we clearly define these thems as follows: **Objective Index :** a feature, such as the correctness, uniqueness, or strength of a rule, calculated by the mathematical analysis. It does not include any human evaluation criteria. **Subjective Index:** The similarity or difference between the interestingness information given beforehand by a human expert and that obtained from a rule. Although it includes some human criteria in its initial state, the similarity or difference is mainly calculated by mathematical analysis. **Real Human Interest:** The interest felt by a human expert based on a rule in his/her mind.

Focusing on interesting rule selection with objective indices, researchers have developed more than forty objective indices based on the number of instances, probability, statistical values, quantity of information, distance of rules or their attributes, and complexity of a rule. [1, 2, 3]. Most of these indices are used to remove meaningless rules rather than to discover interesting ones for a human expert, because they cannot include domain knowledge.

Ohsaki et al.[4] investigated the relation between objective indices and real human interests, through a consideration of real data mining results and their human evaluations. In this work, a comparison showed that it was difficult to exactly predict real human interest with a single objective index. However, their work never showed any concrete method to predict human evaluations with these objective indices.

6.3 Rule Evaluation Support with Rule Evaluation Model Based on Objective Indices

In practical data mining situations, a human expert repeatedly performs costly rule evaluation procedures. In these situations, the useful experiences gained from each evaluation, such as focused attributes, interesting combinations, and valuable facts, is not explicitly used by any rule selection system, but is tacitly stored in the mind of the human expert. For these problems, we have suggested a method using rule evaluation models based on objective rule evaluation indices as a way to explicitly describe the criteria of a human expert, thus re-using the human evaluations.

6.3.1 Constructing a Rule Evaluation Model

We considered the process of modeling the rule evaluations of human experts as the process to clarify the relationships between the human evaluations and the features of input if-then rules. Based on this consideration, we decided that the rule evaluation model construction process could be implemented as a learning task. Fig. 6.1 shows this rule evaluation support method based on the re-use of human evaluations and objective indices for each mined rule as a rule evaluation model.

This method is iteratively carried out its training phase and its prediction phase.

In the training phase, the attributes of a meta-level training data set are obtained by objective indices such as recall, precision and other rule evaluation values. The human evaluations for each rule are combined as classes of each instance. To obtain this data set, a human expert has to evaluate a part or

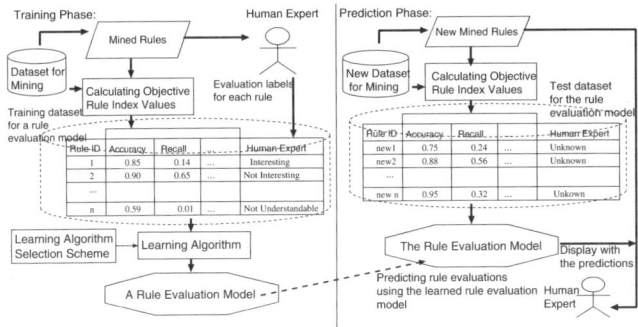

Fig. 6.1. Overview of the rule evaluation support method using rule evaluation models

entire of the input rules at least once. After obtaining the training data set, a rule evaluation model is obtained by using a learning algorithm.

In the prediction phase, a human expert receives predictions for the rest of a new rules based on their objective index values.

Since rule evaluation models are used for predictions, we needed to choose a learning algorithm with high accuracy similar to the current classification problems.

6.3.2   Learning Algorithm Selection with Meta-learning

To enhance a classification task, people often use meta-learning algorithms. One of the approaches integrates prepared base-level learning algorithms with a meta-strategy such as voting, selecting and meta-level learning. We call this approach "selective meta-learning". In addition, we developed another meta-learning scheme, which constructs a proper learning algorithm for a given dataset using de-composed base-level learning algorithms. This approach is called "constructive meta-learning".

In the field of meta-learning, there have been many studies on selective meta-learning algorithms. There are two approaches for selective meta-learning scheme. One includes bagging [5] and boosting [6], combining base-level classifiers from multiple training data with different distributions. In these

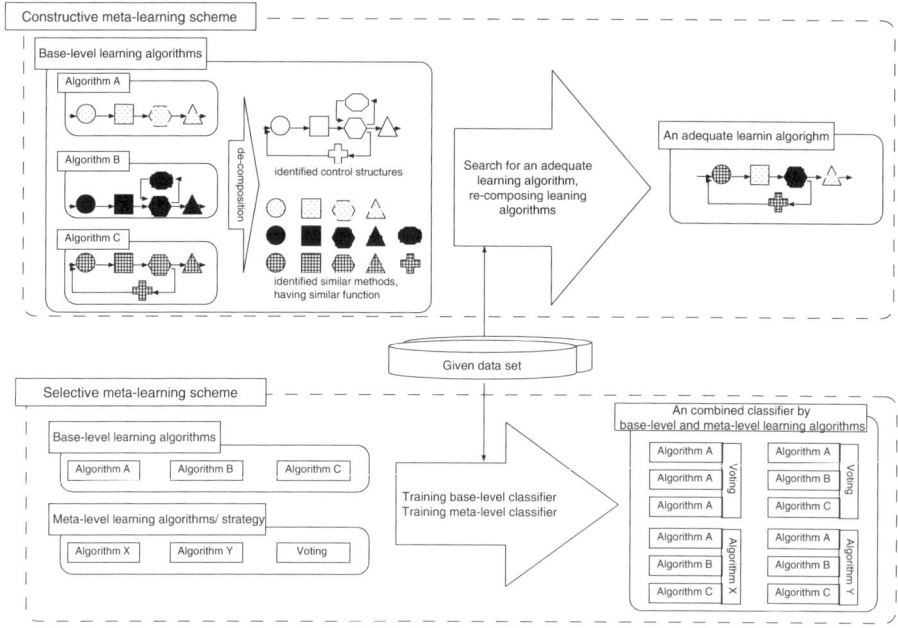

Fig. 6.2. Overview of constructive and selective meta-level processing scheme

meta-learning algorithms, just one learning algorithm is selected to learn base-level classifiers. The other approach includes voting, stacking [7] and cascading [8], which combines base-level classifiers from different learning algorithms. METAL [9] and IDA [10] are also selective meta-learning approaches, selecting a proper learning algorithm for a given data set with a heuristic score, which is called meta-knowledge.

The constructive meta-level processing scheme [11] takes a meta-learning approach, where the objective process is controlled with meta-knowledge, as shown in the upper part of Fig. 6.2. In this scheme, we construct meta-knowledge represented by method repositories. The meta-knowledge consists of information about functional parts, restrictions on the combination of these functional parts, and ways to re-construct object algorithms with the these functional parts.

6.4 Performance Comparisons of Learning Algorithms for Rule Model Construction

To more accurately predict the human evaluation labels for a new rule based on objective indices, we had to obtain a rule evaluation model with a higher predictive accuracy in our rule evaluation support method.

In this section, we first present the results of empirical evaluations of a dataset obtained from the result of meningitis data mining [12] and that of the eight rule sets from eight UCI benchmark datasets [13]. Based on the experimental results, we discuss the followings: the accuracy of rule evaluation models, the learning curves of the learning algorithms, and the contents of the learned rule evaluation models.

For evaluating the accuracy of the rule evaluation models, we compared the predictive accuracies on the entire dataset and Leave-One-Out validation. The accuracy of a validation dataset D is calculated with correctly predicted instances $Correct(D)$ as $Acc(D) = (Correct(D)/|D|) \times 100$, where $|D|$ is the size of the dataset. The recalls of class i on a validation dataset are calculated using correctly predicted instances about the class $Correct(D_i)$ as $Recall(D_i) = (Correct(D_i)/|D_i|) \times 100$, where $|D_i|$ is the size of instances of class i. Further, the precision of class i is calculated using the size of instances which are predicted i as $Precision(D_i) = (Correct(D_i)/Predicted(D_i)) \times 100$.

With regard to the learning curves, we obtained curves for the accuracies of learning algorithms on the entire training dataset to evaluate whether each learning algorithm could perform in the early stage of the rule evaluation process. The accuracies of randomly sub-sampled training datasets were averaged with 10 trials on each percentage of the subset.

By observing the elements of the rule evaluation models on the meningitis data mining results, we considered the characteristics of the objective indices, which are used in these rule evaluation models.

In order to construct a dataset to learn a rule evaluation model, the values of the objective indices were calculated for each rule by considering 39 objective indices as shown in Table 6.1. Thus, each dataset for each rule set has the same

Table 6.1. Objective rule evaluation indices for classification rules used in this research. **P:** Probability of the antecedent and/or consequent of a rule. **S:** Statistical variable based on P. **I:** Information of the antecedent and/or consequent of a rule. **N:** Number of instances included in the antecedent and/or consequent of a rule. **D:** Distance of a rule from the others based on rule attributes.

Theory	Index Name (**Abbreviation**) [Reference Number of Literature]
P	Coverage (**Coverage**), Prevalence (**Prevalence**)
	Precision (**Precision**), Recall (**Recall**)
	Support (**Support**), Specificity (**Specificity**)
	Accuracy (**Accuracy**), Lift (**Lift**)
	Leverage (**Leverage**), Added Value (**Added Value**)[2]
	Klösgen's Interestingness (**KI**)[14], Relative Risk (**RR**)[15]
	Brin's Interest (**BI**)[16], Brin's Conviction (**BC**)[16]
	Certainty Factor (**CF**)[2], Jaccard Coefficient (**Jaccard**)[2]
	F-Measure (**F-M**)[17], Odds Ratio (**OR**)[2]
	Yule's Q (**YuleQ**)[2], Yule's Y (**YuleY**)[2]
	Kappa (**Kappa**)[2], Collective Strength (**CST**)[2]
	Gray and Orlowska's Interestingness weighting Dependency (**GOI**)[18]
	Gini Gain (**Gini**)[2], Credibility (**Credibility**)[19]
S	χ^2 Measure for One Quadrant (χ^2-**M1**)[20]
	χ^2 Measure for Four Quadrant (χ^2-**M4**)[20]
I	J-Measure (**J-M**)[21], K-Measure (**K-M**)[4]
	Mutual Information (**MI**)[2]
	Yao and Liu's Interestingness 1 based on one-way support (**YLI1**)[3]
	Yao and Liu's Interestingness 2 based on two-way support (**YLI2**)[3]
	Yao and Zhong's Interestingness (**YZI**)[3]
N	Cosine Similarity (**CSI**)[2], Laplace Correction (**LC**)[2]
	ϕ Coefficient (ϕ)[2], Piatetsky-Shapiro's Interestingness (**PSI**)[22]
D	Gago and Bento's Interestingness (**GBI**)[23]
	Peculiarity (**Peculiarity**)[24]

number of instances as the rule set. Each instance has 40 attributes, including those of the class.

We applied five learning algorithms to these datasets to compare their performances as rule evaluation model learning methods. We used the following learning algorithms from Weka [25]: C4.5 decision tree learner [26] called J4.8, neural network learner with back propagation (BPNN)[27], support vector machines (SVM)[1][28], classification via linear regressions (CLR)[2][29], and OneR [30]. In addition, we also used the following selective meta-learning algorithms: Bagging [5], Boosting [6] and Stacking[3] [7].

6.4.1 Constructing Rule Evaluation Models for the Meningitis Data Mining Result

In this case study, we considered 244 rules, which had been mined from six datasets about six types of diagnostic problems as shown in Table 6.2. In these

---

[1] A polynomial kernel function was used.

[2] We set up the elimination of collinear attributes and the model selection with a greedy search based on the Akaike information metric.

[3] This stacking took the other seven learning algorithms as base-level learner and J4.8 as a meta-level learner.

Table 6.2. Description of the meningitis datasets and the results of data mining

Dataset	#Attributes	#Class	#Mined rules	#'I'	#'NI'	#'NU'
Diag	29	6	53	15	38	0
C_Course	40	12	22	3	18	1
Culture+diag	31	12	57	7	48	2
Diag2	29	2	35	8	27	0
Course	40	2	53	12	38	3
Cult_find	29	2	24	3	18	3
TOTAL	—	—	244	48	187	9

datasets, some appearances of meningitis patients were considered to be attributes and the diagnosis of each patient was considered as a class. Each rule set had been mined using appropriate rule induction algorithms composed by a constructive meta-learning system called CAMLET [12]. We labeled each rule with one of three evaluations (I: Interesting, NI: Not-Interesting, NU: Not-Understandable) based on evaluation comments provided by a medical expert.

Constructing a proper learning algorithm to construct the meningitis rule evaluation model. We developed a constructive meta-learning system called CAMLET [11] to choose an appropriate learning algorithm for a given dataset using a machine learning method repository. To implement the method repository, we first identified each functional part, called method, from the following eight learning algorithms: Version Space [31], AQ15 [32], Classifier Systems [33], Neural Network, ID3 [34], C4.5, Bagging and Boosting. With the method repository, CAMLET constructs an appropriate learning algorithm for a given dataset by searching through the possible learning algorithm specification space obtained by the method repository for the best one, using a Genetic Algorithm.

After the initial population was set up $\tau = 4$ and a number of refinement were made $N = 100$, CAMLET searched through up to 400 learning algorithms, from 6000 possible learning algorithms, for the best one. Fig.6.3 shows the algorithm constructed by CAMLET for the dataset of the meningitis data mining result.

This algorithm iterates boosting of a C4.5 decision tree for randomly split training datasets. Each classifier set generated by the C4.5 decision tree learner is reinforced with a method from Classifier Systems. Then the learned committee aggregates with weighted voting from boosting.

Comparison of the classification performances. In this section, we present the results of accuracy comparison over the entire dataset, the recall of each class label, and their precisions. Since Leave-One-Out uses just one test instance and the remainders are used repeatedly as the training dataset for each instance of a given dataset, we could evaluate the performance of a learning algorithm for a new dataset without any ambiguity.

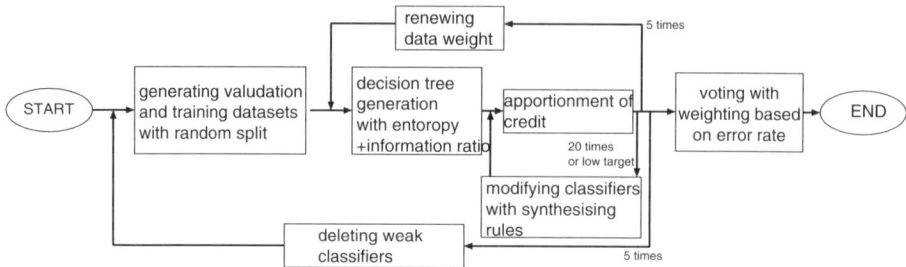

Fig. 6.3. The learning algorithm constructed by CAMLET for the dataset of the meningitis data mining result

Table 6.3. Accuracies (%), Recalls (%), and Precisions (%) of the five learning algorithms

Learning Algorithms	Acc.	Evaluation on the training dataset					
		Recall			Precision		
		I	NI	NU	I	NI	NU
CAMLET	89.4	70.8	97.9	11.1	85.0	90.2	100.0
Stacking	81.1	37.5	96.3	0.0	72.0	87.0	0.0
Boosted J4.8	99.2	97.9	99.5	100.0	97.9	99.5	100.0
Bagged J4.8	87.3	62.5	97.9	0.0	81.1	88.4	0.0
J4.8	85.7	41.7	97.9	66.7	80.0	86.3	85.7
BPNN	86.9	81.3	89.8	55.6	65.0	94.9	71.4
SVM	81.6	35.4	97.3	0.0	68.0	83.5	0.0
CLR	82.8	41.7	97.3	0.0	71.4	84.3	0.0
OneR	82.0	56.3	92.5	0.0	57.4	87.8	0.0

Learning Algorithms	Acc.	Leave—One—Out(LOO)					
		Recall			Precision		
		I	NI	NU	I	NI	NU
CAMLET	80.3	7.4	73.0	0.0	7.4	73.0	0.0
Stacking	81.1	37.5	96.3	0.0	72.0	87.0	0.0
Boosted J4.8	74.2	37.5	87.2	0.0	39.1	84.0	0.0
Bagged J4.8	77.9	31.3	93.6	0.0	50.0	81.8	0.0
J4.8	79.1	29.2	95.7	0.0	63.6	82.5	0.0
BPNN	77.5	39.6	90.9	0.0	50.0	85.9	0.0
SVM	81.6	35.4	97.3	0.0	68.0	83.5	0.0
CLR	80.3	35.4	95.7	0.0	60.7	82.9	0.0
OneR	75.8	27.1	92.0	0.0	37.1	82.3	0.0

The results of the performances of the five learning algorithms for the entire training dataset and the results of Leave-One-Out are shown in Table 6.3. All the Accuracies, Recalls of I and NI, and Precisions of I and NI are higher than those of the predicting majority labels.

As compared to the accuracy of OneR, the other learning algorithms achieve equal or higher performances using combinations of multiple objective indices, rather than by sorting with a single objective index. With regard to the Recall values for class I, BPNN achieved the highest performance. The other algorithms exhibit lower performance than OneR, because they tended to learn classification patterns for the majority class NI.

The accuracy of Leave-One-Out demonstrates the robustness of each learning algorithm. The Accuracy (%) of these learning algorithms ranged from 75.8% to 81.9%. However, these learning algorithms were not able to classify the instances of class NU, because it is difficult to predict a minor class label in this dataset.

The learning algorithm constructed by CAMLET showed the second highest accuracy for the entire training dataset, compared with the other learning algorithms. Although boosted J4.8 outperformed CAMLET on the entire training dataset, it had lower accuracy on LOO. This means boosted J4.8 had the problem of overfitting. Thus, CAMLET showed higher adaptability than the other selective meta-learning algorithms.

Learning curves of the learning algorithms. Since the rule evaluation model construction method required that the mined rules be evaluated by a human expert, we investigated the learning curves of each learning algorithm to estimate a minimum training subset to obtain a valid rule evaluation model. The upper table in Fig. 6.4 shows the accuracies of the entire training dataset for each subset of training dataset. The percentage for the achievements of each learning algorithm compared with its accuracy over the entire dataset are shown in the lower section of Fig. 6.4.

As observed in these results, SVM and CLR, which learn hype-planes, obtained achievement ratios greater than 95% using less than 10% of the training subset. Although decision tree learner, boosted J4.8, and BPNN could learn to

%training sample	10	20	30	40	50	60	70	80	90	100
CAMLET	76.7	78.4	80.8	81.6	81.7	82.6	82.8	84.8	84.6	89.3
Stacking	69.6	77.8	75.3	77.9	72.2	82.2	75.4	83.4	86.5	81.1
Boosted J4.8	74.8	77.8	79.6	82.8	83.6	85.5	86.8	88.0	89.7	99.2
Bagged J4.8	77.5	79.5	80.5	81.4	81.8	82.1	83.2	83.2	84.1	87.3
J4.8	73.4	74.7	79.8	78.6	72.8	83.2	83.7	84.5	85.7	85.7
BPNN	74.8	78.1	80.6	81.1	82.7	83.7	85.3	86.1	87.2	86.9
SMO	78.1	78.6	79.8	79.8	79.8	80.0	79.9	80.2	80.4	81.6
CLR	76.6	78.5	80.3	80.2	80.3	80.7	80.9	81.4	81.0	82.8
OneR	75.2	73.4	77.5	78.0	77.7	77.5	79.0	77.8	78.9	82.4

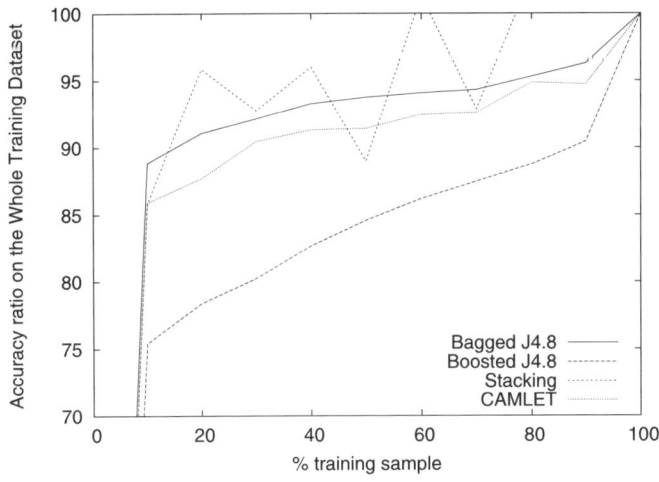

Fig. 6.4. Learning curves of Accuracies (%) on the learning algorithms over subsampled training dataset

be better classifier for the entire dataset than these hyper-plane learners and bagged J4.8, they needed more training instances to become accurate classifiers. Looking at the result of learning algorithm constructed by CAMLET, this algorithm achieves almost the same performance as bagged J4.8, with smaller training subset. However, it can outperform bagged J4.8 with larger training subsets. Although the constructed algorithm was based on boosting, the combination of a reinforcement method from Classifier Systems and the outer loop was able to overcome the disadvantage of boosting for a smaller training subset.

Rule evaluation models for the meningitis data mining result dataset. In this section, we present rule evaluation models for the entire dataset learned using CAMLET, OneR, J4.8 and CLR. This is because they are represented as explicit models such as a rule set, a decision tree, and a linear model set.

As shown in Fig. 6.5, the indices used in the learned rule evaluation models are taken, not only from a group of indices that increases with the correctness of a rule, but also from different groups of indices. Indices such as YLI1, Laplace Correction, Accuracy, Precision, Recall, Coverage, PSI and Gini Gain are indices that were formerly used for models. Later indices include GBI and Peculiarity, which sums up the difference in antecedents between one rule and the other rules in the same rule set. This corresponds to a comment made by the human expert. He said that he evaluated these rules not only according to their correctness but also their interestingness based on his expertise

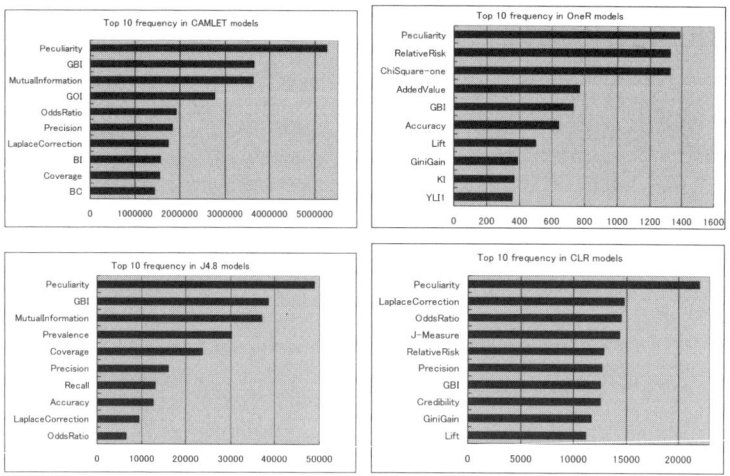

Fig. 6.5. Top 10 frequencies of the indices used by the models of each learning algorithm with 10000 bootstrap samples of the meningitis datamining result dataset and executions

6.4.2 Constructing Rule Evaluation Models on Artificial Evaluation Labels

We also evaluated our rule evaluation model construction method using rule sets obtained from five datasets of the UCI machine learning repository to confirm the lower limit performances on probabilistic class distributions.

We selected the following five datasets: anneal, audiology, autos, balance-scale, breast-cancer, breast-w, colic, and credit-a. With these datasets, we obtained rule sets with bagged PART, which repeatedly executes PART [35] on the bootstrapped training datasets.

For these rule sets, we calculated 39 objective indices as attributes of each rule. With regard to the classes of these datasets, we used three class distributions with multinomial distributions. Table 6.4 shows us a process flow diagram for obtaining these datasets and their descriptions, with three different class

Table 6.4. The datasets of the rule sets learned from the UCI benchmark datasets

	#Mined Rules	#Class labels L1	L2	L3	%Def. class
Distribution I		(0.30)	(0.35)	(0.35)	
anneal	95	33	39	23	41.1
audiology	149	44	58	47	38.9
autos	141	30	48	63	44.7
balance-scale	281	76	102	103	36.7
breast-	122	41	34	47	38.5
breast-w	79	29	26	24	36.7
colic	61	19	18	24	39.3
credit-a	230	78	73	79	34.3
Distribution II		(0.30)	(0.50)	(0.20)	
anneal	95	26	47	22	49.5
audiology	149	44	69	36	46.3
autos	141	40	72	29	51.1
balance-scale	281	76	140	65	49.8
breast-	122	40	62	20	50.8
breast-w	79	29	36	14	45.6
colic	61	19	35	7	57.4
credit-a	230	78	110	42	47.8
Distribution III		(0.30)	(0.65)	(0.05)	
anneal	95	26	63	6	66.3
audiology	149	49	91	9	61.1
autos	141	41	95	5	67.4
balance-scale	281	90	178	13	63.3
breast-	122	42	78	2	63.9
breast-w	79	22	55	2	69.6
colic	61	22	36	3	59.0
credit-a	230	69	150	11	65.2

Table 6.5. Overview of constructed learning algorithms by CAMLET to the datasets of the rule sets learned from the UCI benchmark datasets

	Distribution I			Distribution II			Distribution III		
	original classifier set	overall control structure	final eval. method	original classifier set	overall control structure	final eval. method	original classifier set	overall control structure	final eval. method
anneal	C4.5 tree	Win+Boost+CS	Weighted Voting	C4.5 tree	Boost+CS	Weighted Voting	C4.5 tree	Boost+CS	Weighted Voting
audiology	ID3 tree	Boost	Voting	Random Rules	CS+GA	Weighted Voting	Random Rules	Simple Iteration	Best Select.
autos	Random Rules	Win+Iteration	Weighted Voting	ID3 tree	Boost+Iteration	Weighted Voting	Random Rules	Boost	Weighted Voting
balance-scale	Random Rules	Boost	Voting	Random Rules	Boost+CS	Weighted Voting	Random Rules	CS+GA	Voting
breast-cancer	Random Rules	GA+Iteration	Voting	ID3 tree	Boost+CS +Iteration	Weighted Voting	Random Rules	Win+Iteration	Weighted Voting
breast-w	ID3 tree	Win	Weighted Voting	ID3 tree	Iteration	Best Select.	ID3 tree	CS+Iteration	Weighted Voting
colic	Random Rules	CS+Win	Voting	ID3 tree	Win+Iteration	Best Select.	ID3 tree	Win+Iteration	Voting
credit-a	C4.5 tree	Win+Iteration	Voting	Random Rules	Win+Iteration	Best Select.	ID3 tree	CS+Boost+Iteration	Best Select.

CS means including reinfoecement of classifier set from Classifiser Systems
Boost means including methods and control structure from Boosting

Win means including methods and control structure from Window Strat
GA means including reinforcement of classifier set with Genetic Algori

distributions. The class distribution for "Distribution I" is $P = (0.35, 0.3, 0.3)$ where p_i is the probability of class i. Thus, the number of class i instances in each dataset D_j become $p_i D_j$. Similarly, the probability vector of "Distribution II" is $P = (0.3, 0.5, 0.2)$ and that of "Distribution III" is $P = (0.3, 0.65, 0.05)$.

Constructing proper learning algorithms for rule sets from UCI datasets. In the same way as the construction of an appropriate learning algorithm for the meningitis data mining result, we constructed appropriate learning algorithms for the datasets of rule sets from the eight UCI datasets. Table 6.5 shows an overview of the constructed learning algorithms for each dataset, which had three different class distributions.

For these datasets, CAMLET constructed various learning algorithms based on 'random rule set generation', ID3 decision tree, and C4.5 decision tree. Therefore, these learning algorithms consisted of new combinations of methods that had previously never been seen in learning algorithms. Most of the learning algorithms include 'Voting' from bagging or 'Weighted Voting' from boosting. With regard to these results, CAMLET constructed selective meta-learning algorithms for the datasets with the three different class distributions.

Accuracy Comparison on Classification Performances. For the above mentioned datasets, we used the five learning algorithms to estimate whether their classification results reached or exceeded the accuracies when just predicting each majority class. Table 6.6 shows the accuracies of the nine learning algorithms applied to each class distribution of the three datasets. The learning algorithms constructed by CAMLET, boosted J4.8, bagged J4.8, J4.8, and BPNN always performed better than just predicting the majority class of each dataset. In particular, Bagged J4.8 and Boosted J4.8 outperformed J4.8 and BPNN for almost all datasets. However, their performances were suffered from probabilistic class distributions for larger datasets, such as balance-scale and credit-a.

Table 6.6. Accuracies (%) on entire training datasets labeled with three different distributions

	J4.8	BPNN	SVM	CLR	OneR	Bagged J4.8	Boosted J4.8	Stacking	CAMLET
						Distribution I			
anneal	74.7	71.6	47.4	56.8	55.8	87.4	100.0	27.4	77.9
audiology	47.0	51.7	40.3	45.6	52.3	87.2	47.0	21.5	63.1
autos	66.7	63.8	46.8	46.1	56.0	89.4	66.7	29.8	53.2
balance-scale	58.0	59.4	39.5	43.4	53.0	83.3	58.0	39.5	39.5
breast-cancer	55.7	61.5	40.2	50.8	59.0	88.5	70.5	23.8	41.0
breast-w	86.1	91.1	38.0	46.8	54.4	96.2	100.0	34.2	77.2
colic	91.8	82.0	42.6	60.7	55.7	88.5	100.0	29.5	67.2
credit-a	57.4	48.7	35.7	39.1	54.8	91.3	57.4	26.5	55.7
						Distribution II			
anneal	68.4	66.3	56.8	60.0	56.8	85.3	87.4	49.5	67.4
audiology	60.4	61.1	43.6	55.0	56.4	87.2	69.8	50.3	
autos	63.1	64.5	52.5	53.2	57.4	90.8	100.0	39.0	67.4
balance-scale	61.6	57.7	49.8	55.2	58.0	80.4	61.6	45.6	41.9
breast-cancer	68.0	70.5	47.5	58.2	59.8	77.9	96.7	33.6	64.8
breast-w	89.9	93.7	49.4	58.2	62.0	98.7	100.0	59.5	78.5
colic	77.0	78.7	57.4	62.3	67.2	85.2	100.0	29.5	88.5
credit-a	61.3	59.1	41.3	52.6	56.1	89.6	62.2	47.4	53.5
						Distribution III			
anneal	74.7	70.5	67.4	70.5	73.7	84.2	94.7	67.4	66.3
audiology	65.8	67.8	63.8	64.4	67.1	83.2	67.1	59.7	65.1
autos	85.1	73.8	68.1	70.2	73.8	87.9	100.0	66.7	67.4
balance-scale	70.5	69.8	64.8	65.8	69.8	80.1	85.8	62.6	63.0
breast-cancer	71.3	77.0	66.4	65.6	77.9	86.9	79.5	73.0	73.0
breast-w	74.7	86.1	73.4	68.4	74.7	87.3	100.0	63.3	70.9
colic	70.5	77.0	65.6	60.7	73.8	85.2	100.0	49.2	60.7
credit-a	70.9	70.0	65.2	65.2	71.3	85.7	87.8	61.7	65.2

SVM, CLR and Stacking affected the class distribution differences. Their performances were sometimes lower the percentage for each majority class. Although Stacking is a kind of selective meta-learning algorithm, it performed worse than the other two selective meta-learning algorithms, because it included SVM and CLR at the same time and failed to control the predictive results of these worse learning algorithms.

Evaluation of Learning Curves. Similar to the evaluations of the learning curves on the meningitis rule set, we estimated the minimum training subsets for a valid model, which works better than just predicting the majority class of the datasets.

Table 6.7 shows the sizes of the minimum training subsets, which can help construct more accurate rule evaluation models than the percentages of the majority class formed by each learning algorithm. For datasets with balanced

Table 6.7. Number of minimum training subsamples for outperforming the Accuracy (%) of default class

	Distribution I								
	J4.8	BPNN	SVM	CLR	OneR	Bagged J4.8	Boosted J4.8	Stacking	CAMLET
anneal	20	14	17	29	29	16	14	36	20
audiology	21	18	65	64	41	21	14	56	27
autos	38	28	76	77	70	28	28	77	31
balance−scale	12	14	15	15	32	14	9	51	128
breast−cancer	16	17	22	41	22	14	14	41	36
breast−w	7	10	10	18	14	10	6	19	11
colic	8	8	9	22	14	8	8	24	8
credit−a	9	12	16	30	28	9	8	51	19
	Distribution II								
	J4.8	BPNN	SVM	CLR	OneR	Bagged J4.8	Boosted J4.8	Stacking	CAMLET
anneal	29	20	16	42	46	26	21	46	29
audiology	36	45	−	61	67	27	30	67	
autos	49	39	49	123	88	44	34	74	44
balance−scale	81	84	69	221	168	60	64	135	−
breast−cancer	31	28	102	40	46	28	28	62	28
breast−w	14	11	23	30	26	11	10	31	19
colic	24	20	36	42	36	15	18	37	22
credit−a	51	74	−	134	109	49	42	105	78
	Distribution III								
	J4.8	BPNN	SVM	CLR	OneR	Bagged J4.8	Boosted J4.8	Stacking	CAMLET
anneal	54	58	64	76	−	42	38	64	46
audiology	64	73	45	76	107	50	50	103	84
autos	66	102	84	121	98	45	39	76	76
balance−scale	118	103	133	162	156	86	92	132	−
breast−cancer	50	31	80	92	80	38	36	60	41
breast−w	44	36	31	48	71	34	34	52	53
colic	28	24	46	30	42	28	22	48	54
credit−a	118	159	−	−	173	76	76	120	109

class distribution (Distribution I), these learning algorithms were able to learn valid models with less than 20% of the given training datasets. However, for the datasets with imbalanced distributions (Distribution II & III), they needed more training subsets to construct valid models, because their performances with the entire training datasets fell to the percentages of the majority class of each dataset, as shown in Table 6.6.

Comparison of results of the meta-learning algorithms. Comparing Stacking and CAMLET, CAMLET achieved a higher accuracy than Stacking, as shown in Table 6.6. This shows that the approach of CAMLET, the decomposition and re-construction of learning algorithms, is better than just combining prepared learning algorithms. Although CAMLET can construct boosted and bagged C4.5, which outperformed than the learning algorithms constructed by CAMLET, CAMLET could not search for these algorithms as the appropriate learning algorithms for these datasets. We need to improve the search method used by CAMLET to allow it to construct a more appropriate learning algorithm for a given dataset.

6.5 Conclusion

In this paper, we described the evaluation of the nine learning algorithms for a rule evaluation support method using rule evaluation models to predict evaluations for an if-then rule based on objective indices by re-using the evaluations made by a human expert.

Based on a performance comparison of the nine learning algorithms for the dataset from the result of meningitis data mining, the rule evaluation models achieved higher accuracies than just predicting the majority class. For this dataset, the learning algorithm constructed by CAMLET presented higher accuracy with higher reliability than the other eight learning algorithms, including three selective meta-learning algorithms. For the datasets of rule sets obtained from eight UCI datasets, although committee type learners such as SVM and CLR, and Stacking failed to reach the percentage of the majority class of some datasets, the other learning algorithms were able to go to or beyond the percentages of the majority class of each dataset with smaller than 50% of each training dataset. Thus, our constructive meta-learning scheme has shown its higher flexibility for different class distributions based on various criteria.

Considering the difference between the actual evaluation labeling and the artificial evaluation labeling, it was shown that the evaluation of the medical expert considered the particular relations between an antecedent and a class, or another antecedent, in each rule. These results indicated that our approach could detect human criteria differences as several performance differences of rule evaluation models.

In the future, we will improve CAMLET's method repository to construct suitable learning algorithms for rule evaluation models. We will also apply this rule evaluation support method to other datasets from various domains.

References

1. Hilderman, R.J., Hamilton, H.J.: Knowledge Discovery and Measure of Interest. Kluwer Academic Publishers, Dordrecht (2001)
2. Tan, P.N., Kumar, V., Srivastava, J.: Selecting the Right Interestingness Measure for Association Patterns. In: Proceeding of International Conference on Knowledge Discovery and Data Mining KDD 2002, pp. 32–41 (2002)
3. Yao, Y.Y., Zhong, N.: An Analysis of Quantitative Measures Associated with Rules. In: Proceeding of Pacific-Asia Conference on Knowledge Discovery and Data Mining PAKDD 1999, pp. 479–488 (1999)
4. Ohsaki, M., Kitaguchi, S., Kume, S., Yokoi, H., Yamaguchi, T.: Evaluation of Rule Interestingness Measures with a Clinical Dataset on Hepatitis. In: Boulicaut, J.-F., Esposito, F., Giannotti, F., Pedreschi, D. (eds.) PKDD 2004. LNCS (LNAI), vol. 3202, pp. 362–373. Springer, Heidelberg (2004)
5. Breiman, L.: Bagging predictors. Machine Learning 24(2), 123–140 (1996)

6. Freund, Y., Schapire, R.E.: Experiments with a new boosting algorithm. In: Proc. of Thirteenth International Conference on Machine Learning, pp. 148–156 (1996)
7. Wolpert, D.: Stacked Generalization. Neural Network 5(2), 241–260 (1992)
8. Gama, J., Brazdil, P.: Cascade generalization. Machine Learning 41(3), 315–343 (2000)
9. Metal (2002), http://www.metal-kdd.org/
10. Bernstein, A., Provost, F.: An intelligent assistant for knowledge discovery process. In: Proceeding IJICAI 2001 Workshop on Wrappers for Performance Enhancement in KDD (2001)
11. Abe, H., Yamaguchi, T.: Constructive Meta-Learning with Machine Learning Method Repositories. In: Orchard, B., Yang, C., Ali, M. (eds.) IEA/AIE 2004. LNCS (LNAI), vol. 3029, pp. 502–511. Springer, Heidelberg (2004)
12. Hatazawa, H., Negishi, N., Suyama, A., Tsumoto, S., Yamaguchi, T.: Knowledge Discovery Support from a Meningoencephalitis Database Using an Automatic Composition Tool for Inductive Applications. In: Proc. of KDD Challenge, in conjunction with PAKDD 2000, pp. 28–33 (2000)
13. Hettich, S., Blake, C.L., Merz, C.J.: UCI Repository of machine learning databases, Department of Information and Computer Science, University of California, Irvine (1998), http://www.ics.uci.edu/mlearn/MLRepository.html
14. Fayyad, U.M., Piatetsky-Shapiro, G., Smyth, P., Uthurusamy, R. (eds.): Explora: A Multipattern and Multistrategy Discovery Assistant. In: Advances in Knowledge Discovery and Data Mining, pp. 249–271. AAAI/MIT Press (1996)
15. Ali, K., Manganaris, S., Srikant, R.: Partial classification using association rules. In: Proceedings of the International Conference on Knowledge Discovery and Data Mining KDD 1997, pp. 115–118 (1997)
16. Brin, S., Motwani, R., Ullman, J., Tsur, S.: Dynamic itemset counting and implication rules for market basket data. In: Proc. of ACM SIGMOD Int. Conf. on Management of Data, pp. 255–264 (1997)
17. Rijsbergen, C.: Information Retrieval, Ch. 7, Butterworths, London (1979), http://www.dcs.gla.ac.uk/Keith/Chapter.7/Ch.7.html
18. Gray, B., Orlowska, M.E.: CCAIIA: Clustering Categorical Attributes into Interesting Association Rules. In: Pacific-Asia Conf. on Knowledge Discovery and Data Mining PAKDD 1998, pp. 132–143 (1998)
19. Hamilton, H.J., Shan, N., Ziarko, W.: Machine Learning of Credible Classifications. In: Proc. of Australian Conf. on Artificial Intelligence AI 1997, pp. 330–339 (1997)
20. Goodman, L.A., Kruskal, W.H.: Measures of association for cross classifications. Springer Series in Statistics, vol. 1. Springer, Heidelberg (1979)
21. Smyth, P., Goodman, R.M.: Rule Induction using Information Theory. In: Piatetsky-Shapiro, G., Frawley, W.J. (eds.) Knowledge Discovery in Databases, pp. 159–176. AAAI/MIT Press (1991)
22. Piatetsky-Shapiro, G.: Discovery, Analysis and Presentation of Strong Rules. In: Piatetsky-Shapiro, G., Frawley, W.J. (eds.) Knowledge Discovery in Databases, pp. 229–248. AAAI/MIT Press (1991)
23. Gago, P., Bento, C.: A Metric for Selection of the Most Promising Rules. In: European Conference on the Principles of Data Mining and Knowledge Discovery PKDD 1998, pp. 19–27 (1998)
24. Zhong, N., Yao, Y.Y., Ohshima, M.: Peculiarity Oriented Multi-Database Mining. IEEE Transaction on Knowledge and Data Engineering 15(4), 952–960 (2003)
25. Witten, I.H., Frank, E.: DataMining: Practical Machine Learning Tools and Techniques with Java Implementations. Morgan Kaufmann, San Francisco (2000)

26. Quinlan, J.R.: Programs for Machine Learning. Morgan Kaufmann Publishers, San Francisco (1993)
27. Hinton, G.E.: Learning distributed representations of concepts. In: Proceedings of 8th Annual Conference of the Cognitive Science Society (1986)
28. Platt, J.: Fast Training of Support Vector Machines using Sequential Minimal Optimization. In: Schölkopf, B., Burges, C., Smola, A. (eds.) Advances in Kernel Methods - Support Vector Learning, pp. 185–208. MIT Press, Cambridge (1999)
29. Frank, E., Wang, Y., Inglis, S., Holmes, G., Witten, I.H.: Using model trees for classification. Machine Learning 32(1), 63–76 (1998)
30. R.C.H.: Very simple classification rules perform well on most commonly used datasets. Machine Learning 11, 63–91 (1993)
31. Mitchell, T.M.: Generalization as Search. Artificial Intelligence 18(2), 203–226 (1982)
32. Michalski, R., Mozetic, I., Hong, J., Lavrac, N.: The AQ15 Inductive Learning System: An Over View and Experiments, Reports of Machine Learning and Inference Laboratory, No.MLI-86-6, George Mason University (1986)
33. Booker, L.B., Holland, J.H., Goldberg, D.E.: Classifier Systems and Genetic Algorithms. Artificail Inteligence 40, 235–282 (1989)
34. Quinlan, J.R.: Induction of Decision Tree. Machine Learning 1, 81–106 (1986)
35. Frank, E., Witten, I.H.: Generating accurate rule sets without global optimization. In: Proc. of the Fifteenth International Conference on Machine Learning, pp. 144–151 (1998)

7

Mining Statistical Association Rules to Select the Most Relevant Medical Image Features

Marcela X. Ribeiro[1], Andre G.R. Balan[1], Joaquim C. Felipe[2],
Agma J.M. Traina[1], and Caetano Traina Jr.[1]

[1] Department of Computer Science,
   University of São Paulo at São Carlos, Brazil
[2] Department of Physics and Mathematics,
   University of São Paulo, at Ribeirão Preto, Brazil

Abstract. In this chapter we discuss how to take advantage of association rule mining to promote feature selection from low-level image features. Feature selection can significantly improve the precision of content-based queries in image databases by removing noisy and redundant features. A new algorithm named StARMiner is presented. StARMiner aims at finding association rules relating low-level image features to high-level knowledge about the images. Such rules are employed to select the most relevant features. We present a case study in order to highlight how the proposed algorithm performs in different situations, regarding its ability to select the most relevant features that properly distinguish the images. We compare the StARMiner algorithm with other well-known feature selection algorithms, showing that StARMiner reaches higher precision rates. The results obtained corroborate the assumption that association rule mining can effectively support dimensionality reduction in image databases.

7.1 Introduction

Nowadays, computational applications often need to deal with complex data, such as images, video, time series, fingerprints, and DNA sequences. Focusing on one of the most studied type of complex data - images, and more particularly on medical images - two kinds of systems are now widely used: the Picture Archiving and Communication Systems (PACS) and the Computer-Aided Diagnosis (CAD) systems.

The CAD research community has been seeking, for several years, efficient and precise algorithms to correctly classify medical images into relevant categories, aiming at supporting medical diagnosis. The development of PACS broadens the effective use of images on diagnosing, as well as in medicine teaching. However, in order to be effectively useful, the processing of image retrieval in PACS and CAD systems must also be fast and consistent with the judgment of specialists.

The volume of images generated on medical exams grows exponentially, demanding efficient and effective methods of image retrieval and analysis. In fact, the number of images generated in hospitals and medical centers corresponds to several Terabytes per day in a medium size hospital, what demands efficient

D.A. Zighed et al. (Eds.): Mining Complex Data, SCI 165, pp. 113–131.

mechanisms to store and retrieve them. Therefore, content-based image retrieval (CBIR) techniques have been intensively investigated in the last years [16].

CBIR techniques rely on image processing algorithms to extract relevant characteristics (features) from the images. The characteristics are grouped into feature vectors, which are stored and organized by indexing structures aiming at achieving fast and efficient image retrieval. Generally, CBIR techniques use intrinsic visual features of images, such as color, shape and texture [13] yielding vectors with hundreds or even thousands of features. Unlike one would think, having a large number of features actually represents a problem. As the number of the extracted features grows, the process of storing, indexing, retrieving, and comparing them becomes more and more time consuming. Moreover, in several situations, many features are correlated, meaning that they bring redundant information about the images that can deteriorate the ability of the system to correctly distinguish them. The large number of features leads CBIR systems to face the problem known as the "dimensionality curse" [17]. Beyer [7] has proved, as the number of features increases, the significance of each feature tends to diminish. Hence, it is important to keep the number of features as low as possible, establishing a tradeoff between the representation power and the feature vector size.

Image features are also commonly employed in the classification task. A significant example is the classification of tumor masses detected in mammograms as benign or malignant. Initially, the radiologist classifies the images based on the shape of the lesion. Malignant tumors generally infiltrate the surrounding tissue, resulting in an irregular or hardly-distinguishable contour, while benign masses have a smooth contour. Figure 7.1 illustrates two examples of tumor masses.

This chapter discusses how to apply techniques of mining statistical association rules to improve content-based image retrieval in medical domain. We present a new algorithm (the StARMiner - **S**tatistical **A**ssociation **R**ule **Min**er) to determine a minimal set of representative features. The algorithm uses statistical measurements, which describe the behavior of the features considering the image categories, to find representative rules. We compare the efficacy of StARMiner and other well- known feature selection algorithms, Relief-F and DTM (Decision Tree Method) in the task of feature selection using a case study.

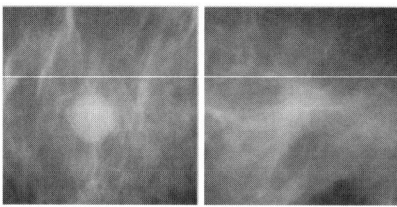

Fig. 7.1. Typical breast tumor masses: benign (left) and malignant (right)

7.2   Background

Image mining has been the target of many researches in the field of data mining and information retrieval in current years. A major challenge of the image mining field is to effectively relate low-level features (automatically extracted from image pixels) to high-level semantics based on the human perception.

According to [11], researches in image mining can be generally classified into two main directions: *domain-specific* and *general-purpose* directions. The *domain-specific* direction focuses on image processing techniques, where the goal is to process the image and to extract the features that best contribute to differentiate images from different types. The *general-purpose* direction focuses on mining algorithms that aim at reducing the semantic gap between high-level human perception of images and low-level image feature representation. Indeed, general-purpose techniques work improving the accuracy of specific-domain techniques, working in a complementary way.

Mining images demands the extraction of their main features regarding specific criteria. After extracted, the feature vector and the image descriptions are submitted to the mining process.

When working with image databases, high-level data manually supplied by domain experts can also be employed together with low-level features in image mining processes. However, with the growing of large-scale image repositories, manual annotation of images has become unfeasible because of its inherent problems of subjectivity, non-scalability, and non-uniformity of vocabulary. CBIR systems are proposed to overcome these limitations, where the most similar images of a given one are retrieved based on comparisons of visual features (automatically extracted from images). The retrieved images can be employed to label a new one or, in case of medical images, to help the decision making process of diagnosing a new image.

In this section, two tasks of data mining are discussed: feature selection and association rule mining. We concentrate on applying these techniques to extract patterns from images and to improve content-based search in medical image databases. In this section, we also describe how to evaluate the results of a CBIR system using the precision versus recall curves (P&R). We employed P&R graphs to evaluate our experiments.

7.2.1   Feature Selection

Dimensionality reduction (also called dimension reduction) is the process of reducing the number of features (attributes) used to represent a dataset under consideration. Dimensionality reduction approaches diminish the feature vector size by removing redundant, correlated and noisy data. In most cases, dimensionality reduction speeds up the processing of data mining algorithms and improves their accuracy.

The dimensionality reduction approaches are also classified into *feature selection* and *feature transformation* approaches. There is no agreement about the nomenclature of *feature transformation* in literature. In fact, *feature transforma-tion* approaches are also called *feature extraction* or *feature reduction* approaches. The key difference between *feature selection* and *feature transformation* is that the former consists in selecting a subset of the original features while the last generates a completely new set of features to represent a given dataset.

In this chapter, we focus on *feature selection* approaches. *Feature selection* ap-proaches do not transform the original features; they only remove the subset of redundant and irrelevant features, preserving the semantical meaning of original data. Feature selection techniques can be divided into *binary* and *continuous* techniques. *Continuous* feature selection techniques assign continuous weights to each feature, while the *binary* approach assigns binary weights to each fea-ture. Feature selection techniques can also follow either the *filter* or the *wrapper* model. In the *filter* model, the feature selection process is performed before the learning phase and works as a pre-processing step of the learning algorithm. In the *wrapper* model, the feature selection algorithm uses the learning algorithm as a subroutine. The main disadvantage of the wrapper model is the huge com-putational effort employed by the learning algorithm to evaluate each feature subset [18].

A vast amount of feature selection algorithms has been presented in litera-ture. One of the former algorithms was presented in [19], where the property of monotonicity is employed to prune the search space, and divergence is employed to evaluate features.

One of the most well-known feature selection algorithms is Relief [14]. The general principle of Relief is to measure the quality of features according to how their values distinguish instances of different classes. Given a randomly selected instance S from a dataset R, with k features (attributes), Relief searches the dataset for the nearest neighbor of the same class, which is called nearest hit H, and Relief also searches the dataset for the nearest neighbor of the different class, called nearest miss M. It updates the quality estimator $W[f_i]$ of all features f_i, depending on the difference between the feature values of the instances S, H e M. This process repeats n times, where n is a parameter specified by the user. The time complexity of Relief is $O(nkN)$, where N is the number of instances of the dataset and k is the number of features. Relief returns a rank of attributes ordered according to their relevance, but it does not indicate the number of features that should be selected. One limitation of the Relief algorithm is that it works only for datasets with binary classes. This limitation is overcome by Relief-F [15] that also tackles datasets with multi-valued classes.

Another well-known feature selection technique is the Decision Tree Method (DTM) [8]. DTM adopts a forward search to generate feature subsets, using the entropy criterion to evaluate them. DTM runs C4.5 [21], an algorithm that builds a decision tree. Since a decision tree is a sequence of attributes that defines the state of an instance, DTM selects the features that appear in the pruned decision tree as the best subset, i.e., the features appearing in the path to any leaf node in the pruned tree are selected as the best subset.

In [12], the chi-square distribution is used to infer data distribution and to promote feature selection. Indeed, other statistical tests can be used to infer data distribution. In this chapter, as we will describe in the next section, we use a statistical test to generate association rules. A performance comparison among feature selection methods can be found in [22].

7.2.2 Association Rules

In this chapter we show how to use association rules for feature selection. Association rule mining is a descriptive task of data mining, where the goal is to find relevant relationships among data items. It was initially motivated by business applications, such as catalog design, store layout, and customer categorization [3]. However, finding associations has also been widely used in many other applications such as data classification and summarization [20, 24].

The problem of mining association rules was first stated in [1] as follows. Let $I = \{i_1, \ldots, i_n\}$ be a set of literals called items. A set $X \in I$ is called an itemset. Let R be a table with transactions t involving elements that are subsets of I. An association rule is an expression of the form $X \rightarrow Y$, where X and Y are itemsets. X is called body or antecedent of the rule, and Y is called head or consequent of the rule.

Let $|R|$ be the number of transactions in relation R. Let $|Z|$ be the total number of occurrences of the itemset Z in transactions of relation R. *Support* and *confidence* measures (Equations 7.1 and 7.2) are used to determine the rules returned by the mining process.

$$Support = \frac{|X \cup Y|}{|R|} \tag{7.1}$$

$$Confidence = \frac{|X \cup Y|}{|X|} \tag{7.2}$$

The problem of mining association rules, as it was first stated, involves finding rules that satisfy the restrictions of minimum support and minimum confidence specified by the user.

Apriori [2] is one of the first and widely used association rule mining algorithm. One drawback of the Apriori algorithm is its low performance because of the successive dataset scans carried out by the algorithm. Some algorithms were developed for speeding up the association rule mining. Examples of such algorithms include Partition [26], FP-Growth [10] and Eclat [28]. These algorithms were developed to mine the first and the simplest type of association rules, the Boolean association rules, which are rules that correlate categorical (nominal) data items. In [4, 23, 27] procedures for mining quantitative association rules are presented. Quantitative association rules relate continuous-valued attributes. In [29, 30] the problem of generating association rules correlating items from multiple databases was dealt with. In [25] an algorithm for mining association rules in data warehouses was presented.

Mining association rules in image datasets has been a great challenge. Procedures of association rule mining do not produce interesting results by themselves. Images should be previously pre-processed by image processing algorithms to produce the image data that is submitted to the mining processes.

7.2.3   Content-Based Retrieval Evaluation

When working with content-based retrieval, performing exact searches on image datasets are not useful, since searching for the same data already under analysis has very few applications. Therefore, the retrieval of complex data is mainly performed regarding similarity. The most well-known and useful types of similarity queries are the k-nearest neighbor (for instance: "given the Thorax-XRay of John Doe, find the five images most similar to it from the image database"), and range queries (for instance: "given the Thorax-XRay of John Doe, find the images that differ from it up to three units"). Similarity search is performed comparing the feature vectors using a distance function to quantify how close (or similar) each pair of vectors is.

This chapter is focused on medical images, more specifically on the feature vectors employed to compare and retrieve the images by similarity. The motivation is to reduce the usually large number of extracted features, because for PACS and CAD systems, it is usual to gather as many image characteristics as possible, leading to high-dimensional feature vectors, which encompasses much redundant information. Consequently, it is necessary to sift the features that keep the most meaningful information. Notice that the proposed approach can be straightforwardly extended to work on other types of complex data beyond images, since similarity queries are generally the most suitable for complex data.

In this chapter, we present a technique that uses association rules to improve the content-based image retrieval on medical domain. One important issue related to CBIR systems consists on how to evaluate their efficacy. A standard approach to evaluate the accuracy of the similarity queries is the precision and recall (P&R) graph [5]. Precision and recall are defined in Equation 7.3 and Equation 7.4.

$$Precision = \frac{TRS}{TS} \tag{7.3}$$

$$Recall = \frac{TRS}{TR} \tag{7.4}$$

In Equations 7.3 and 7.4, TR is the total number of relevant images for a given query; TRS is the number of relevant images actually returned in the query, and TS is the total number of images returned in the query. In our experiments we use precision and recall (P&R) curves in order to analyze our proposed algorithm StARMiner.

7.3 Feature Selection through Statistical Association Rule Mining

A statistical association rule is a type of rule that shows an interesting relationship among subsets of data based on the distribution of the quantitative values. The term statistical association rule is given to any association rule whose generation process uses statistical tests to confirm its validity. The goal of working with statistical association rules is that they do not require data discretization. A discretization process often leads to a loss of information and can distort the results of a mining algorithm.

Feature vectors describe the images quantitatively. Hence, a suitable approach to find association rules should consider quantitative data. In this section we present StARMiner (Statistical Association Rule Miner), a new algorithm for statistical association rule mining. The goal of StARMiner is to find statistical association rules to select a minimal set of features that preserves the ability of discerning image according to their types. The method proposed here extends the techniques of statistical association rule mining proposed in [4].

Let x_j be a category of an image f_i an image feature (attribute). The rules returned by the StARMiner algorithm have the format $x_j \rightarrow f_i$. StARMiner only returns rules that satisfy the Condition 1 and Condition 2, as follows.

Condition 1. The feature f_i must have a behavior in images from category x_j different from its behavior in images from all the other categories.
Condition 2. The feature f_i must present a uniform behavior in every image from category x_j.

Conditions 1 and 2 are implemented in the StARMiner algorithm incorporating restrictions of interest in the mining process, in the way described as follows. Let T be a dataset of medical images, x_j an image category, $T_{x_j} \in T$ the subset of images of category x_j, f_i the i_{th} feature of the feature vector F, and f_{i_k} the value of feature f_i in the image k. Let $\mu_{f_i}(Z)$ and $\sigma_{f_i}(Z)$ be, respectively, the mean and standard deviation of the values of feature f_i in the subset of images Z. The algorithm uses three thresholds defined by the user: $\Delta\mu_{min}$ - the minimum allowed difference between the average of the feature f_i in images from category x_j and the average of f_i in the remaining dataset; σ_{max} - the maximum standard deviation of f_i values allowed in a category and; γ_{min} - the minimum confidence to reject the hypothesis $H0$. StARMiner mines rules of the form $x_j \rightarrow f_i$, if the conditions given in Equations 7.7, 7.8 and 7.9 are satisfied.

$$\mu_{f_i}(V) = \frac{\sum_{k \in V}(f_{i_k})}{|V|} \tag{7.5}$$

$$\sigma_{f_i}(V) = \sqrt{\left(\frac{\sum_{k \in V}(f_{i_k} - \mu_{f_i}(V))^2}{|V|}\right)} \tag{7.6}$$

$$\mu_{f_i}(T_{x_j}) - \mu_{f_i}(T - T_{x_j}) \geq \Delta\mu_{min} \tag{7.7}$$

$$\sigma_{f_i}(T_{x_j}) \leq \sigma_{max} \tag{7.8}$$

$$H_0 : \mu_{f_i}(T_{x_j}) = \mu_{f_i}(T - T_{x_j}) \tag{7.9}$$

In Equation 7.9, H_0 should be rejected with a confidence equal to or greater than γ_{min}, in favor of the hypothesis that the means $\mu_{f_i}(T_{x_j})$ and $\mu_{f_i}(T - T_{x_j})$ are statistically different. To reject H_0 with confidence γ_{min}, the Z value, calculated using Equation 7.10, must be in the rejection region illustrated in Figure 7.2. The critical Z values Z_1 and Z_2 depend on the γ_{min} value as shown in Table 7.1:

$$Z = \frac{\mu_{f_i}(T_{x_j}) - \mu_{f_i}(T - T_{x_j})}{\frac{\sigma_{f_i}(T_{x_j})}{\sqrt{(|T_x|)}}} \tag{7.10}$$

A rule $x_j \rightarrow f_i$, returned by the algorithm, relates a feature f_i with a category x_j, where the values of f_i have a statistically different behavior in images of category x_j. This property indicates that f_i is an interesting feature to distinguish the images of category x_j from the remaining images. The StARMiner algorithm also gives information about the feature behavior in the mined rules. A rule mined by StARMiner, on its complete form is:

$$x_j \rightarrow f_i, \mu_{f_i}(T_{x_j}), \mu_{f_i}(T - T_{x_j}), \sigma_{f_i}(T_{x_j}), \sigma_{f_i}(T - T_{x_j})$$

where, $\mu_{f_i}(T_x)$ and $\sigma_{f_i}(T_x)$ are, respectively, the mean and the standard deviation of f_i values in the images from category x_j; $\mu_{f_i}(T_{x_j})$ and $\sigma_{f_i}(T - T_{x_j})$ are, respectively, the mean and the standard deviation of f_i values in the images that are not from category x_j. Algorithm 1 presents a description of StARMiner.

To perform StARMiner, the dataset under analysis is scanned twice. The first scan calculates the mean of each feature (lines 1 to 6). The second dataset scan

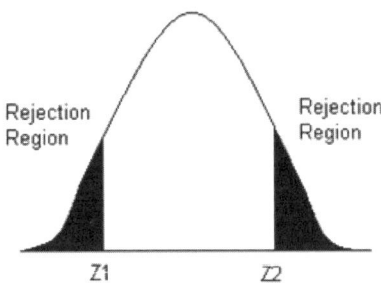

Fig. 7.2. Illustration of the rejection regions of a hypothesis test

Table 7.1. Critic Z values

γ_{min}	0.9	0.95	0.99
Z_1	-1.64	-1.96	-2.58
Z_2	1.64	1.96	2.58

Algorithm 1. StARMiner Algorithm

Input: Dataset T of image transactions structured as $\{x, f_1, f_2, f_n\}$ where x represents the image category and f_i an image feature; thresholds $\Delta\mu_{min}$, σ_{max}, γ_{min},

Output: The mined rules

1. Scan dataset T
2. **for** each feature f_i **do**
3.   **for** each category x_j **do**
4.   calculate $\mu_{f_i}(T_{x_j})$ and $\mu_{f_i}(T - T_{x_j})$
5.   **end for**
6. **end for**
7. Scan dataset T
8. **for** each feature f_i **do**
9.   **for** each category x_j **do**
10.   calculate $\sigma_{f_i}(T_{x_j})$ and $sigma_{f_i}(T - T_{x_j})$
11.   calculate Z value
12.   **if** $(\mu_{f_i}(T_{x_j}) - \mu_{f_i}(T - T_{x_j})) \geq \Delta\mu_{min}$ **And** $\sigma_{f_i}(T_{x_j}) \leq \sigma_{max}$ **And** $(Z < Z_1$ or $Z > Z_2)$ **then**
13.   write $x_j \rightarrow f_i, \mu_{f_i}(T_{x_j}), \mu_{f_i}(T - T_{x_j}), \sigma_{f_i}(T_{x_j}), \sigma_{f_i}(T - T_{x_j})$
14.   **end if**
15.   **end for**
16. **end for**

(lines 7 to 16) calculates the standard deviation for each feature and the Z value, used in the hypotheses test. The restrictions of interest are processed in lines 11 and 12. A rule is returned only if it satisfies the input thresholds ($\Delta\mu_{min}$, σ_{max}, γ_{min}). The complexity of StARMiner is $\Theta(ckN)$, where N is the number of instances of the dataset, k is the number of features, and c is the number of categories. **Criterion 1** is employed to perform feature selection using the StARMiner results.

Criterion 1. The features in the set of rules returned by StARMiner algorithm are selected as the most relevant ones.

The StARMiner algorithm makes possible to find rules that properly categorizes the images. That is, the algorithm spots the features with high power on differentiating image categories, since they have a particular and uniform behavior in images of a given category. This is important, because the features (or attributes) that present a uniform behavior to every image in the dataset, independently of the image category, do not contribute to categorize them, and should be eliminated.

To validate the StARMiner algorithm we used a procedure composed of three steps illustrated in Figure 7.3. These steps are detailed as follows.

Since StARMiner is a supervised feature selection algorithm, the image dataset is divided in the training set and the test set. The training set is submitted to the feature selection algorithm, where the test set is used to evaluate the method in the content-based queries.

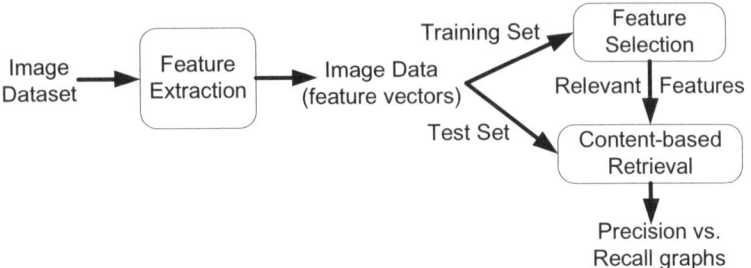

Fig. 7.3. Steps of the procedure used to validate the StARMiner algorithm

Step 1 - Feature Extraction. The image dataset is submitted to a feature extractor, generating a feature vector for each image.

Step 2 - Feature Selection. The feature vectors, together with the categories of the training images, are submitted to the feature selection algorithm. Thereafter, the reduced feature vectors are used to index the images from the test set in content-based searches, reducing the computational cost to execute similarity queries.

Step 3 - Content-Based Retrieval. For each image in the test dataset, one k-nearest neighbor query is performed. The measurements of precision and recall (P&R) are computed for each query result and an average P&R curve is computed for both original vectors and reduced vectors. As a rule of thumb on analyzing P&R curves, the closer the curve is to the top of the graph, the better the retrieval technique is.

In the next section, a case study is presented. StARMiner is applied to mine statistical association rules in a real medical image dataset, selecting relevant features and promoting dimensionality reduction of the dataset. The case study show the applicability of statistical association rule mining to find patterns, relating low-level features automatically extracted from images, with high-level information about image categories. The mined patterns can help image processing researchers to better understand low-level feature behavior and to find which low-level feature is really important to describe the image content. The mined rules also detail the behavior of features in subsets of images. In the case study, the feature selection performed by StARMiner is also compared to other well-known feature selection algorithms. The results show that StARMiner reaches higher values of precision when processing similarity queries.

7.4   Case Study

In this section, we describe a case study that exemplify the applicability of the StARMiner algorithm. The case study uses the SegMRF dataset. The SegMRF dataset consists of 704 medical images (angiogram and magnetic resonance - MR - images) obtained from the Clinical Hospital at Ribeirao Preto of the University

Table 7.2. Summary of image categories from the SegMRF dataset

Image Category	Number of Images
Angiogram	36
MR Axial Pelvis	86
MR Axial Head	155
MR Sagittal Head	258
MR Coronal Abdomen	23
MR Sagittal Spine	59
MR Axial Abdomen	51
MR Coronal Head	36

of Sao Paulo. The SegMRF dataset is classified in eight categories detailed in Table 7.2.

In **Step 1**, the images from SegMRF were segmented using Markov Random Fields (MRFs), which has been proved to be a suitable segmentation model for textured images [9]. MRF segments an image using local features, assigning each pixel to a region based on its relationship to the neighboring pixels. The final segmentation is achieved by minimizing the expected value of the number of misclassified pixels. The segmentation algorithm employed is the same presented in [6] which is an improved version of the EM/MPM method [9].

Since MRFs express only local properties of images, it is also important to extract global properties to discriminate them well. The global description is achieved by estimating the fractal properties of each segmented region. For each region segmented based on texture, six features were extracted: the mass (m); the centroid coordinates $(xo$ and $yo)$; the average gray level (a); the Fractal dimension (D); and the linear coefficient used to estimate D (b). Therefore, when an image is segmented in L regions, the feature vector has $L \times 6$ elements. In this experiment, we segmented the images in five regions. Figure 7.4 illustrates the feature vector described. It is important to stress that considering just five regions (as illustrated in Figure 7.4), the feature vector generated is quite compact. The feature vector can discriminate the images well, but even so, StARMiner demonstrated that it still has superfluous information that does not need to be stored.

The SegMRF dataset was divided in: training set, composed of 176 images; and, test set, composed of 528 images. In Step 2, StARMiner was run over the feature vectors of the training images, generating 21 rules. We evaluated various threshold values, and the best results were achieved using $\Delta\mu_{min} = 0.2$, $\sigma_{max} = 0.13$, $\gamma_{min} = 0.98$. An example of a rule obtained is:

$$angiogram \rightarrow region\ 2\ gray\ level\ average\ (a_2)$$
$$\mu_{a_2}(angiogram\ images) = 0.1$$
$$\mu_{a_2}(non\text{-}angiogram\ images) = 0.43$$
$$\sigma_{a_2}(angiogram\ images) = 0.07$$
$$\sigma_{a_2}(non\text{-}angiogram\ images) = 0.18$$

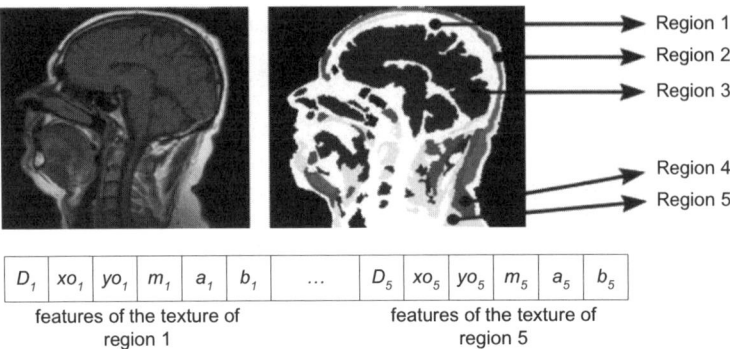

| D_1 | xo_1 | yo_1 | m_1 | a_1 | b_1 | ... | D_5 | xo_5 | yo_5 | m_5 | a_5 | b_5 |

features of the texture of   features of the texture of
region 1   region 5

Fig. 7.4. Segmentation regions and the feature vector

This rule indicates the importance of the average gray level in region 2 (a_2) to identify images from angiogram category. The rule shows that, the mean of a_2 in images of angiogram (0.1) is different from the mean of the same feature for the remaining images (0.43). The standard deviation of a_2 in images from angiogram is small (0.07) comparing with the standard deviation of the values of a_2 for the remaining images (0.18). These values indicates that the feature a_2 (region 2 gray level average) has a particular behavior in images of angiogram, contrasting with its behavior in images of other categories, evidencing that it is a relevant feature to distinguish images from angiograms.

Table 7.3 shows the attributes selected by StARMiner. The attributes selected by StARMiner are underlined. The attributes that are not underlined should be eliminated from the feature vector.

The results indicate that the fractal dimension D feature has a very low contribution to distinguish the images for this dataset (see Table 7.3). The results also indicate that features of average gray level a are the most relevant to distinguish the images in categories, because none feature of this type was removed by the feature selection process.

Table 7.3. Attributes selected by StARMiner. The attributes selected by StARMiner are underlined. The attributes that are not underlined should be eliminated from the feature vector.

Region 1	Region 2	Region 3	Region 4	Region 5
_D_	D	D	D	D
_xo_	xo	**_xo_**	**_x0_**	**_xo_**
yo	**_yo_**	**_yo_**	**_yo_**	**_yo_**
_m_	**_m_**	**_m_**	m	m
_a_	**_a_**	**_a_**	**_a_**	**_a_**
b	**_b_**	**_b_**	**_b_**	**_b_**

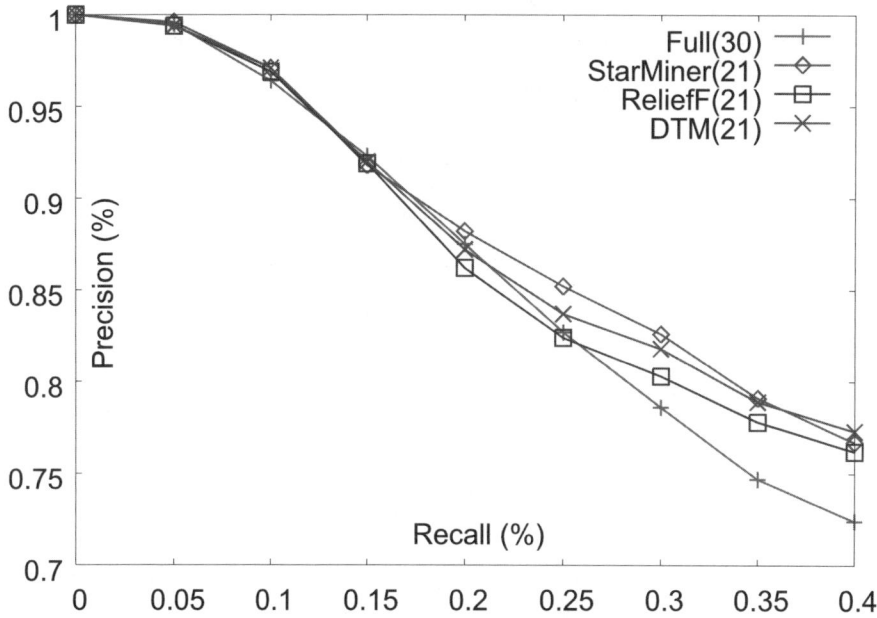

Fig. 7.5. P&R graph built using the SegMRF test set represented by: 30 original features, 21 selected by StARMiner, 21 selected by Relief-F and 21 selected by DTM

In Step 2, we measured the effectiveness of StARMiner algorithm in the task of feature selection. To perform it, we also applied Relief-F to the training images. The 21 most relevant features returned by Relief-F were also taken to compose a feature vector. In addition, DTM was also applied to the training images and the 21 most relevant features selected were also placed in a feature vector. The StARMiner algorithm took 0.25 seconds to select the features, Relief-F took 0.72 seconds, and DTM took 0.85 seconds.

To build the Precision vs. Recall graphs, we considered four cases of feature vectors used to represent the images: (a) using the 30 original features; (b) using the 21 features selected by StARMiner; (c) using the 21 features selected by Relief-F; (d) using the 21 features selected by DTM. Similarity queries were executed over the test set and the P&R graphs were drawn. Figure 7.5 shows the P&R graph obtained.

The graph in Figure 7.5 shows that the results obtained with 21 features are quite better than the results gotten with all 30 features. Thus, although using approximately 70% of the processing effort originally required, the precision of content-based queries is improved (the computational effort of a similarity query is proportional to the feature vector size).

To guarantee that we have selected the minimum set of relevant features that maintain the precision results, we also executed the same k-nearest neighbor

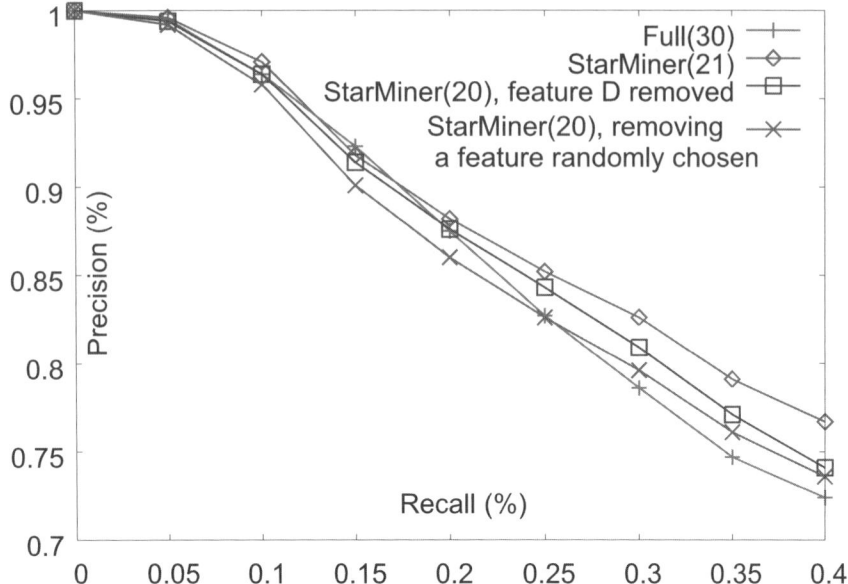

Fig. 7.6. Precision vs. Recall obtained using 30 features, 21 features selected by StARMiner, 20 features obtained by removing the feature D from those selected by StARMiner, and 20 features obtained by randomly removing one feature from those selected by StARMiner.

queries using 20 features obtained by randomly removing one feature from the 21 features selected by StARMiner, and, as a result of it, the P&R graph always worsened. Figure 7.6 shows a P&R graph comparing the precision of the queries using 30 features, 21 features selected by StARMiner, and 20 features obtained by randomly removing one feature from those selected by StARMiner. The selected features shown in Table 7.3 points out that the feature D has a very low contribution to represent the images. We redid the similarity queries removing the feature D of region 1 from the set of 21 selected features to check if it drops the precision values. The results of this test are also shown in Figure 7.6.

Comparing the curves from Figure 7.6, it is possible to note that removing the feature D from the 21 features selected by StARMiner, the precision values reduce much less than removing another feature, since the contribution of the feature D is small. For Region 1, the feature D is meaningful, but among the selected features, it is the one that brings the smallest contribution to differentiate the image types.

It is interesting to visualize the images returned when applying similarity queries using the whole set of features and the reduced set selected by StARMiner. We asked for the 15-nearest neighbors over the same image center shown on the left side of the screen of Figures 7.7 and 7.8.

Figure 7.7 shows the results when using the 30 original features, while Figure 7.7 shows the results using only the 21 selected features. The features

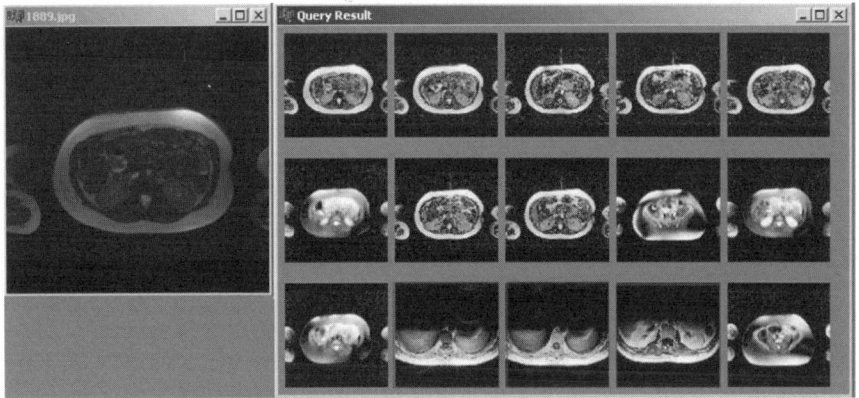

Fig. 7.7. Example of query using the 30 original features

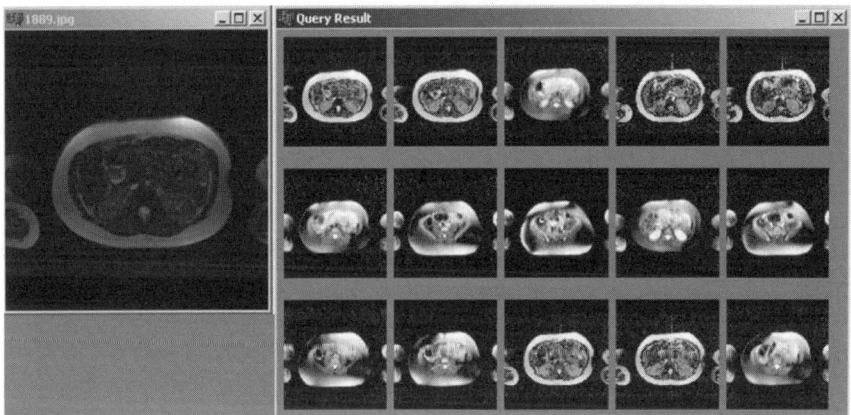

Fig. 7.8. Example of query using the 21 features selected by StARMiner

selected by StARMiner works better than the 30 original features for the executed query. The query performed using the original 30 features (Figure 7.7) achieved a precision of 80%, while the query performed using the 21 selected features achieved a precision of 100% (Figure 7.8). This happened in several situations, corroborating previous claims that allowing correlated attributes to be in the feature vector, instead of improving, it actually worsens the retrieval process.

7.5 Future Trends

The field of mining complex data has a wide variety of subjects still not explored, especially when leading with image analysis. Techniques to effectively reduce the semantic gap between low-level features representation and the high-level human interpretation of images should be further developed. Some improvements have been achieved using relevance feedback techniques. However, these techniques are still limited, and demand to gather information from sequences of relevance feedback iterations from different users. Relevance feedback approaches saturate the provided gain within a few number of iterations and are not automatic. It makes this approach susceptible to the problem of subjectivity and inconsistency caused by the typical human restrictions of time, interest or tiredness. However, the use of relevance feedback brings the human to be part of the process, and if well implemented, it can really diminish the semantic gap.

On the other hand, development in the field of association rule mining can effectively help to reduce the semantic gap automatically, where patterns relating semantic meaning to low-level representation can be found. Advances in feature selection techniques can also help to reduce the semantic gap by determining the minimal subset of features that effectively represents semantic information embedded in complex data.

Another question that should be answered is: how to make computers to recognize objects in a scene (image) in the same way as humans do? Researchers from several areas, including medicine, computer science, physics and electrical engineering, are working together to answer such question. In the future, the result of this effort might facilitate the human life in several aspects. For instance, small computer devices that reproduce the electrical stimuli occurred in the brain when a person is looking to a scene could be developed. Such devices could then be implanted in blind people's brains allowing them to see a scene, and even having the perception of colors and depth.

In the future, the image mining field, which is a special case of a complex data mining field, should follow two main directions:

- **Mining for patterns:** mining of images aimed at finding interesting patterns in a image, scene or sequence of scenes;
- **Mining for search:** mining to support content-based retrieval of images.

The development in the *mining for patterns* direction can lead to the discovery of other types of data mining tasks. For example, a new future task of data mining can be the *mapping task*. *Mapping task* can be a future field of data mining specialized in mapping data, such as space, time and behavior, in trends, objects and evolution lines. In addition, progress in *mining for patterns* direction will improve in medical care, agriculture, climate forecast, and space exploration. A recent demand for the *mining for patterns* research is regarding the improvement of surveillance and security.

With the development of the Internet, improvements in the *mining for search* direction are mandatory. Nowadays, a small potential of the Internet is effectively explored. There are very few and limited mechanisms of content-based search available in the Internet. The users desire to search for images, scenes or movies having a given subject. Today, well-known problems forbid them to perform such searches: (a) inconsistence in image descriptions; and (b) low-level features are not associated to their semantic meaning. New mining techniques should be developed to solve such problems. Such techniques might be available in the Internet, spreading information and knowledge.

It is expected that the progress in the field of mining complex data makes computational devices and programs to go a step further in the task of reproducing human brain functions. In fact, a human brain analyzes millions of complex data in a fraction of seconds, bringing to conscience just a small fraction of such data. This data fraction is the information that is stored in memory. Thus, the human brain is the perfect machine of mining complex data and, in future, scientists might be reproducing such machine. The selection of the most representative features extracted from images, based on association rules, is one of the most promising approaches that can turn these developments into reality.

7.6 Conclusions

This chapter details two issues related to mining complex data: feature selection and association rule mining. A new approach to select the most relevant image features in image datasets has been presented, consequently allowing dimensionality reduction of medical image features. The presented method uses statistical association rules to select the most relevant features. The presented mining algorithm, StARMiner, finds rules involving the attributes that most contribute to differentiate many classes of medical images. The accuracy of the method was verified in several case studies, and one representative was discussed in this chapter. The experiments performed k-nearest neighbor queries to measure the ability of the proposed technique in reducing the number of features needed to perform similarity queries maintaining the accuracy of the results. The results show that a significant reduction in the number of features can be obtained improving the retrieval efficacy of the features, leading to an impressive gain in time. The experiments also indicated that the features selected by StARMiner are more relevant to discriminate images than those selected by Relief-F and DTM algorithms. Furthermore, the results indicate that the mining of statistical association rules to select relevant features is an effective approach for dimensionality reduction in medical image datasets. Future trends in the field of image mining were discussed and an optimistic description of possible future scenery of the field was presented.

Acknowledgments

This research was supported by the Sao Paulo State Research Foundation (FAPESP) and by the Brazilian National Research Council (CNPq).

References

1. Agrawal, R., Imielinski, T., Swami, A.N.: Mining association rules between sets of items in large databases. In: ACM SIGMOD Intl. Conf. on Management of Data, Washington, D.C, pp. 207–216 (1993)

2. Agrawal, R., Srikant, R.: Fast algorithms for mining association rules. In: Intl. Conf. on Very Large Databases (VLDB), Santiago, Chile, pp. 487–499 (1994)
3. Apte, C., Liu, B., Pednault, E.P.D., Smyth, P.: Business applications of data mining. Communications of the ACM (CACM) 45(8), 49–53 (2002)
4. Aumann, Y., Lindell, Y.: A statistical theory for quantitative association rules. In: The fifth ACM SIGKDD Intl. Conf. on Knowledge discovery and data mining, San Diego, California, United States, pp. 261–270 (1999)
5. Baeza-Yates, R.A., Ribeiro-Neto, B.A.: Modern Information Retrieval. Addison-Wesley, Wokingham
6. Balan, A.G.R., Traina, A.J.M., Traina Jr., C.,, P.M.: d. A. Marques. Fractal analysis of image textures for indexing and retrieval by content. In: 18th IEEE Intl. Symposium on Computer-Based Medical Systems - CBMS, Dublin, Ireland, pp. 581–586 (2005)
7. Beyer, K., Godstein, J., Ramakrishnan, R., Shaft, U.: When is "nearest neighbor" meaningful? In: Beeri, C., Bruneman, P. (eds.) ICDT 1999. LNCS, vol. 1540, pp. 217–235. Springer, Heidelberg (1998)
8. Cardie, C.: Using decision trees to improve case-based learning. In: 10th Intl. Conf. on Machine Learning, pp. 25–32 (1993)
9. Comer, M.L., Delp, E.J.: The em/mpm algorithm for segmentation of textured images: Analysis and further experimental results. IEEE Trans. on Image Processing 9(10), 1731–1744 (2000)
10. Han, J., Pei, J., Yin, Y.: Mining frequent patterns without candidate generation. In: Int'l Conf. on Management of Data, Dallas, Texas, USA (2000)
11. Hsu, W., Lee, M.L., Zhang, J.: Image mining: Trends and developments. Journal of Intelligent Information Systems 19(1), 7–23 (2002)
12. Huang, S.H.: Dimensionality reduction in automatic knowledge acquisition: A simple greedy search approach. IEEE Trans. on Knowledge and Data Engineering (TKDE) 15(6), 1364–1373 (2003)
13. Kinoshita, S.K., de Azevedo-Marques, P.M., Pereira Jr., R.R., Heisinger Rodrigues, J.A.: Content-based retrieval of mammograms using visual features related to breast density patterns. Journal of Digital Imaging 20(2), 172–190 (2007)
14. Kira, K., Rendell, L.A.: A practical approach for feature selection. In: 9th Intl. Conf. on Machine Learning, Aberdeen, Scotland, pp. 249–256 (1992)
15. Kononenko, I.: Estimating attributes: Analysis and extension of relief. In: European Conf. on Machine Learning, Catania, Italy, pp. 171–182 (1994)
16. Liu, Y., Zhang, D., Lu, G., Ma, W.-Y.: A survey of content-based image retrieval with high-level semantics. Pattern Recognition Letters 40, 262–282 (2007)
17. Malcok, M., Aslandogan, Y., Yesildirek, A.: Fractal dimension and similarity search in high-dimensional spatial databases. In: IEEE Intl. Conf. on Information Reuse and Integration, Waikoloa, Hawaii, USA. pp. 380–384 (2006)
18. Molina, L.C., Belanche, L., Nebot, A.: Feature selection algorithms: A survey and experimental evaluation. In: IEEE Intl. Conf. on Data Mining 2002 (ICDM 2002), Washington, DC, USA, pp. 306–404 (2002)
19. Narendra, P.M., Fukunaga, K.: A branch and bound algorithm for feature subset selection. IEEE Trans. On Computer 26(9), 917–922 (1977)
20. Ordonez, C., Ezquerra, N., Santana, C.A.: Constraining and summarizing association rules in medical data. Knowledge and Information Systems 9(3), 259–283 (2006)
21. Quinlan, R.: C4.5: Programs for Machine Learning, San Mateo, CA (1993)

22. Refaeilzadeh, P., Tang, L., Liu, H.: On comparison of feature selection algorithms. In: AAAI 2007 Workshop on Evaluation Methods for Machine Learning II, Vancouver, Canada, pp. 1–6 (2007)
23. Ribeiro, M.X., Balan, A.G.R., Felipe, J.C., Traina, A.J.M., Traina Jr., C.: Mining statistical association rules to select the most relevant medical image features. In: 1st Intl. Workshop on Mining Complex Data (IEEE MCD 2005), Houston, USA, pp. 91–98 (2005)
24. Ribeiro, M.X., Marques, J., Traina, A.J.M., Traina-Jr, C.: Statistical association rules and relevance feedback: Powerful allies to improve the retrieval of medical images. In: 19th IEEE Intl. Symposium on Computer-Based Medical Systems, Salt Lake City, USA, pp. 887–892 (2006)
25. Ribeiro, M.X., Vieira, M.T.P.: A new approach for mining association rules in data warehouses. In: Christiansen, H., Hacid, M.-S., Andreasen, T., Larsen, H.L. (eds.) FQAS 2004. LNCS (LNAI), vol. 3055, pp. 28–110. Springer, Heidelberg (2004)
26. Savarese, A., Omiecinski, E., Navathe, S.: An efficient algorithm for mining association rules in large databases. In: 21st Conf. on Very Large Databases (VLDB 1995) (1995)
27. Srikant, R., Agrawal, R.: Mining quantitative association rules in large relational tables. In: ACM SIGMOD Intl. Conf. on Management of Data, Montreal, Canada, pp. 1–12 (1996)
28. Zaki, M.J., Parthasarathy, S., Ogihara, M., Li, W.: New algorithms for fast discovery of association rules. In: ACM SIGKDD Int'l Conf. on Knowledge Discovery and Data Mining, Newport Beach, USA (1997)
29. Zhang, S., Wu, X., Zhang, C.: Multi-database mining. IEEE Computational Intelligence Bulletin 2(1), 5–13 (2003)
30. Zhong, N., Ohshima, M., Yao, Y.Y., Ohsuga, S.: Interestingness, peculiarity, and multi-database mining. In: IEEE Intl. Conf. on Data Mining, pp. 566–573 (2001)

8

From Sequence Mining to Multidimensional Sequence Mining

Karine Zeitouni

PRISM Laboratory, University of Versailles-St-Quentin
45 Avenue des Etats-Unis, 78035 Versailles, France
Karine.Zeitouni@prism.uvsq.fr

Abstract. Sequential pattern mining has been broadly studied and many algorithms have been proposed. The first part of this chapter proposes a new algorithm for mining frequent sequences. This algorithm processes only one scan of the database thanks to an indexed structure associated to a bit map representation. Thus, it allows a fast data access and a compact storage in main memory. Experiments have been conducted using real and synthetic datasets. The experimental results show the efficiency of our method compared to existing algorithms. Beyond mining plain sequences, taking into account multidimensional information associated to sequential data is for a great interest for many applications. In the second part, we propose a characterization based multidimensional sequential patterns mining. This method first groups sequences by similarity; then characterizes each cluster using multidimensional properties describing the sequences. The clusters are built around the frequent sequential patterns. Thus, the whole process results in rules characterizing sequential patterns using multidimensional information. This method has been experimented towards a survey on population daily activity and mobility in order to analyze the profile of the population having typical activity sequences. The extracted rules show our method effectiveness.

Keywords: Sequential data mining, data structures, algorithms, optimization.

8.1 Introduction

The problem of mining sequential patterns was first introduced in the context of market basket analysis [2]. It aims to retrieve frequent patterns in the sequences of products purchased by customers through time ordered transactions. Several algorithms have been proposed in order to improve the performances and to reduce required space in memory [20] [26] [11] [13] [10]. Other works have concerned mining frequent sequences in DNA [9] or Web Usage Mining [19] [22]. In general, it could apply to any database containing a collection of item sequences. Our target application was based on a time-use survey, more precisely the database reports the daily activities and displacements carried out by each surveyed person in a household.

After having tested the most cited algorithms, we have observed their weakness to scale with large size datasets. Indeed, most of them perform multiple scans of the database, which is the main bottleneck in mining. Others require too large memory space to load the data when the database size increases.

D.A. Zighed et al. (Eds.): Mining Complex Data, SCI 165, pp. 133–152.
springerlink.com   © Springer-Verlag Berlin Heidelberg 2009

This leads us to propose, in the first part of this chapter, a new algorithm which aims at enhancing the performances of mining sequential association rules while reducing resource consumption. We make the following contributions:

1. This algorithm only makes one scan of the database;
2. It is based on a highly compact main memory data structure, saving the required storage resources;
3. It allows a fast access to the data thanks to index structure;
4. The experimental results show that our algorithm outperforms existing ones.

Mining *sequential patterns* has many interesting applications as it is. In addition to performance issue, many works have proposed new features, such as incremental sequential pattern mining [5] [12], restriction by constraints [14] or dealing with new types of data, such as query plans [26]. Among interesting extensions, *multidimensional sequence mining* is a major issue [16]. In fact, it allows discovering rules that links between sequences (e.g. transaction history) and regular attributes data (such as those in client file). Such rules may describe customer profiles, e.g. to which category of individuals a given purchase (or a given path traversal pattern) corresponds, or discover to which category of individuals correspond a given path traversal pattern. This is the subject of the second part of this chapter.

Our approach consists in mining individual profiles - based on attributes - for the most frequent sequential patterns. At this end, we propose a characterization based approach where a whole sequence is considered as a complex attribute. Thus, it makes sense to integrate reasoning on sequences (frequent patterns, similarity, grouping) while other dimensions are considered as descriptive of each sequence group. Briefly, our approach is based on two steps. The first gathers all database sequences around the most similar *sequential pattern* in order to derive classes of sequences represented by their *sequential patterns*. The second step describes these classes (and their sequential patterns) by their *multidimensional* attributes values characterizing them.

The characteristic rules express which attribute properties are typical to frequent sequential patterns. The sequential patterns should fulfill a given support threshold, and the rule should be satisfied with a given confidence threshold. The extraction of such rules raises three main questions:

1. How to determine that a sequence or a subsequence is similar to another?
2. How to group multidimensional sequences with a given sequential pattern?
3. How to determine the most characteristic properties for a group of sequences?

We have adopted different solutions that we detail afterward.

Both methods have been experimented using a real dataset related to population daily activity and mobility survey. It aims at mining frequent patterns of activity sequences, then at analyzing the profile of the population having those typical activity sequences. In addition, other experiments have been conducted to test the scalability of the sequential pattern mining algorithm, and use synthetic data and public available data widely used.

This chapter combines and extends two previously published papers, namely [17] [18]. It is organized as follows: a background section will provide an overview of the state of the art, before stating the concepts and definitions used further, and finally, it introduces the use case. Our contributions are then presented in two parts; the first is related to sequential pattern mining, then the second proposes an approach for multidimensional sequences mining. Each part details the implementation and the experimental results. The conclusion outlines the main contribution of the paper and discusses its perspectives.

8.2 Background

8.2.1 State of the Art

Most works related to mining frequent sequences are in the field of customer transaction analysis. Early work on frequent patterns -*Apriori* algorithm- only considered transactions, not sequence of transactions [1]. This algorithm is costly because it carries out multiple scans of the database to determine frequent subsets of items. Three algorithms dealing with sequence of transactions are presented and compared in [2], and [20]: *AprioriAll*, *AprioriSome* and *DynamicSome*. *AprioriAll* algorithm is an adaptation of *Apriori* to sequences where candidate generation and support are computed differently. *AprioriAll*, and *AprioriSome* only compute maximal frequent sequences. Their principle is to jump to candidates of size k+next(k) in the next scan, where next(k)>1. Maximum frequent sequences of lower size that have not been calculated are given in the backward phase. The value of next(k) increases with Pk = |Lk|/|Ck|, where Lk stands for frequent sequences of size k, and Ck the whole generated candidates of size k. *DynamicSome* algorithm is based on *AprioriSome* but uses a jump by a multiple of user defined step. *SPAM* algorithm [10] uses a bitmap representation of transaction sequences once the entire database has been loaded in a lexicographic tree. But this algorithm considers that the entire database and all used data structures should completely fit into main memory, and then do not adapt for large datasets. *GSP* algorithm [20] utilizes the property that all contiguous subsequences of a frequent sequence also have to be frequent. As *Apriori*, it generates frequent sequences, then candidate sequences by adding one or more items. *PrefixSPAN* [14] first finds the frequent items after scanning the database once. The sequence database is then projected, according to the frequent items, into several smaller databases. Finally, all sequential patterns are found by recursively growing subsequence fragments in each projected database. Employing a divide-and-conquer strategy with the *PatternGrowth* methodology, *PrefixSPAN* efficiently mines the complete set of patterns.

As for *multidimensional sequences mining*, it has been studied recently and partially. Main works deal with frequent patterns that mix sequence items and dimensions. The main contribution is from [16]. They have defined three algorithms for mining *multidimensional sequential patterns*: *UniSeq* uses *PrefixSPAN* [14] to mine *multidimensional sequential patterns* with sequences extended with *dimensional information*. *Dim-Seq* uses the BUC-like algorithm [4] to first mine *multidimensional patterns*, then *PrefixSPAN* is used to mine the *sequential patterns* associated to the *multidimensional patterns*. *Seq-Dim* first mines *multidimensional patterns* using BUC-like algorithm, then uses *PrefixSPAN* to mine the associated *sequential patterns*. Notice that the three algorithms produce the same result. However, they do not make a real distinction between *multidimensional values* and *sequential items*. Moreover, mining *multidimensional sequential patterns* do not allow extracting characteristic rules. Finally, they do not use specific methods to select the sequences while some multidimensional patterns are likely to be shared by groups of similar sequences.

8.2.2 Concepts and Definitions

In the proposed approach, we consider a database composed of sequences s and attributes A_i (with value a_i) describing the sequences (table 8.1).

Definition 1: Let I be a set of items. A sequence $s = <s_1, s_2, ..., s_i, ..., s_n>$ is defined as an ordered list of elements $s_i \in I$. s_i is also called a sequence element.

More general definition was initially proposed in [2], where each sequence element is rather an item-set. However, we argue that basic sequences composed of single items are sufficient for many applications, and adopt this definition.

Definition 2: A *sequences database* S is composed of a set of tuples (*sid, s*) where *sid* denotes the identity of the sequence *s*.

Definition 3: A sequence $s = <s_1, s_2, ..., s_m>$ is called a subsequence of another sequence $t = <t_1, t_2, ..., t_n>$, denoted $s \subseteq t$ if and only if there exist integers $1 \leq j1 < j2 < ... < jn \leq m$ such that $s_1 = t_{j1}, s_2 = t_{j2}, ..., s_n = t_{jn}$.

In other words, a sequence s is *included* in a sequence t if the ordered list of elements of s is included in the ordered list of elements of t.

Definition 4: The support of a sequence α in a sequence database S is the number of tuples in the database containing α i.e., support(α) | {<sid, s> | (<sid, s>$\in$ S) $\wedge$ ($\alpha \subseteq$ s)}.

Definition 5: Given a positive integer *min_support*, a sequence α is called a *sequential pattern* in sequence database S if support(α) $\geq$ min_support.

Another concept is the similarity of sequences in order to compare them. Sequence similarity is a well known problem in the fields of bio-informatics. It aims at determining if a DNA sequence is similar or not to another. The most popular algorithms are BLAST [3], FASTA [15] and LCSS (*Longest Common Subsequence*) [6]. We use this last method in our definition of sequence similarity. It measures the minimum number of insertions and deletions to transform s_1 into s_2. This method is widely implemented and used for this purpose.

Definition 6: Given two sequences $s = <s_1, s_2, ..., s_i, ..., s_m>$ and $t = <t_1, t_2, ..., t_j, ..., t_n>$ such that ($i \in [1,m]$, $j \in [1,n]$). Let lcs (s,t) the size of the longest common subsequence. The *dissimilarity distance* between s and t is defined as: d(s, t) = n + m - 2*lcs (s, t).

Definition 7: Given a positive integer DT called *dissimilarity threshold*, sequence s is said similar to a sequence t if their dissimilarity distance is lower than DT.

Definition 8: A *multidimensional sequences database* is of schema (*RID,S*, $A_1, ..., A_m$), where *RID* is a primary key, $A_1, ..., A_m$ are dimensions, and *S* the domain of sequences. A *multidimensional sequence* is defined as (*rid, s*, $a_1, ..., a_m$) where $a_i \in$ Ai, for $1 \leq i \leq$ m and *s* a sequence (see table 1).

Definition 9: We define a *Class of a Sequence* S, denoted S_C, as the cluster composed of multidimensional sequences whose sequences are similar to S, where S is a *sequential pattern*.

In order to define characteristic rules, we adopt and formalize the definition given in [9]. They define *characterization* of a sub-set as the property descriptions specific to this sub-set, comparing to all objects in the database.

Definition 10: We denote *se* a subset of the database DB, *prop* a multidimensional property $(a_{i1}, ..., a_{ik})$, $freq^{se}(prop)$ the number of objects in *se* that meet the property *prop*; and card(*se*) the cardinality of *se*. The *significance* of *prop* in the subset *se* is defined as: $F^{DB}_{se}(prop) = (freq^{se}(prop)/card(se)) / (freq^{DB}(prop)/card(DB))$

Definition 11: Given a real R standing for the significance threshold. *prop* is said characteristic of *se*, and denoted as: *prop* ➔ *se [significance]*, if and only if: $F^{DB}_{se}(prop) = significance \geq R$.

Definition 12: Let S_c be the class of a *sequential pattern* SP_c. We define a *multidimensional sequential rule* as: *prop* ➔ SP_c *[significance]*.

This *multidimensional sequential rule* means that the multidimensional property *prop* is characteristic of the *sequential pattern* SP_c with the computed *significance*. The example of table 8.1 shows a *multidimensional sequence database*. The tuple $(1, <s_1,s_2, .., s_n>, a_1, ..., a_m)$ stands for a *multidimensional sequence* of the database.

Table 8.1. A Multidimensional Sequence Database

RID	S	A_1	A_2	...	A_i	...	A_m
1	$<s_1, s_2, .., s_i, .., s_n>$	a_1	a_2		a_i		a_m
k	...						

8.2.3 Description of the Use Case and Datasets

The target application is related to population time-use analysis and more precisely their daily activities and displacements. This dataset describes daily activities and displacements carried out by each person of a surveyed household at the scale of a whole urban area. It can be seen as a sequence of activities, also called *activity program* [24]. For example, during a day, an individual can leave home, drive children to school, go to work, pick children up from school and come back to home. This sequence can be described as (Home, School, Work, School, Home). In order to simplify the notations, we represent each activity by a specific character, e. g. H for Home, W for Work, and S for School. Other activities are Market (denoted M), Restaurant (R), Leisure (L), etc. This alphabet can be as long as necessary. Then, by removing the comma separators, a sequence could be simplified to a character string, e.g. HSWSH for the previous sequence. Although we have used activity programs as an example in our experiments, the analysis is also relevant for other sequences, such as the transport mode used for displacements, the departure time, and so on.

Activity programs of most individuals may be the same or be similar. Each activity program could be seen as a sequence of single value, making it possible to discover frequent activity sequences that characterise groups of the surveyed individuals. This allows analyzing the mobility of this urban population. Likewise, when considering transport mode, schedules or duration sequences, it would be possible to determine a typology of used transport modes, schedules, and so on. Besides, the survey holds other information about individuals as their age, gender, profession, and so on. Then, each group having similar activity pattern is likely to have some characteristic attribute values. Hence, it is very relevant to characterize those groups and their corresponding sequential patterns. Those data have been used in the second part yielding characteristic rules for the groups of the surveyed individuals that share approximately those activity patterns. Moreover, other datasets have been used mainly to test the scalability and to compare our algorithms in the most widely used contexts, such as public datasets[1].

8.3   A Sequential Pattern Mining Algorithm

This section proposes an algorithm of *sequential pattern* mining. We focus on the specific case where the considered sequences composed of single items instead of item-sets. We argue that this is the case if the most popular sequences, such as DNA [9], documents, web-logs, or activity program sequences. Our algorithm is compared to *PrefixSPAN*, and SPAN, ones of the most efficient mining algorithms.

8.3.1   Algorithm Overview

The proposed algorithm is two phases. The first stage is the data encoding into a memory resident data structures. The second one is the frequent generation that in turn is composed of candidate generation, and candidate support checking. This algorithm only makes one scan of the database during which the total number of distinct sequences, the frequency of these sequences and the number of sequences by size are computed. This allows computing the support of each generated sequence.

The backbone of our approach is its main memory data structure, called *IBM* as Indexed Bit Map. It is composed of four elements: (i) a Bitmap: a matrix that represents the distinct sequences of the database, (ii) *SV*: a sequence vector that encodes all the ordered combinations of sequences, (iii) INDEX: an index on the Bitmap that allows a direct access to sequences according to their size, (iv) NB: a table associated to the Bitmap which informs about the frequency of each distinct sequence (Fig. 8.1). Only distinct sequences are stored in the Bitmap, they are classified by decreasing size. An index by size allows a direct access to sequences according to their size, which optimizes the candidate generation and counting phase of the algorithm. In the example of Fig. 8.1, IBM encodes the whole distinct sequences of the database. Notice that Index and the Bitmap are numbered down-up. Here, there are 6 distinct sequences of size 1 to 5: H, W, HSR, HSH, HSMH and HSRWH. Each cell of the INDEX indicates the first line where the corresponding size of sequence is stored. For

---

[1] http://kdd.ics.uci.edu

example, the cell number 5 (with value 6) corresponds to the line number 6 of the first sequence of size 5 encoded in the Bitmap. The table NB stores the frequency of each distinct sequence in the database. Thus the sequence HSMH occurs 20 times in the database.

In the first stage of the algorithm, INDEX, *SV*, NB and the Bitmap are built on the fly during one pass. At each insertion of a sequence, the Bitmap matrix may become larger, and a set of shifting operations are applied to the bit values stored in this table.

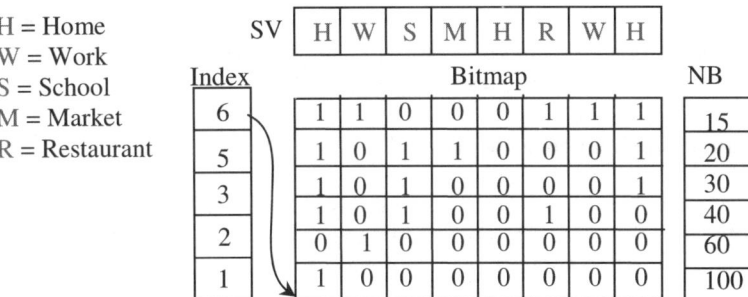

H = Home
W = Work
S = School
M = Market
R = Restaurant

SV	H	W	S	M	H	R	W	H

Index	Bitmap								NB
6	1	1	0	0	0	1	1	1	15
5	1	0	1	1	0	0	0	1	20
3	1	0	1	0	0	0	0	1	30
	1	0	1	0	0	1	0	0	40
2	0	1	0	0	0	0	0	0	60
1	1	0	0	0	0	0	0	0	100

Fig. 8.1. The data structure

IBM (sequence database *DB*, threshold *t*)
01 *SV* = φ   -- *SV* empty at the beginning
02 **For** each sequence *s* in *DB*
03   Merge-sequence-vector (*s*) -- merge *s* with *SV*
04   Update *NB*
05   If s ∉ *IBM*
06   Encode and Insert *s* in the *IBM*
07   Update *Index*   Shift the IBM table according to the new *SV*
09 *k* = 1
10 Gen-frequent-sequences (*t*) -- check the frequent of size 1
11 **While** there exist a frequent sequence of size *k*
12   *k* = *k*+1
13   Generate *Ck*
14   Gen-frequent-sequences (*t*)

Fig. 8.2. IBM algorithm

Fig. 8.2 shows the general IBM algorithm that takes as parameters: the sequence database *DB* and a threshold *t*. This value (*t*) stands for the minimum subsequences frequency taken into account for the generation of the candidates. Then for each sequence *s* that it reads from the database during the scan, *SV* (linc 03) is generated using a merging process (as detailed in Fig. 8.3). Three situations may hold. The first is when the sequence is already encoded in *the Bitmap* which only requires the update of *NB* table (line 04): the line corresponding to this sequence in *NB* is incremented to maintain its frequency counting. The second case is when the sequence itself is not present in *the Bitmap*, but may be represented in *SV*. In this case, *s* will be encoded and inserted in *the Bitmap* as a new line, involving the update of *Index* and *NB* accordingly (lines 06 and 07). The last case is when the sequence cannot be represented in *SV*, i.e. it is affected by the Merge-sequence-vector(*s*) function (Fig. 8.4). In this case, the algorithm adds the process of shifting the *Bitmap* table in order to adapt existing sequences coding to the new *SV* (line 08). Once all the data have been encoded in this structure, new candidates (line 10) are generated (see candidates generation section) and compared to the data stored in the *Bitmap* (line 11) with a fast access thanks to the *Index* (see support counting section).

8.3.2 Generation of the Sequence Vector

The sequence vector is generated during the unique scan of the database according to the algorithm of Fig. 8.3, which depicts the Merge-sequence-vector function which builds the *SV* vector according to the sequences present in the database.

Merge-sequence-vector (sequence *s*)
01 **For** each item *a* of *s*   -- *SV* is fetched by a position pointer initially set to 0
02   **If** *a* ∉ *SV suffix*   -- *after the current position*
03   **If** (∃ *b* ∈ *s* suffix such that *b* ∈ *SV* suffix)   insert *a* before *b*
04   **Else** insert *a* at the end of *SV*
05   update the current position of *SV* to the position of *a*

Fig. 8.3. Generation of the Sequence Vector

It takes as parameter a sequence s of the database. It fetches the sequence items one by one checking their presence in SV in the right order (line 01). If a given item a is not presents in *SV* after the current position (line 02), the function checks if their exist an item *b* in *SV* such that *b* in located after *a* in *s* and also located after the current position in *SV*. If those two conditions hold, item *a* will be inserted before item *b* in *SV* (line 03). If only the first condition holds, *a* will be positioned at the end of *SV* (line 04). The process will continue for the next item starting by the position of item *a* in *SV* (line 05). Notice that SV will stay unchanged when the sequence items belong to SV suffix. This holds when the current SV is sufficient to encode the new sequence. In the example of Fig. 8.4, the arrow with value 1 shows an insertion of a sequence HSH. This operation does not change SV. Since this sequence already exists in the Bitmap, only the corresponding frequency in the NB table is incremented. The arrow with value 2 shows an insertion of the sequence HWMWH. This operation modifies the SV vector; then, shifting operations are applied to the Bitmap in order to preserve the existing encoded sequences, and a new line is added in the Bitmap for the new sequence. Finally, the frequency of this new sequence is set to one.

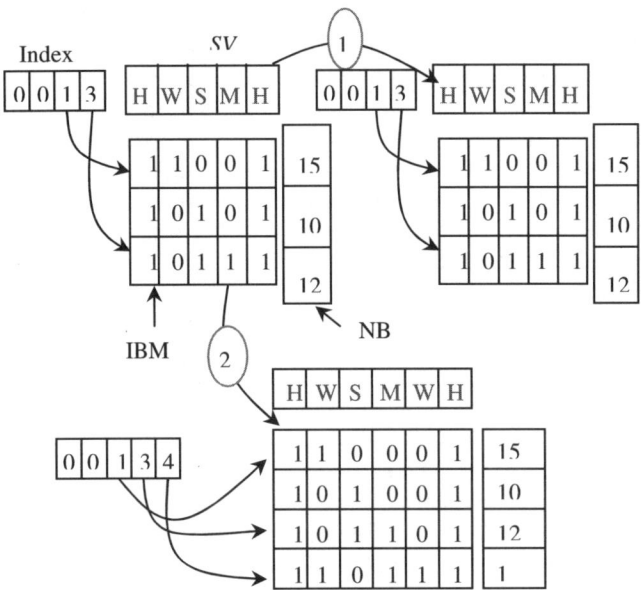

Fig. 8.4. IBM data structure generation process

8.3.3 Candidate Generation and Support Counting

The second phase is based on the *APRIORI* principle, with the difference that it operates in a main memory data structure instead of scanning the database. Using the Index table in IBM, the process starts by the sequences candidates of size 1. Their support is counted (as explained later). They are retained as frequent if the support fulfils the support threshold. Then, bigger size candidates are generated from these frequent items using the fusion process (joining phase) as in the GSP algorithm [20]. The frequencies of the candidates are counted again, and the process is repeated.

The fusion process consists to merge two candidates having a common contiguous subsequence of size n-2 in one sequence *s* of size n. For example, consider the two candidates c = MMH and c' = MHM of size 3. MH is a common contiguous subsequence of c and c', and of size 2. Therefore, the candidate s = MMHM is generated from c and c'.

More formally, given two sequences $c=c_1c_2...c_{n-1}$ and $c'=c'_1c'_2...c'_{n-1}$ of size n-1, a sequence of size n $s=s_1s_2...s_n$ may be generated from c and c' as follows:

(i) if $c = c_1$ and $c' = c'_1$, then $s = c_1c'_1$.
(ii) if n > 2 and $\forall\, i \in [2..n-1]$ $c_i = c'_{i-1}$, then $s= c_1c_2...c_{n-1}c'_{n-1}$

As for the support counting of the generated candidates, it is facilitated by the data structure. For a given candidate C of size S, the algorithm (see Fig. 8.5) first looks in the cell number S of the Index where the first sequence of size S is encoded. Then, this line is accessed. For each line starting from this line to the last line of the Bitmap table, the algorithm determines using the *SV* vector if C is contained in each line of

IBM. If so, the corresponding frequency of this sequence stored in the NB table, is added to the frequency of the candidate. After the comparison with each line until the last one, the support of C is computed.

Fig. 8.5 shows the Gen-frequent-sequences algorithm that determines all frequent candidates of size k. It takes as input a threshold t. All frequent candidates of size k are put in the set Lk. The function scans each generated candidate Ck of size k (line 02) and checks if it is included in the distinct sequences of size greater or equal than k in *IBM* (line 03). *Index*[k] points to the first sequence of size k and *max* stands for the sequence of greatest size encoded in *IBM*. Then if a candidate is a subsequence of a given sequence encoded in *IBM* (line 05), his support is incremented by its frequency (*NB*[k]) in the database. At the end of this process, if the support of a given candidate of size k is greater then the threshold t, the candidate is placed into the set Lk.

Gen-frequent-sequences (Threshold t)
01 $Lk = \varnothing$ -- Set of frequent sequences of size k
02 **For** all sequences $s \in Ck$ -- Candidates of size k
03   **For** all lines l in IBM from line *Index*[k] to max
04   **If** $s \subset l$ s.count = s.count + *NB*[k] -- frequence of s
05   **If** s.count $\geq t$ $Lk = Lk \cup s$

Fig. 8.5. Generation of Frequent Sequences of Size k

Suppose the example of Fig. 8.1 and a generated candidate C=HSH of size S=3. Then the algorithm will access the cell number 3 of the Index which pin points to the line 3 of the IBM table, where the first sequence of size 3 starts. This sequence does not contain C, but those in line 4 to 6 contain C. So the frequency of C is computed as 30+20+15=65. The support of C is equal to 65/(100+60+40+30+20+15)=0.245. If the support threshold is equal to 0.4, C candidate will not be retained as frequent pattern.

8.4 Implementation and Performance Study

The experiments aimed to validate our approach and to compare it to other methods. This comparison focuses on processing performances, storage costs, and scalability. The tests were performed on a 2.5 Ghz Pentium IV with 1 GB of memory running Microsoft Windows XP Professional. They aimed at showing: (i) our method effectiveness by applying it to a real dataset, and (ii) its scalability while increasing the data size.

Indeed, IBM has been performed on real data related to daily activity programs of the population of a north French urban area. In this application, the number of items is about 10. The database contains about 10,800 sequences among which 3,429 distinct sequences. The sequence size varies between 2 and 34 with an average size 6. The application aims at discovering frequent activity patterns in order to derive some population profiles. Interesting and previously unknown patterns have been produced which allow the decision makers better understanding of the daily activity and mobility for the reported population. For instance: (Home, Leisure, Home) is the most frequent with a support of 49%; (Home, Work, Home) is surprisingly less frequent with 37%, among which 9% correspond to (Home, Work, Home, Work, Home); (Home, School, Leisure, Home) appears in 11% of the sequences; while (Home, Shopping, Home, Leisure, Home) appears in 8% of them.

As for the scalability test, we generated and tested three different sizes of datasets with respectively: 100,000; 300,000; 600,000; and 1,000,000 rows. Items and the size of the sequences have been randomly generated for most experiments. The size of sequences was randomly generated from 2 to 60, and the number of distinct items was about 10 (from 0 to 9). This number has been pushed to 20, 35 and 75 distinct items, notably by using the public dataset from UCI KDD archive. For our experimentations, we have used the packages PrefixSPAN-0.4.tar.gz[2] and Spam.1.3.1.tar.gz[3]. Moreover, we have augmented the activity dataset by randomly generating arbitrary sequences. Finally, we have used the public datasets provided in UCI KDD archive chess.dat[4] that has a larger alphabet than in the activity survey dataset. We have measured two features: the overall response time from one hand, and the storage cost from the other.

Moreover, we have studied two variants of the IBM algorithm at the implementation level. The first one, called IBM2, is based on the observation that the binary matrix manipulation necessitates shifting operations. In order to avoid these superfluous and costly computations, we have proposed the variant algorithm IBM2 where the Bitmap is replaced by a matrix of Boolean types. This type takes 8 bits in languages like Java or C++. This solution requires more space in memory, but it performs better since accessing each cell becomes direct. The gain in performances of IBM2 has been confirmed by experimental results bellow. The second variant, called *IBM_OPT* (respectively *IBM2_OPT* when combined with *IBM2*), uses the fact that the structure of IBM in independent from the support threshold value. The idea is to serialize it as a *Java object* and stored in a file, which makes it available for later use after a simple load in the main memory. Thus the cost of pre-processing can be totally

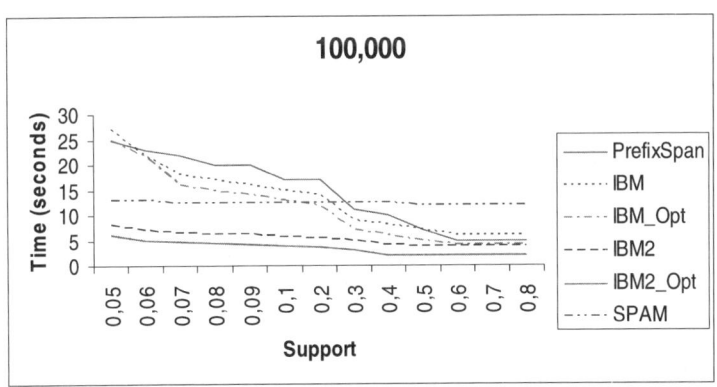

Fig. 8.6. Performances with 100,000 rows

[2] http://chasen.org/~taku/software/*PrefixSPAN*/
[3] http://himalaya-tools.sourceforge.net/Spam/#download
[4] http://kdd.ics.uci.edu

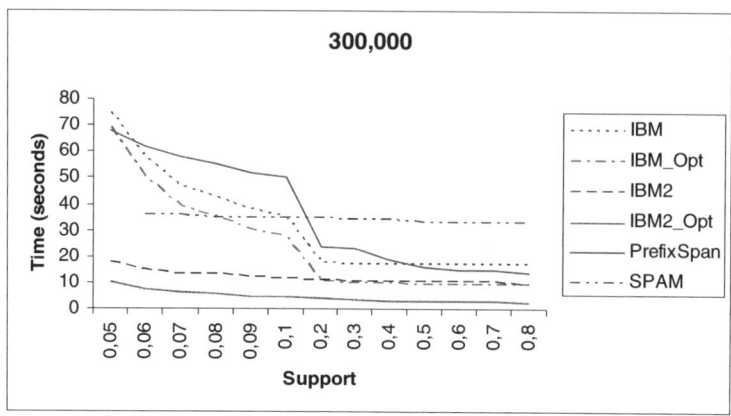

Fig. 8.7. Performances with 300,000 rows

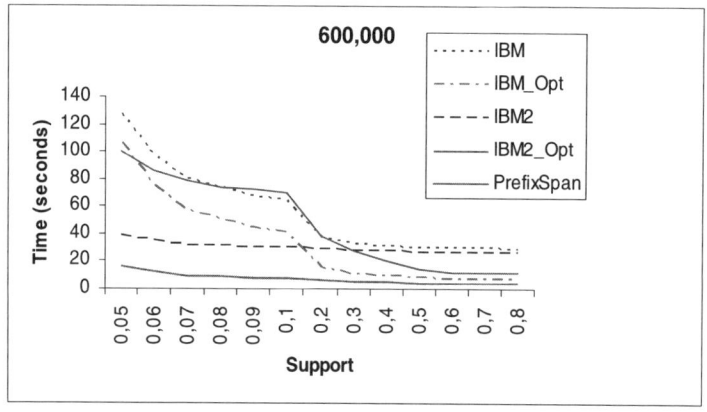

Fig. 8.8. Performances with 600,000 rows

avoided. This is particularly suitable to data mining process where the user interactively tries different support parameters before getting a satisfactory result. This optimization improves the performance in processing time, especially for large and very large databases as shown in the next section.

Fig. 8.6 shows all proposed algorithms, namely *IBM, IBM2, IBM_Opt* and *IBM2_Opt,* outperform *PrefixSPAN* in the most cases for the dataset having 100,000 sequences. *SPAM* has approximately the same response time and outperforms *IBM* when the support decreases, but *IBM* remains faster. The size of *SV* is about 173 items, the generation time for the structure is equal to 2 seconds and the number of distinct items in the dataset is around 17,000 sequences. This experiment also confirms that *IBM* outperforms *IBM*, and shows the performance gain increases as the support decreases. Indeed, a lower support increases the number of generated candidates. The more the candidates are generated, the more the number of comparisons in the structure increases and the more the number of shifting operations increases.

Finally, this experiment shows that *IBM_OPT* (respectively *IBM2_OPT*) outperform *IBM* (respectively *IBM2*). The performance gain (materialized by the gap between the corresponding curves) remains constant. Indeed, this difference corresponds to the saved time of the data structure construction and it is independent of the support threshold. For the dataset composed of 300,000 sequences, we observe the same behavior than in the previous experiment (Fig. 8.7). The size of *SV* is about 219 with 50,000 distinct sequences in the dataset and the structure is generated in 7.5 seconds. For a dataset composed of 600,000 sequences (Fig. 8.8), *SPAM* algorithm produced a memory overflow and failed. The size of *SV* is composed of 265 items with 90,000 distinct sequences in the dataset and the time to generate the structure is about 22.6 seconds. We observe that *PrefixSPAN* shows better performances than *IBM*. *IBM_OPT* and *IBM2_OPT* still outperform *PrefixSPAN*. *IBM2* remains better than *IBM* and than *PrefixSPAN*, especially for a support bellow 0.3.

For a dataset composed of 1,000,000 sequences, the size of *SV* is about 370 with 160,000 distinct sequences in the dataset and the time to generate the structure is about 80 seconds. *PrefixSPAN* consolidates its performance compared to *IBM*. *IBM_OPT* always outperforms *PrefixSPAN*. In general, *IBM* shows better performances than *PrefixSPAN*, unless the support is high. Conversely, *IBM_OPT* outperforms *PrefixSPAN* when the support is greater than 0.1. In order to measure the impact of the alphabet size on the performances, we have generated datasets using a larger alphabet (i.e. with more than 20 distinct items). Fig.. 8.10 and 8.11 show the results for 20 and 35 distinct items for dataset composed of respectively 130,000 and 100,000 sequences. For 20 distinct items, *SV* is composed of 192 items with 26,000 distinct sequences in the dataset. For the dataset composed of 35 distinct items, *SV* is composed of 214 items with 32,000 distinct sequences. We observe that *PrefixSPAN* outperforms *IBM_OPT* and IBM, but *IBM* and *IBM_OPT* win *PrefixSPAN*. Until 35 distinct items, *IBM_OPT* wins *PrefixSPAN*, with a support lower than 0.2. But, *PrefixSPAN* outperforms *IBM* due to the generation of the structure.

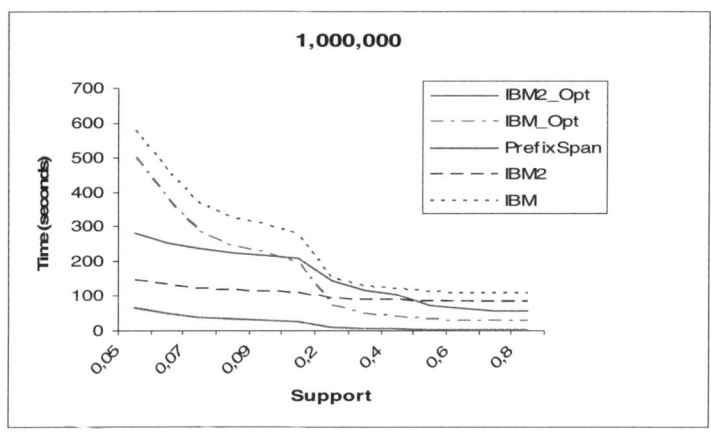

Fig. 8.9. Performances with 1,000,000 rows

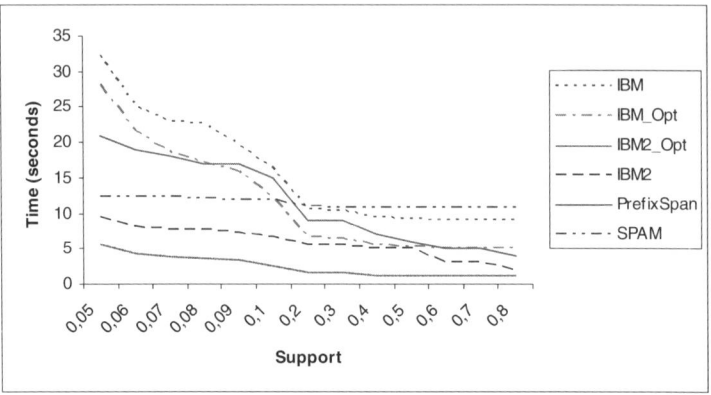

Fig. 8.10. Performances for a dataset with 130,000 rows and 20 distinct items

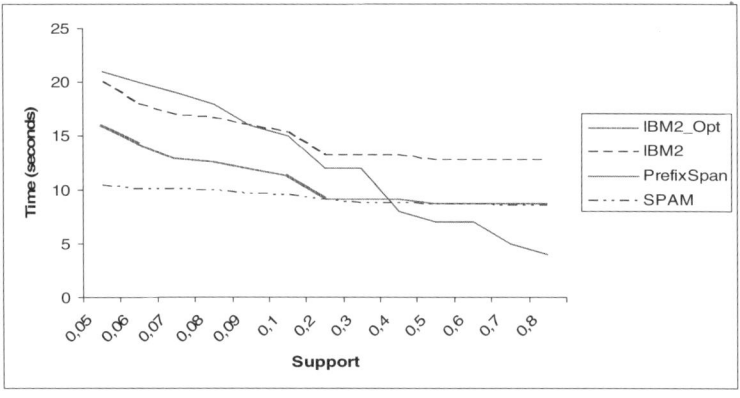

Fig. 8.11. Performances for a dataset with 100,000 rows and 35 distinct items

For more than 35 distinct items, *PrefixSPAN* becomes the faster, but this is true only for this large database size. Indeed, an interesting result has been obtained on a smaller database which has a larger alphabet (of 75 items).

This test (see Fig. 8.12) used the public dataset *chess.dat*, provided notably in the UCI KDD archive. The dataset is composed of 75 distinct items with 3,196 sequences of size between 36 to 37 items. The generation time of the structure is insignificant. Although the number of distinct items is greater than 35 (75 here), IBM and *IBM* outperform *PrefixSPAN* (see Fig. 8.12). The main reason is the size of sequences, which is larger (from 36 to 37) compared to the synthetic datasets. Thus height of the tree representing the projected database in *PrefixSPAN* is greater than for the synthetic data. Therefore, traversing this projected data in *PrefixSPAN* is slower than scanning our IBM data structure The memory space required by IBM, *IBM* and *PrefixSPAN* are respectively equal to 2.3 MB, 0.29 MB and 9 MB. Nevertheless, the *SV* is only composed of 91 items. Concerning the storage cost, we have measured the memory resource consumption for each algorithm for each tested dataset.

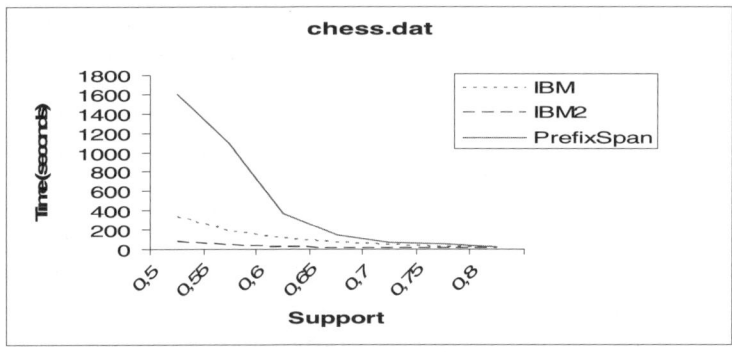

Fig. 8.12. Performances with *chess.dat* file

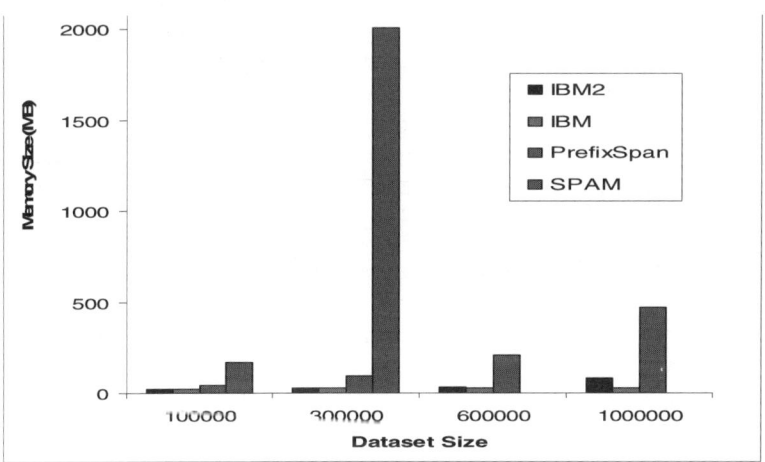

Fig. 8.13. Memory consumption

Fig. 8.13 shows the total memory consumption (in MB) used by *IBM* and *IBM2*, *SPAM*, and *PrefixSPAN*. For instance, with a database composed of 600,000 rows, *SV* contains about 265 values for 90,000 distinct rows. The size of the Bitmap in *IBM2* is then equal to: 265*90,000 = 23.85 MB. As *IBM* is 8 times more compact, the size of the binary Bitmap is less than 3 MB. With 1,000,000 rows (Fig. 8.9), *SV* contains 370 elements for 160,000 distinct rows. Then, the size of the Bitmap reaches 59.2 MB in *IBM2*, whereas the size of the Bitmap fits in 7.5 MB in *IBM* algorithm. These results show that *IBM* is more appropriate than *IBM2* for very large databases, due to data compression. However, *IBM2* runs faster than *IBM*. This is due to the costs of shifting operations necessary to access target values, whereas *IBM2* directly accesses the target sequences. As we can see in Fig. 8.13, the difference between memory costs in *IBM* and in *IBM2* is insignificant compared to memory costs in *SPAM* and in *Prefix-SPAN*. For example with 1,000,000 rows, the total memory size for *IBM* is equal to 28 MB whereas for *PrefixSPAN*, it is about 468 MB.

The size of the Bitmap also depends on the size of *SV*, which also increases with the number of distinct sequences. Notice that *SV* size does not depend on the size of the database itself. In fact, it only increases when the encountered sequence can not be encoded using the current *SV*. Moreover, since the probability to find common ordered items between *SV* and the current sequence becomes high as the building process advances, *SV* size becomes stable regardless of the size of the database.

In order to prove the efficiency of IBM in extreme cases, we have performed a test with a sequence c_T of the following form: HWHWHW...HW of size 200, composed of repeated series of H and W items. This type of sequence is likely to increase the size of the data structure. This experiment aims first to demonstrate that this situation does not affect the processing costs, and second to evaluate the loss of storage performance.

Tests have been done with datasets composed of 600,000 and 1,000,000 sequences. We observe no variation of processing costs. This is because, according to the sequence vector generation algorithm (see the general algorithm), the repeated items H and W that are not located in other sequences are put at the end of *SV*. Therefore, unless repeated series are actually frequent in the database, the probability to have long repeated series of HW in the middle of *SV* is very low. Then, using the data structure for candidate generation and frequent patterns will not be affected, because the items H and W put at the end of the structure would never been accessed.

8.5   Characterization Based Multidimensional Sequence Mining

In order to extract relevant knowledge from a multidimensional sequence database, we propose a characterization based approach. This process combines: *sequential patterns mining*, sequence similarity, and characterization. It produces *sequential multidimensional rules* as defined in Background section.

8.5.1 Method Overview

Our approach is sequence centered. Indeed, it aims to characterize the most frequent sequences according to the other dimensions. It is divided into three steps:

1. first, *sequential patterns* are mined ;
2. then these subsequences are used to represent classes composed of subset of database sequences. In order to determine to which class a sequence belongs, an algorithm of similarity between sequences is used. The sequences of database which are the most similar to a given *sequential pattern* are gathered in the same class. At the end of this process, a set of classes composed of database sequences and represented by their *sequential pattern* are defined. In order to reduce the intersections between classes, only *maximal sequential patterns* are considered i.e. *sequential patterns* that are not included in another one ;
3. once the classes have been built, a characterization algorithm is used to discover the dimension values characterizing them.

8.5.2 General Algorithm

Based on the three steps previously described, the general algorithm for mining multidimensional sequential patterns is depicted in Fig. 8.14. The input parameters of this algorithm are:

- A *multidimensional sequences* database DB.
- A support FT for mining *sequential patterns*.
- A *dissimilarity* threshold DT for mining the most similar sequence to a given *sequential pattern*.
- A *characterization* threshold CT and a set of attributes and associated values E(Ai., Vj).

The algorithm first mines *maximal sequential patterns* and places the result in D (line 03). Each *pattern* becomes a *model* of a class of sequences. Individuals whom sequences are the most similar to a given *maximal sequential pattern* are placed in the corresponding class S_c (line 05). The characterization algorithm (line 07) determines which attribute values characterize the subset S_c among E(Ai, Vj).

Mining_MSP(DB, FT, DT, CT, E(Ai,Vj))
01   D = ∅
02   R = ∅
03   D = Maximal_Sequential_Patterns (DB, FT)
04   **For each** S ∈ D
05 S_c = Cluster (DB, S, DT)
06   **For Each** S_c
07   R = Characterization (S_c, DB, CS, E(Ai., Vj))

Fig. 8.14. General Algorithm

Characterization (class *se*, database *DB*, real S, set E of (attribute A, value V))
00 *se*.characterization = ∅ ;
01 compute freq^{DB}(prop) for all properties *prop* = (attributes, values) ;
02 For each attribute A
03   **For** each value V of A
04   compute freq^{se}(*prop*) for the property *prop* = (attribute, value) ;
05   **If** F^{DB}_{se}(*prop*) ≥ S   Add (*se*.characterization, *prop*) ;

Fig. 8.15. Characterization Algorithm

The first step performs *maximal sequential pattern mining*. As seen before, *IBM* algorithm shows the best performances. Therefore, we use this algorithm at this step of the process. Frequent patterns are likely to have much more similar sequences in the database than non frequent ones. Therefore, our approach of class generation starts from those frequent patterns to determine classes' centers. We call them *models*.

However, not all *sequential patterns* can be considered as *models*, because some are very similar to each others, and then do not maximize the dissimilarity between different classes. Especially, as subsequences of frequent patterns are all frequents, their similarity is high. Therefore, *models* are restricted to maximal frequent patterns. The second step assigns the sequences of the database to the classes that are represented by those *models*. This is performed based on similarity computation between the database sequences and those *models*. However, most algorithms rather compute the *dissimilarity distance*. Depending on the selected similarity algorithm, the more the *dissimilarity distance* is small, the more the candidates are similar i.e. have common ordered items. We adopt LCSS algorithm because the presence of each item in the sequence is here more relevant than its occurrence probability (see section 2). A *dissimilarity* threshold allows the user to adjust the *overlapping* between classes as well as the class coverage. The more this threshold will be high, the more the *overlapping* probability will be high. Inversely, if this threshold is too small, too many database sequences may not be clustered.

The final step is to discover the main characteristics of individuals (profiles) characterizing each *class* i.e. the dimension values specific to a given *class*. In order to extract such knowledge, we propose an algorithm based on the definition of the characterization proposed by Han and. Kamber [9]. The algorithm is detailed in Fig. 15 above.

8.5.3 Experimentation

This experimentation has been done using the real dataset provided by *household activity survey*. The dataset contains 10840 individuals, with the dimensions: gender, age (10 classes of age) and work types (about 9). The total number of distinct sequence activities is about 3429. The tests have been realized using the following threshold values: *support*=0.1 ; *dissimilarity*=1 ; *characterization*= 8.

The algorithm has found 17 classes. For instance, the sequential pattern HMRH has been mined with a support equal to 0.14. Its similar sequences are: HWMRH, HMHR, HMRH, HMMRH, HRH, HLMRH, HMH, and HMRRH. This set of sequences composes the class of HMRH. The following rules have been mined for this class:

(gender = F) ∧ (30 < age < 40) ∧ (job category = TE8) ➔ HMRH [9.064].
(gender = F) ∧ (50 < age < 60) ∧ (job category = TE7) ➔ HMRH [10.413].
(gender = F) ∧ (50 < age < 60) ∧ (job category = TE5) ➔ HMRH [12.466].
(gender = H) ∧ (60 < age < 70) ∧ (job category = TE6) ➔ HMRH [10.407].

The three first rules mean that individuals having an activity sequence similar to HMRH standing for (Home, Market, Restaurant, Home) are frequently (significance between 9.064 and 12.466) women between 30 and 40 years old working as liberal profession (here TE8), or women working as clerk (TE5) or without profession (TE7) between 50 and 60 years old.

8.6 Conclusion and Future Work

This chapter has proposed novel approaches for mining sequential patterns and multidimensional characterization. The first part has presented a new algorithm *IBM* and its variants *IBM2*, *IBM_OPT* and *IBM2_OPT*. The aim of this algorithm is to mine frequent sequences in item sequences. *IBM* only makes one scan of the database and provides efficient data structure that optimizes memory space as well as access costs.

It has been applied to discover all frequent activity sequences in a time use survey database within an urban area. The experiments have shown that *IBM* and its variants provide better performances than existing algorithms in most cases: a better response time is reached, while a very low memory is required. Experimental results have also shown that *IBM2* and *IBM2_OPT* outperforms *IBM* and *IBM_OPT*, which in turns outperforms *SPAM* and *PrefixSPAN* for large and very large databases and for a limited number of distinct items.

Notice that the proposed data structure for *IBM* and *IBM2* algorithms, and especially the *SV* vector, could be used for other purposes as similarity search between sequences and sequence clustering. Another perspective is to apply it to different application contexts, as the analysis of query plans and genomic sequences. We have already experimented many various datasets: activity sequences, chess play sequences, and the web pages sequences *msnbc.dat* from the UCI KDD archive. In the context of the activity-mobility survey, we will explore the mining of spatial sequences, such as trajectories [7]. This is still a challenging research issue.

This chapter also describes a sequence centered approach for mining *multidimensional sequential rules*. It characterizes the main sequences and performs in three steps. First, it mines *sequential patterns* and sets them as class *models*. In the second step, a similarity algorithm is used to gather each *model* with similar sequences of the database to form the classes. Then in the third step, a characterization algorithm is used to extract attribute values characterizing each class. Compared to existing works, our approach: (i) allows extracting rules of the form $(a_1, a_2, a_i,..., a_n)$ ➔ $<s_1, s_2, s_i,..., s_n>$ [R] where attribute values $(a_1, a_2, a_i ..., a_n)$ are characteristics of objects whom sequences are similar to $<s_1, s_2, s_i,..., s_n>$ with a significance R; ii) considers similar sequences rather than their exact values, producing more relevant knowledge for a better decision–making. In perspective, this approach will be extended to multidimensional spatial and temporal sequences in order to characterize the trajectories of moving objects. At this end, similarity search of trajectories may use the approach of [23] which is also based on LCSS.

References

1. Agrawal, R., Srikant, R.: Fast Algorithms for Mining Association Rules. In: Proc. of the 20th Int. Conf. Very Large Data Bases (VLDB), Santiago, Chile (September 1994)
2. Agrawal, R., Srikant, R.: Mining sequential patterns. In: Proc. of the 11th Int'l Conference on Data Engineering, Taipei, Taiwan (March 1995)
3. Altschul, S.F., Gish, W., Miller, W., Myers, E.W., Lipman, D.J.: Basic Local Alignment Search Tool. J. Mol. Biol. 215, 403–410 (1990)
4. Beyer, K., Ramakrishnan, R.: Bottom-up computation of sparse and iceberg cubes. In: Proc. 1999 ACM-SIGMOD Int. Conf. Management of Data (SIGMOD 1999), Philadelphia, PA, pp. 359–370 (June 1999)
5. Cheng, H., Yan, X., Han, J.: IncSpan: Incremental mining of sequential patterns in large database. In: Proc. 2004 ACM SIGKDD Int. Conf. Knowledge Discovery in Databases (KDD 2004), Seattle, WA, pp. 527–532 (August 2004)
6. Freschi, V., Bogliolo, A.: Longest Common Subsequence between Run-Length-Encoded Strings: a New Algorithm with Improved Parallelism. Elsevier Information Processing Letters 90(4), 167–173 (2004)

7. Gidófalvi, G., Pedersen, T.B.: Mining Long Sharable Patterns in Trajectories of Moving Objects. In: Proc. of STDBM, pp. 49–58 (2006)
8. Han, J., Jamil, H.M., Lu, Y., Chen, L., Liao, Y., Pei, J.: DNA Miner: A system prototype for mining DNA sequences. In: The proc. of the ACM SIGMOD International Conference on the management of data, Day 21-24, Santa Barbara, CA, USA (2001)
9. Han, J., Kamber, M.: Data Mining Concepts and Techniques. Morgan Kaufmann Publishers, San Francisco (2001)
10. Jay, A., Johannes, G., Tomi, Y., Jason, F.: Sequential Pattern Mining using A Bitmap Representation. In: SIGMOD, Edmonton, Alberta, Canada, pp. 429–435 (July 2002)
11. Masseglia, F., Cathala, F., Poncelet, P.: The PSP approach for mining sequential patterns. In: Żytkow, J.M. (ed.) PKDD 1998. LNCS, vol. 1510, pp. 176–184. Springer, Heidelberg (1998)
12. Masseglia, F., Poncelet, P., Teisseire, M.: Incremental mining of sequential patterns in large databases. In: Data & Knowledge Engineering (DKE), vol. 46, pp. 97–121 (July 2003)
13. Pei, J., Han, J., Mortazavi-Asl, B., Pinto, H.: PrefixSPAN: Mining sequential patterns efficiency by prefix-projected pattern growth. In: Proc. of the International Conference on Data Engineering (ICDE), pp. 215–224 (2001)
14. Pei, J., Han, J., Wang, W.: Constraint-Based Sequential Pattern Mining: The Pattern-Growth Methods. Journal of Intelligent Information Systems 28(2), 133–160 (2007)
15. Pearson, W., Lipman, D.: Improved Tools for Biological Sequence Analysis. Proceedings of National Academic Science 85, 2444–2448 (1988)
16. Pinto, H., Han, J., Pei, J., Wang, K., Chen, Q., Dayal, U.: Multi-dimensionnal sequential pattern mining. In: ACM CIKM, pp. 81–88 (2001)
17. Savary, L., Zeitouni, K.: Indexed Bit Map (IBM) for Mining Frequent Sequences. In: Jorge, A.M., Torgo, L., Brazdil, P.B., Camacho, R., Gama, J. (eds.) PKDD 2005. LNCS (LNAI), vol. 3721, pp. 659–666. Springer, Heidelberg (2005)
18. Savary, L., Zeitouni, K.: Mining Multidimensional Sequential Rules - A Characterization Approach. In: First International Workshop on Mining Complex Data in conjunction with ICDM 2005 (IEEE MCD 2005), vol. 27, Houston, Texas, USA, pp. 99–102 (November 2005)
19. Spiliopoulou, M., Faulstich Lukas, C., Winkler, K.: A data miner analyzing the navigational behaviour of web users. In: Proc. Of the Workshop on Machine Learning in User Modelling of the ACAI 1999 Int. Conf., Creta, Greece (July 1999)
20. Srikant, R., Agrawal, R.: Mining Sequential Patterns: Generalizations and Performance Improvements. In: Proc. 5th EDBT, Mars 25-29, Avignon, France, pp. 3–17 (1996)
21. Srikant, R.: Fast Algorithms for Mining Association Rules. In: Proc. of the 20th Int. Conf. Very Large Data Bases (VLDB), Santiago, Chile (September 1994)
22. Srivastava, J., Cooley, R., Deshpande, M., Tan, P.-N.: Web Usage Mining: Discovery and Applications of Usage Patterns from Web Data. ACM SIGKDD Explorations, vol. 1(2), pp. 12–23 (2000)
23. Vlachos, M., Kollios, G., Gunopulos, D.: Discovering Similar Multidimensional Trajectories. In: Proc. of 18th International Conference on Data Engineering (ICDE), San Jose, CA, pp. 673–684 (2002)
24. Wang, D., Cheng, T.: A spatio-temporal data model for activity-based transport demand modeling. International Journal of Geographical Information Science 15(6), 561–585 (2001)
25. Zaki, M.J.: Efficient Enumeration of Frequent Sequences. Int. Conference on Information and Knowledge Management, Washington DC (November 1998)
26. Zaki, M.J., Lesh, N., Ogihara, M.: PLANMINE: Sequence mining for plan failures. In: Proc. 1998 Int. Conf. Knowledge Discovery and Data Mining (KDD 1998), New York, NY, pp. 369–373 (August 1998)

Tree-Based Algorithms for Action Rules Discovery

Zbigniew W. Raś[1,2], Li-Shiang Tsay[3], and Agnieszka Dardzińska[4]

[1] Univ. of North Carolina, Charlotte, Dept. of Computer Science,
   9201 Univ. City Blvd., Charlotte, NC 28223, USA
[2] Polish-Japanese Institute of Information Technology,
   Koszykowa 86, 02-008 Warsaw, Poland
[3] North Carolina A&T State Univ., School of Technology,
   Greensboro, NC 27411, USA
[4] Bialystok Technical Univ., Dept. of Computer Science,
   15-351 Bialystok, Poland

Abstract. One of the main goals in Knowledge Discovery is to find interesting associations between values of attributes, those that are meaningful in a domain of interest. The most effective way to reduce the amount of discovered patterns is to apply two interestingness measures, subjective and objective. Subjective measures are based on the subjectivity and understandability of users examining the patterns. They are divided into actionable, unexpected, and novel. Because classical knowledge discovery algorithms are unable to determine if a rule is truly actionable for a given user [1], we focus on a new class of rules [15], called E-action rules, that can be used not only for automatic analysis of discovered classification rules but also for hints of how to reclassify some objects in a data set from one state into another more desired one. Actionability is closely linked with the availability of flexible attributes [18] used to describe data and with the feasibility and cost [23] of desired re-classifications. Some of them are easy to achieve. Some, initially seen as impossible within constraints set up by a user, still can be successfully achieved if additional attributes are available. For instance, if a system is distributed and collaborating sites agree on the ontology [5], [6] of their common attributes, the availability of additional data from remote sites can help to achieve certain re-classifications of objects at a server site [23]. Action tree algorithm, presented in this paper, requires prior extraction of classification rules similarly as the algorithms proposed in [15] and [17] but it guarantees a faster and more effective process of E-action rules discovery. It was implemented as system $DEAR_2.2$ and tested on several public domain databases. Support and confidence of E-action rules is introduced and used to prune a large number of generated candidates which are irrelevant, spurious, and insignificant.

9.1 Introduction

Finding useful rules is an important task of knowledge discovery in data. Most of the researchers focused on techniques for generating patterns, such as classification rules, association rules...etc, from a data set. They assume that it is users responsibility to analyze these patterns and infer actionable solutions for specific

D.A. Zighed et al. (Eds.): Mining Complex Data, SCI 165, pp. 153–163.
springerlink.com   © Springer-Verlag Berlin Heidelberg 2009

problems within a given domain. The classical knowledge discovery algorithms have the potential to identify enormous number of significant patterns from data. Therefore, people are overwhelmed by a large number of uninteresting patterns which are very difficult to analyze and use to form timely solutions. So, there is a need to look for new techniques and tools with the ability to assist people in identifying rules with useful knowledge.

There are two types of interestingness measure: objective and subjective (see [10], [1], [19], [20]). Subjective interestingness measures include unexpectedness [19] and actionability [1]. When a rule contradicts the user's prior belief about the domain, uncovers new knowledge, or surprises him, it is classified as unexpected. A rule is deemed actionable, if the user can take action to gain an advantage based on this rule. Domain experts basically look at a rule and say that this rule can be converted into an appropriate action.

E-action rules mining is a technique that intelligently and automatically assists humans in acquiring useful information from data. This information can be turned into actions which can benefit users. The approach gives suggestions about how to change certain attribute values of a given set of objects in order to reclassify them according to a user wish.

There are two frameworks for mining actionable knowledge: loosely coupled and tightly coupled [9]. In the tightly coupled framework, action rules are extracted directly from a database [7], [8], [22]. In the loosely coupled framework, proposed in [15], the extraction of actionable knowledge is preceded by classification rules discovery. It is further subdivided into:

- strategies generating action rules from certain pairs of classification rules [18], [21], [23],
- strategies generating action rules from single classification rules [16], [24].

This paper relates to a loosely coupled framework. In most of the algorithms for action rules mining, there is no guarantee that the discovered patterns in the first step will lead to actionable knowledge that is capable of maximizing profits. One way to approach this problem is to assign a cost function to all changes of attribute values [24]. If changes of attribute values in the classification part of an action rule are too costly, then they can be replaced by composing this rule with other action rules, as proposed in [23]. Each composition of these rules uniquely defines a new action rule. Objects supporting each new action rule, let's say r, are the same as objects supporting the action rule replaced by r but the cost of reclassifying them is lower for the new rule.

E-action rule models the actionability concept in a better way than action rule [15] by introducing a notion of its supporting class of objects. E-action rules are constructed from certain pairs of classification rules. They can be used not only for evaluating discovered patterns but also for reclassifying some objects in a dataset from one state into a new more desired state. For example, classification rules found from a bank's data can be very useful to describe who is a good client (whom to offer some additional services) and who is a bad client (whom to watch carefully to minimize the bank loses). However, if bank managers need to improve their understanding of customers and seek for specific actions to improve the services, mere classification rules are not sufficient. In this paper, we propose to use classification rules to build a new strategy of action based on their condition features in order to get a desired effect on their decision feature. Going back to the bank example, the strategy of action would consist of modifying some condition features in order to improve our understanding of customers behavior and then improve the services. E-action rules are useful in many other fields, including medical diagnosis. In medical diagnosis, classification rules can explain the relationships between symptoms and sickness and in predicting the diagnosis of a new patient. E-action rules are useful in providing a hint to a doctor what symptoms have to be modified in order to recover a certain group of patients from a given illness.

Action Tree algorithm is presented for generating E-action rules and it is implemented as System $DEAR\_2.2$. The algorithm follows a top-down strategy that searches for a solution in a part of the search space. It is seeking at each stage for a stable attribute that has a least number of values. Then, the set of rules is split recursively using that attribute. When all stable attributes are processed, the final subsets are split further based on a decision attribute. This strategy generates an action tree which is used to construct E-action rules from the leaf nodes of the same parent.

9.2   Information System and E-Action Rules

An information system is used for representing knowledge. Its definition, presented here, is due to Pawlak [12].

By an information system we mean a pair $S = (U, A)$, where:

- U is a nonempty, finite set of objects,
- A is a nonempty, finite set of attributes i.e. $a : U \longrightarrow V_a$ is a function for any $a \in A$, where V_a is called the domain of a.

Elements of U are called objects. In this paper, for the purpose of clarity, objects are interpreted as customers. Attributes are interpreted as features such as, offers made by a bank, characteristic conditions etc.

We consider a special case of information systems called decision tables [12]. In any decision table together with the set of attributes a partition of that set into conditions and decisions is given. Additionally, we assume that the set of conditions is partitioned into stable conditions and flexible conditions. For simplicity reason, we assume that there is only one decision attribute. *Date of birth* is an example of a stable attribute. The *interest rate* on any customer account is an example of a flexible attribute as the bank can adjust rates. We adopt the following definition of a decision table:

By a decision table we mean any information system $S = (U, A_{St} \cup A_{Fl} \cup \{d\})$, where $d \notin A_{St} \cup A_{Fl}$ is a distinguished attribute called the decision. The elements of A_{St} are called stable conditions, whereas the elements of A_{Fl} are called flexible conditions.

Table 9.1. Decision System

	a	b	c	d
x_1	2	1	2	L
x_2	2	1	2	L
x_3	1	1	0	H
x_4	1	1	0	H
x_5	2	3	2	H
x_6	2	3	2	H
x_7	2	1	1	L
x_8	2	1	1	L
x_9	2	2	1	L
x_{10}	2	3	0	L
x_{11}	1	1	2	H
x_{12}	1	1	1	H

As an example of a decision table we take $S = (\{x_1, x_2, x_3, x_4, x_5, x_6, x_7, x_8,$ $x_9, x_{10}, x_{11}, x_{12}\}, \{a, c\} \cup \{b\} \cup \{d\})$ represented by Table 9.1. The set $\{a, c\}$ lists stable attributes, b is a flexible attribute and d is a decision attribute. Also, we assume that H denotes a *high* profit and L denotes a *low* one.

In order to induce rules in which the THEN part consists of the decision attribute d and the IF part consists of attributes belonging to $A_{St} \cup A_{Fl}$, for instance system $LERS$ [4] can be used for rules extraction.

Alternatively, we can extract rules from sub-tables $(U, B \cup \{d\})$ of S, where B is a d-reduct (see [11]) of S, to improve efficiency of the algorithm when the number of attributes is large. The set B is called d-reduct in S if there is no proper subset C of B such that d depends on C. The concept of d-reduct in S was introduced to induce efficiently rules from S describing values of the attribute d depending on minimal subsets of $A_{St} \cup A_{Fl}$.

By $L(r)$ we mean all attributes listed in the IF part of a rule r. For example, if $r_1 = [(a_1, 2) \wedge (a_2, 1) \wedge (a_3, 4) \longrightarrow (d, 8)]$ is a rule then $L(r_1) = \{a_1, a_2, a_3\}$.

By $d(r_1)$ we denote the decision value of that rule. In our example $d(r_1) = 8$. If r_1, r_2 are rules and $B \subseteq A_{St} \cup A_{Fl}$ is a set of attributes, then $r_1/B = r_2/B$ means that the conditional parts of rules r_1, r_2 restricted to attributes B are the same. For example if $r_2 = [(a_2, 1) * (a_3, 4) \longrightarrow (d, 1)]$, then $r_1/\{a_2, a_3\} = r_2/\{a_2, a_3\}$.

In our example, we get the following certain rules with support greater or equal to 2:

$(b, 3) * (c, 2) \longrightarrow (d, H)$, $(a, 1) * (b, 1) \longrightarrow (d, L)$,
$(a, 1) * (c, 1) \longrightarrow (d, L)$, $(b, 1) * (c, 0) \longrightarrow (d, H)$,
$(a, 1) \longrightarrow (d, H)$

Now, let us assume that $(a, v \longrightarrow w)$ denotes the fact that the value of attribute a has been changed from v to w. Similarly, the term $(a, v \longrightarrow w)(x)$

means that $a(x) = v$ has been changed to $a(x) = w$. Saying another words, the property (a, v) of object x has been changed to property (a, w).

Let $S = (U, A_{St} \cup A_{Fl} \cup \{d\})$ be a decision table and rules r_1, r_2 are extracted from S. The notion of an extended action rule (E-action rule) was given in [21]. Its definition is given below. We assume here that:

- B_{St} is a maximal subset of A_{St} such that $r_1/B_{St} = r_2/B_{St}$,
- $d(r_1) = k_1$, $d(r_2) = k_2$ and $k_1 \leq k_2$,
- $(\forall a \in [A_{St} \cap L(r_1) \cap L(r_2)])[a(r_1) = a(r_2)]$,
- $(\forall i \leq q)(\forall e_i \in [A_{St} \cap [L(r_2) - L(r_1)]])[e_i(r_2) = u_i]$,
- $(\forall i \leq r)(\forall c_i \in [A_{Fl} \cap [L(r_2) - L(r_1)]])[c_i(r_2) = t_i]$,
- $(\forall i \in p)(\forall b_i \in [A_{Fl} \cap L(r_1) \cap L(r_2)])[[b_i(r_1) = v_i]\&[b_i(r_2) = w_i]]$.

Let $A_{St} \cap L(r_1) \cap L(r_2) = B$. By (r_1, r_2) -E-action rule on $x \in U$ we mean the expression r:

$$[\prod\{a = a(r_1) : a \in B\} \wedge (e_1 = u_1) \wedge (e_2 = u_2) \wedge ... \wedge (e_q = u_q) \wedge (b_1, v_1 \longrightarrow w_1) \wedge (b_2, v_2 \longrightarrow w_2) \wedge ... \wedge (b_p, v_p \longrightarrow w_p) \wedge (c_1, \longrightarrow t_1) \wedge (c_2, \longrightarrow t_2) \wedge ... \wedge (c_r, \longrightarrow t_r)](x) \Longrightarrow [(d, k_1 \longrightarrow k_2)](x)$$

Object $x \in U$ supports (r_1, r_2)-E-action rule r in $S = (U, A_{St} \cup A_{Fl} \cup \{d\})$, if the following conditions are satisfied:

- $(\forall i \leq p)[b_i \in L(r)][b_i(x) = v_i] \wedge d(x) = k_1$
- $(\forall i \leq p)[b_i \in L(r)][b_i(y) = w_i] \wedge d(y) = k_2$
- $(\forall j \leq p)[a_j \in (A_{St} \cap L(r_2))][a_j(x) = u_j]$
- $(\forall j \leq p)[a_j \in (A_{St} \cap L(r_2))][a_j(y) = u_j]$
- object x supports rule r_1
- object y supports rule r_2

By the support of E-action rule r in S, denoted by $Sup_S(r)$, we mean the set of all objects in S supporting r. Saying another words, this set is defined as:

$$\{x : [a_1(x) = u_1] \wedge [a_2(x) = u_2] \wedge ... \wedge [a_q(x) = u_q] \wedge [b_1(x) = v_1] \wedge [b_2(x) = v_2] \wedge ... \wedge [b_p(x) = v_p] \wedge [d(x) = k_1]\}.$$

By the confidence of r in S, denoted by $Conf_S(r)$, we mean
$[Sup_S(r)/Sup_S(L(r))] \times [Conf(r_2)]$
To find the confidence of (r_1, r_2)-E-action rule in S, we divide the number of objects supporting (r_1, r_2)-action rule in S by the number of objects supporting left hand side of (r_1, r_2)-E-action rule times the confidence of the classification rule r_2 in S.

9.3 Discovering E-Action Rules

In this section we present a new algorithm, called Action-Tree algorithm, for discovering E-action rules. Basically, we partition the set of classification rules R discovered from a decision system $S = (U, A_{St} \cup A_{Fl} \cup \{d\})$, where A_{St} is the set of stable attributes, A_{Fl} is the set of flexible attributes and, $V_d = \{d_1, d_2, ..., d_k\}$

is the set of decision values, into subsets of rules having the same values of stable attributes in their classification parts and defining the same value of the decision attribute. Classification rules can be extracted from S using, for instance, discovery system $LERS$ [2].

Action-tree algorithm for extracting E-Action rules from decision system S is as follows:

i. Build Action-Tree
   a. Partition the set of classification rules R in a way that two rules are in the same class if their stable attributes are the same
      1. Find the cardinality of the domain V_{v_i} for each stable attribute v_i in S.
      2. Take v_i, which $card(V_{v_i})$ is the smallest, as the splitting attribute and divide R into subsets each of which contains rules having the same value of the stable attribute v_i.
      3. For each subset, obtained in step 2, determine if it contains rules of different decision values and different values of flexible attributes. If it does, go to step 2. If it doesn't, there is no need to split the subset further and we place a mark.
   b. Partition each resulting subset into new subsets each of which contains only rules having the same decision value.
   c. Each leaf of the resulting tree represents a set of rules which do not contradict on stable attributes and also it uniquely defines decision value d_i. The path from the root to that leaf gives the description of objects supported by these rules.
ii. Generate E-action rules
   a. Form E-action rules by comparing all unmarked leaf nodes of the same parent.
   b. Calculate support and confidence of each new-formed E-action rule. If support and confidence meet the thresholds set up by user, print the rule.

The algorithm starts at the root node of the tree, called E-action tree, representing all classification rules extracted from S. A stable attribute is selected to partition these rules. For each value of that attribute an outgoing edge from the root node is created, and the corresponding subset of rules that have the attribute value assigned to that edge is moved to the newly created child node. This process is repeated recursively for each child node. When we are done with stable attributes, the last split is based on a decision attribute for each current leaf of E-action tree. If at any time all classification rules representing a node have the same decision value, then we stop constructing that part of the tree. We still have to explain which stable attributes are chosen to split classification rules representing a node of E-action tree. The algorithm selects any stable attribute which has the smallest number of possible values among all the remaining stable attributes. This step is justified by the need to apply a heuristic strategy (see [14]) which will minimize the number of edges in the resulting tree and the same make the time-complexity of the algorithm lower.

We have two types of nodes: a leaf node and a non-leaf node. At a non-leaf node, the set of rules is partitioned along the branches and each child node gets its corresponding subset of rules. Every path to the decision attribute node, one level above the leaf node, represents a subset of the extracted classification rules when the stable attributers have the same value. Each leaf node represents a set of rules, which do not contradict on stable attributes and also define decision value d_i. The path from the root to that leaf gives the description of objects supported by these rules.

Instead of splitting the set of rules R by stable attributes and next by the decision attribute, we can also start the partitioning algorithm from a decision attribute. For instance, if a decision attribute has 3 values, we get 3 initial sub-trees. In the next step of the algorithm, we start splitting these sub-trees by stable attributes following the same strategy as the one presented for E-action trees. This new algorithm is called action-forest algorithm.

Now, let us take Table 9.1 as an example of a decision system S. Attributes a, c are stable and b, d flexible. Assume now that our plan is to re-classify some objects from the class $d^{-1}(\{d_i\})$ into the class $d^{-1}(\{d_j\})$. In our example, we also assume that $d_i = (d, L)$ and $d_j = (d, H)$.

First, we represent the set R of certain rules extracted from S as a table (see Table 9.2). The first column of this table shows objects in S supporting the rules from R (each row represents a rule). For instance, the second row represents the rule $[[(a, 2) \wedge (c, 1)] \Rightarrow (d, L)]$. The construction of an action tree starts with the set R as a table (see Table 9.2) representing the root of the tree (T_1 in Fig. 9.1). The root node selection is based on a stable attribute with the smallest number of values among all stable attributes. The same strategy is used for a child node selection. After labelling the nodes of the tree by all stable attributes, the tree is split based on the value of the decision attribute. Referring back to the example in Table 9.1, we use stable attribute a to split that table into two sub-tables defined by values $\{1, 2\}$ of attribute a. The domain of attribute a is $\{1, 2\}$ and the domain of attribute c is $\{0, 1, 2\}$. Clearly, $card[V_a]$ is less than $card[V_c]$ so we partition the table into two: one table with rules containing $a = 0$ and another with rules containing $a = 2$. Corresponding edges are labelled by values of attribute a. All rules in the sub-table T_2 have the same decision value. So, we

Table 9.2. Set of rules R with supporting objects

Objects	a	b	c	d
$\{x_3, x_4, x_{11}, x_{12}\}$	1			H
$\{x_1, x_2, x_7, x_8\}$	2		1	L
$\{x_7, x_8, x_9\}$	2		0	L
$\{x_3, x_4\}$		1	0	H
$\{x_5, x_6\}$		3	2	H

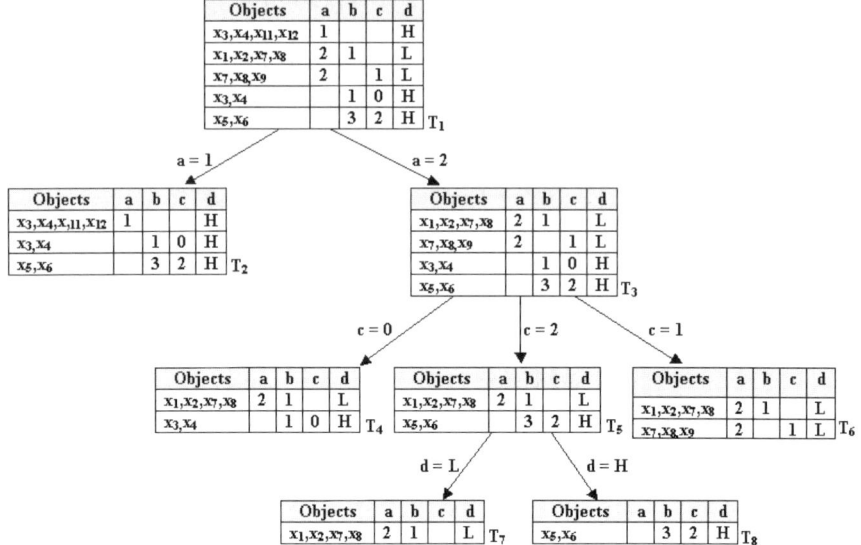

Fig. 9.1. Action tree

can not construct E-action rule from that sub-table which means it is not divided any further. Because rules in the sub-table T_3 contain different decision values and a stable attribute c, T_3 is partitioned into three sub-tables, one with rules containing c=0, one with rules containing c=1, and one with rules containing c=2. Now, rules in each of the sub-tables do not contain any stable attributes. Sub-table T_6 is not split any further for the same reason as sub-table T_2. All objects in sub-table T_4 have the same value of flexible attribute b. There is no way to form a workable strategy from this sub-table so it is not partitioned any further. Sub-table T_5 is divided into two new sub-tables. Each leaf represents a set of rules, which do not contradict on stable attributes and also define decision value d_i.

The path from the root of the tree to that leaf gives the description of objects supported by these rules. Following the path labelled by value $[a = 2]$, $[c = 2]$, and $[d = L]$, we get table T_7. Following the path labelled by value $[a = 2]$, $[c = 2]$, and $[d = H]$, we get table T_8. Because T_7 and T_8 are sibling nodes, we can directly compare pairs of rules belonging to these two tables and construct one E-action rule such as:

$$[[(a, 2) \wedge (b, 1 \rightarrow 3)] \Rightarrow (d, L \rightarrow H)].$$

After the rule is formed, we evaluate it by checking its support and its confidence ($sup = 4$, $conf = 100\%$).

This new algorithm (called DEAR_2.2) was implemented and tested on several data sets from UCI Machine Learning Repository. In all cases, the action tree algorithm was more efficient then the action forest algorithm. The generated E-action rules by both algorithms are the same. The confidence of E-action rules is higher than the confidence action rules.

9.4 Conclusion

E-action rules are structures that represent actionability in an objective way. The strategy used to generate them is data driven and domain independent because it does not depend on domain knowledge. Although the definition of E-action rules is purely objective, we still can not get rid of some degree of subjectivity in determining how actions can be implemented. To build E-action rules, we divide all attributes into two subsets, stable and flexible. Obviously, this partition has to be done by users who decide which attributes are stable and which are flexible. This is a purely subjective decision. A stable attribute has no influence on change, but any flexible attribute may influence changes. Users have to be careful judging which attributes are stable and which are flexible. If we apply E-action rule on objects then it shows how values of their flexible features should be changed in order to achieve their desired re-classification. Stable features always will remain the same. Basically, any E-action rule identifies a class of objects that can be reclassified from an undesired state to a desired state by properly changing some of the values of their flexible features. How to implement these changes often depends on the user. If the attribute is an interest rate on the banking account then banks can take appropriate action as the rule states (i.e., change lower interest rate to 4.75%). In this case, it is a purely objective action. However, if the attribute is a fever then doctors may lower the temperature by following a number of different actions. So, this is a purely subjective concept. Basically, we cannot eliminate some amount of subjectivity in the process of E-action rules construction and implementation.

Acknowledgement

This research was partially supported by the National Science Foundation under grant IIS-0414815.

References

1. Adomavicius, G., Tuzhilin, A.: Discovery of actionable patterns in databases: the action hierarchy approach. In: Proceedings of KDD 1997 Conference, Newport Beach, CA. AAAI Press, Menlo Park (1997)
2. Chmielewski, M.R., Grzymala-Busse, J.W., Peterson, N.W., Than, S.: The rule induction system LERS - a version for personal computers. In: Foundations of Computing and Decision Sciences. vol. 18(3-4), Institute of Computing Science, Technical University of Poznan, Poland, pp. 181–212 (1993)
3. Geffner, H., Wainer, J.: Modeling action, knowledge and control. In: Prade, H. (ed.) ECAI 1998, Proceedings of the 13th European Conference on AI, pp. 532–536. John Wiley & Sons, Chichester (1998)

4. Grzymala-Busse, J.: A new version of the rule induction system LERS. Fundamenta Informaticae 31(1), 27–39 (1997)
5. Guarino, N.: Formal Ontology in Information Systems. IOS Press, Amsterdam (1998)
6. Guarino, N., Giaretta, P.: Ontologies and knowledge bases, towards a terminological clarification. In: Towards Very Large Knowledge Bases: Knowledge Building and Knowledge Sharing. IOS Press, Amsterdam (1995)
7. He, Z., Xu, X., Deng, S., Ma, R.: Mining action rules from scratch. Expert Systems with Applications 29(3), 691–699 (2005)
8. Im, S., Ras, Z.W.: Action rule extraction from a decision table: ARED. In: An, A., et al. (eds.) Foundations of Intelligent Systems, Proceedings of ISMIS 2008, Toronto, Canada, Springer, Heidelberg (2008)
9. Kaur, H.: Actionable rules: issues and new directions. In: Transactions on Engineering, Computing and Technology, pp. 61–64. World Informatica Society (April 2005)
10. Liu, B., Hsu, W., Chen, S.: Using general impressions to analyze discovered classification rules. In: Proceedings of KDD 1997 Conference, Newport Beach, CA. AAAI Press, Menlo Park (1997)
11. Pawlak, Z.: Rough sets-theoretical aspects of reasoning about data. Kluwer, Dordrecht (1991)
12. Pawlak, Z.: Information systems - theoretical foundations. Information Systems Journal 6, 205–218 (1981)
13. Polkowski, L., Skowron, A.: Rough sets in knowledge discovery. Studies in Fuzziness and Soft Computing. Physica-Verlag, Springer (1998)
14. Raś, Z.: Discovering rules in information trees. In: Żytkow, J.M., Rauch, J. (eds.) PKDD 1999. LNCS (LNAI), vol. 1704, pp. 518–523. Springer, Heidelberg (1999)
15. Raś, Z., Wieczorkowska, A.: Action rules: how to increase profit of a company. In: Zighed, D.A., Komorowski, J., Żytkow, J.M. (eds.) PKDD 2000. LNCS (LNAI), vol. 1910, pp. 587–592. Springer, Heidelberg (2000)
16. Raś, Z.W., Wyrzykowska, E., Wasyluk, H.: ARAS: Action rules discovery based on Agglomerative Strategy. In: Raś, Z.W., Tsumoto, S., Zighed, D.A. (eds.) MCD 2007. LNCS (LNAI), vol. 4944, pp. 196–208. Springer, Heidelberg (2008)
17. Raś, Z.W., Tsay, L.-S.: Discovering extended action-rules (System DEAR). In: Intelligent Information Systems 2003, Proceedings of the IIS 2003 Symposium, Zakopane, Poland. Advances in Soft Computing, pp. 293–300. Springer, Heidelberg (2003)
18. Raś, Z.W., Tzacheva, A., Tsay, L.-S.: Action rules. In: Wang, J. (ed.) Encyclopedia of Data Warehousing and Mining, pp. 1–5. Idea Group Inc. (2005)
19. Silberschatz, A., Tuzhilin, A.: On subjective measures of interestingness in knowledge discovery. In: Proceedings of KDD 1995 Conference. AAAI Press, Menlo Park (1995)
20. Silberschatz, A., Tuzhilin, A.: What makes patterns interesting in knowledge discovery systems. IEEE Transactions on Knowledge and Data Engineering 5(6) (1996)
21. Tsay, L.-S., Raś, Z.W.: Action rules discovery: System DEAR2, method and experiments. Journal of Experimental and Theoretical Artificial Intelligence 17(1-2), 119–128 (2005) (special Issue on Knowledge Discovery,)
22. Tsay, L.-S., Raś, Z.W.: Discovering the concise set of actionable patterns. In: An, A., Matwin, S., Raś, Z.W., Ślezak, D. (eds.) Foundations of Intelligent Systems. LNCS (LNAI), vol. 4994. Springer, Heidelberg (2008)

23. Tzacheva, A., Raś, Z.W.: Action rules mining. In: International Journal of Intelligent Systems, vol. 20(7), pp. 719–736. Wiley, Chichester (2005) (special Issue on Knowledge Discovery)
24. Tzacheva, A., Raś, Z.W.: Constraint based action rule discovery with single classification rules. In: An, A., Stefanowski, J., Ramanna, S., Butz, C.J., Pedrycz, W., Wang, G. (eds.) RSFDGrC 2007. LNCS (LNAI), vol. 4482, pp. 322–329. Springer, Heidelberg (2007)

Part III

Graph Data Mining

10

Indexing Structure for Graph-Structured Data

Stanislav Bartoň and Pavel Zezula

Faculty of Informatics, Masaryk University
Botanická 68a, 60200 Brno
Czech Republic
{barton, zezula}@fi.muni.cz

Abstract. An own design of an indexing structure for general graph structured data called ρ-index that allows an effective processing of special path queries is presented. These special queries represent for example a search for all paths lying between two arbitrary vertices limited to a certain path length. The ρ-index is a multilevel balanced tree structure where each node is created with a certain graph transformation and described by modified adjacency matrix. Hence, ρ-index indexes all the paths to a predefined length l inclusive. The search algorithm is then able to find all the paths shorter than or having the length l and some of the paths longer then the predefined l lying between any two vertices in the indexed graph. The designed search algorithm exploits a special graph structure, a transcription graph, to compute the result using the ρ-index . We also present an experimental evaluation of the process of creating the ρ-index on graphs with different sizes and also a complexity evaluation of the search algorithm that uses the ρ-index.

10.1 Introduction

In the context of the Semantic Web, ρ-operators are proposed in [5] as a mean to explore complex relationships [20] between entities. The problem of searching for the complex relationships can be modeled as the process of searching paths in a graph where various entities represent vertices and edges the direct relationships between them. In case of the semantic web the resources or classes and edges the properties between them. The notion of complex relationships can be also identified in bibliographic digital libraries, where entities are publications and the relationship can represent references or direct citations between them.

As proposed in [5], we recognize two kinds of complex relationships. The first one is represented by *a path* lying between two inspected vertices. Speaking in terms of publications this means that one publication indirectly cites or references the other publication – a chain of publications can be built so that one cites another. The second type of complex relationship is *a connection* between two inspected vertices. This symbolizes a fact that the two inspected publications indirectly cite one common publication, see Figure 10.1 for an example of this kind of complex relationship.

D.A. Zighed et al. (Eds.): Mining Complex Data, SCI 165, pp. 167–188.
springerlink.com   © Springer-Verlag Berlin Heidelberg 2009

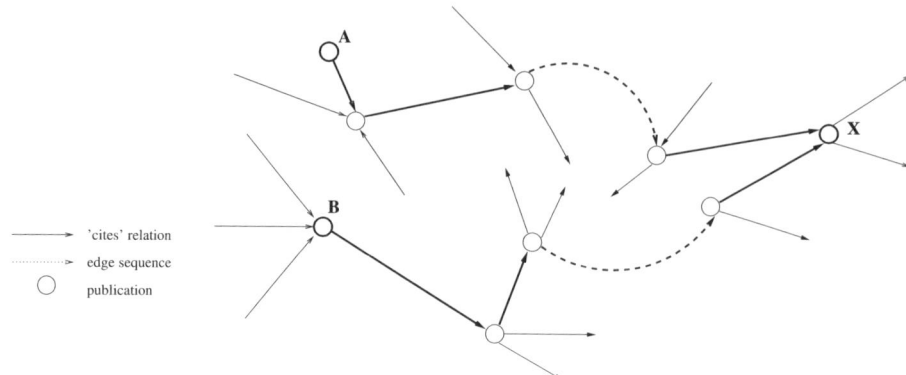

Fig. 10.1. An example of a connection between vertices A and B. Two paths originated in A and B connected in a common vertex X.

The knowledge about complex relationships among publications can be used for example for ranking the result of the search for publications using the complex relationship discovery among entities present in the result and then sorting them according to that information. Another use case can be an automated recommendation of publications based on the preferred set of publications by searching for close connections between the publications from the preferred set. Intuitively, the complex relationship discovery has sense in any other field of interest that incorporates graph structured data. For that reason, this chapter introduces an indexing technique called the ρ-index that enables efficient discovery of all complex relationships between any two inspected entities in large collections of arbitrary graph structured data.

This chapter is then structured as follows, Section 10.2 presents related work in the field of indexing graph structured data, Section 10.3 is a brief insight into the design of the proposed indexing structure. Section 10.4 introduces a search algorithm that is used to discover all paths between any two vertices in the indexed graph using ρ-index. Consequently, the experimental evaluation of the designed indexing structure and the search algorithm is in Section 10.5. Finally, this chapter is concluded and some directions of the future work are proposed in Section 10.6.

10.2   Related Work

The problem of answering various graph queries has two possible solutions. One is through an algorithmic on the fly query answering and the other one is preprocessing some indexing structure that would ease the computational complexity of the query processing.

Firstly we discuss one of the on the fly algorithmic approaches which is Tarjan's algorithmic solution to a *single source path expression problem* from [18, 19]

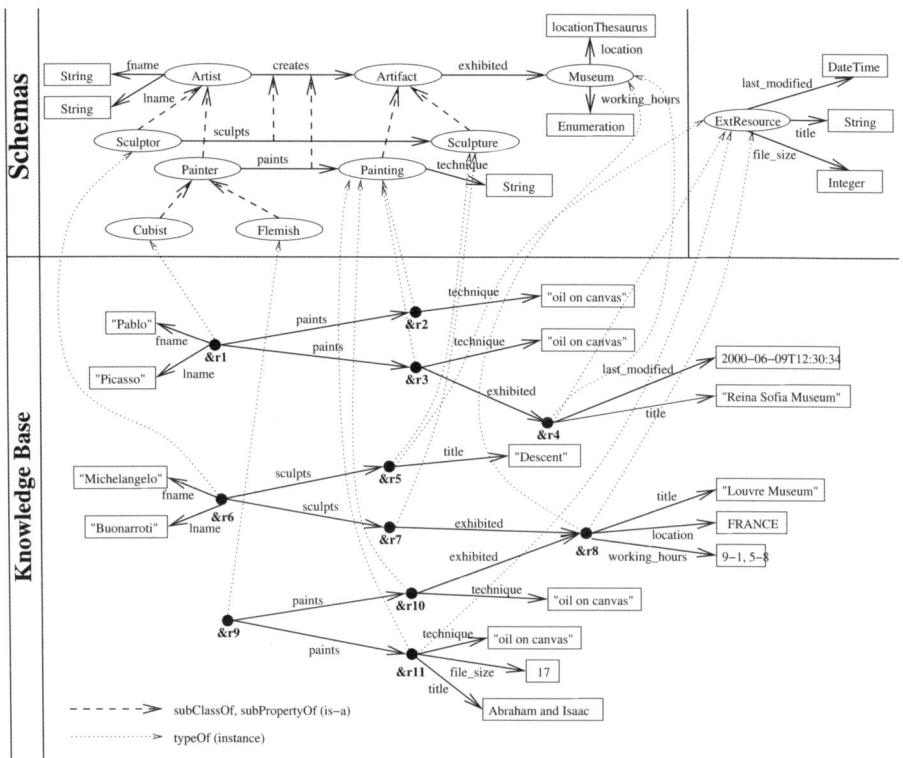

Fig. 10.2. An example of the RDF graph

which can be used to answer the queries for all paths lying between any two vertices in a graph. Hence, given a graph $G = (V, E)$ and a distinguished source
vertex s, for each vertex v find a regular expression $P(s, v)$ which represents
all paths from s to v in G. The problem is that the algorithm is designated
to be used only on directed acyclic graph (DAG). Although, there is a transformation proposed to covert an arbitrary graph to DAG, the computational
complexity of the algorithmic solution is $O(|E|)$ making it infeasible for efficient
query processing.

The indexing structures that can be used for efficient search for all paths
lying between two vertices in a graph were designed for RDF [13] graphs. A
short example of a RDF graph is depicted in Figure 10.2. The first index [5] was
designed directly for the purpose of implementing the ρ-operators that represent
the search for the complex relationships in RDF graphs. Its concept is that it
creates a matrix for each RDF schema [9] that takes part in the indexed RDF
graph where each entry of the matrix represents all paths between the entities
in the schema. This approach indexes only the schema part of the graph due to
the computational and store complexity of the index. When candidate paths are

retrieved from the index the actual existence of their instances in the knowledge base is checked.

The second indexing structure [15] that has been introduced for RDF graphs and can be used to process the queries concerning the complex relationships among vertices in a graph is based on path expressions and suffix arrays [14]. The base idea lies in extracting all possible path expressions from the indexed graph and consequently create all suffix arrays on string representations of the path expressions. The main drawback of this approach lays in its limitation of application to DAGs. Therefore, in this chapter we introduce our own indexing structure for efficient query processing of path oriented queries.

Nonetheless, the search for complex relationships can be reduced to a reachability query answering. Simply, instead of returning all paths lying between two inspected vertices a single boolean value is returned answering a question whether the start vertex can reach the end one. There are numerous algorithmic approaches to solve this problem varying mostly by the structures they use to compute the transitive closure of the relation. They are either a matrix based like [2] that are based on Warshall's algorithm [22] or the graph based algorithms [16, 17] or combining both approaches which results in a algorithm [3]. The indexing structures for efficient reachability query processing are labeling schemes that stem from the XML and tree structure labeling schemes. The most popular labeling schemes are based on either interval labeling scheme [1] or on a structural approach like [10] or again combining both in a hierarchical labeling presented in [23]. Yet, these approaches can be used only to distinguish the existence of a complex relationship between two vertices, further inspection of the complex relationship itself is not possible.

10.3 Design of the Index

The graph theory proved that a very handy representation of a directed graph is its adjacency matrix because using matrix algebra we can comfortably study the graph's properties. For instance, if the adjacency matrix is powered by two, each field in the resulting matrix contains a number of paths of length two lying between each two vertices in the original graph. If the computation continued, the result would contain amounts of all paths of an arbitrary length. Moreover, with a slight modification of the matrix that is introduced later in this section we would get not just the amounts of paths but the paths themselves.

Main difficulty of matrix representation of a graph is that its use is limited to fairly small graphs since the matrix grows in the quadratic space and the multiplication operation on matrices has even cubic time complexity. Therefore, we introduce graph transformation to enable the use of the matrix approach to graphs of arbitrary size.

10.3.1 Graph Segmentation

The graph transformation designed to simplify the graph we used is called *graph segmentation*. It takes the indexed graph and divides it into segments in a way that each vertex is contained in some segment and once assigned to a segment such vertex is not assigned to any other segment. Precise definition can be found in Definition 1. The main difference between subgraph and segment of a graph is that segment can contain edges which's both vertices are not in the same segment.

Definition 1. *Graph segment and a graph segmentation:*

- $LEFT\_V(e) = v_1 \Leftrightarrow e = (v_1, v_2)$
- $RIGHT\_V(e) = v_2 \Leftrightarrow e = (v_1, v_2)$

Segment S in a graph $G : S = (V_S, E_S) : V_S \subseteq V \ \wedge \ E_S = \{e \in E \mid RIGHT\_V(e) \in V_S \ \vee \ LEFT\_V(e) \in V_S\}$

Segmentation $S(G) = \{S | S$ *is a segment of* $G\} \wedge \forall S, S' \in S(G), S \neq S' :$ $V_S \cap V_{S'} = \emptyset \ \wedge \ \bigcup_{S \in S(G)} V_S = V$

Afterward, the vertices and edges between vertices within one segment form a subgraph of the indexed graph. The edges lying between vertices assigned to different segments form edges in the simplified graph. Segments then form the vertices in the simplified graph what we call a segment graph which is defined in Definition 2. By this transformation, multiple edges can appear between vertices in the new graph. Regardless, each multiple edge can be substituted by a single edge since from a path point of view it means a redundant information.

Definition 2. *Segment graph of G:*

$SG(G) = (S(G), \ X), \ X = \{h | h = (S_i, S_j) \Leftrightarrow 1 \leq i, j \leq k \wedge EDGES\_OUT(S_i) \cap EDGES\_IN(S_j) \neq \emptyset\}$
where k *is the number of segments in* $S(G)$.

The segment graph $SG(G)$ has very similar properties as the graph G. Any path followed in the indexed graph can be observed also in the segment graph. Since we left out only the inner edges of each segment. This simplified path in the segment graph we call a sequence of segments just to avoid confusion of terms, see Definition 3. Intuitively, each path in the indexed path can be represented by only one sequence of segments. The method to transform a path into a sequence of segments is to replace each group of vertices and inner edges of each segment by that particular segment and to replace each edge lying between two vertices assigned to different segments by a particular edge from X, with regard to the Definition 2 such edge always exists.

Definition 3. *Sequence of segments:*

$EDGES\_OUT(S) = \{e | e \in E_S \wedge LEFT\_V(e) \in V_S \wedge RIGHT\_V(e) \notin V_S\}$
$EDGES\_IN(S) = \{e | e \in E_S \wedge RIGHT\_V(e) \in V_S \wedge LEFT\_V(e) \notin V_S\}$

Sequence of segments $(S_1 \ldots S_l) : S_1, \ldots S_l \in S(G), \ 1 \leq i \leq l - 1 : EDGES\_OUT(S_i) \cap EDGES\_IN(S_{i+1}) \neq \emptyset$

Thereafter, each of the segments can be represented by its path type adjacency matrix. A path type adjacency matrix is a modification of a usual adjacency matrix known from graph theory. It is designed to represent a graph in a path oriented way. It stores paths in its fields instead of just amounts of those paths. Initially, in each field path type matrix contains a path consisting of a single edge whenever there is an edge between two vertices in the graph. The convenience it presents over the usual adjacency matrix is that after the transitive closure of the path type matrix is computed – the fields contain not just an amount of paths lying between any two vertices, but also the paths themselves. Naturally, the mathematical operations on numbers + and * are replaced by the respective operations on paths – set union and concatenation.

Using the graph segmentation one large graph (G) can be transformed into a smaller simplified graph $(SG(G))$ by identifying certain number of segments and collapsing them into single vertices. The size of the segment by which we mean a number of vertices in the segment can be easily controlled. If the transformed graph is still too big to be described by its path type matrix the whole procedure can be repeatedly applied again taking as an input the already simplified graph. Thus we can acquire a multilevel indexing structure where each vertex represents a graph on the lower level.

Hence, the creation of the ρ-index accompanies a graph segmentation followed by a computation of the path type matrix for each segment. This step is repeated until we get a graph that we are able to describe by its path type adjacency matrix. A size of segments may vary on every particular level. Therefore the maximal sizes of the segments at each level form the parameter settings of the ρ-index. Examples of the parameter settings for ρ-index creation are discussed in Section 10.5. The visual outline of the indexing structure is in Figure 10.3.

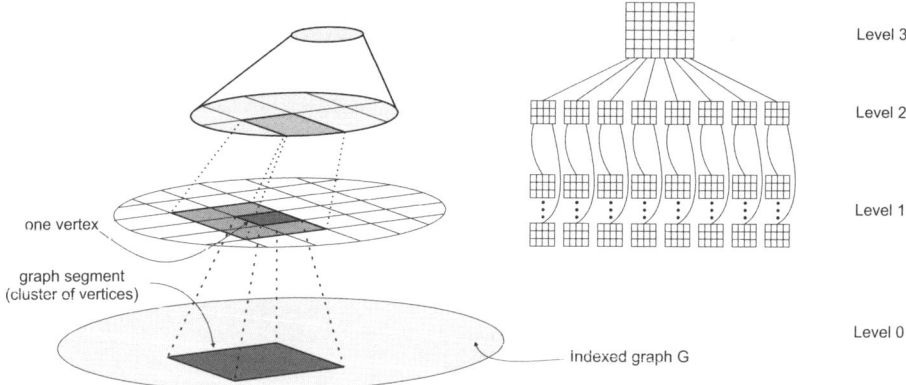

Fig. 10.3. Structure outline of the ρ-index

10.3.2 Graph Segmentation Method and Strategy

Various ways how to assign the vertices to segments have been identified and studied. One of them was a graph to forest of trees transformation which's result is a forest of trees and was proposed in [7]. Combination of vertex clustering and the graph to forest of trees transformation together with its preliminary evaluation can be found in [8]. Further implementation and evaluation showed that the graph to forest of trees makes the resulting indexing structure very tangled and therefore the search algorithm did not present good results.

Therefore, for the experimental evaluation presented in this chapter we have chosen the vertex clustering as a segmentation method. Initially it puts a single vertex into set V_S. Afterwards it incrementally enlarges the segment with vertices to which or from leads an edge to this vertex. Those edges then form the set E_S. This continues until a maximal number of vertices in V_S is reached. For each level the maximal number of vertices in V_S is stated as a parameter.

The nature of the ρ-index tree structure is very dependent on the settings for the maximal number of vertices in V_S at each level. Intuitively, by setting small sizes of the segments a slim and high tree can be created. On the other hand, using a large number at first level a wide and low tree is acquired. The evaluation of different parameter settings and how they affect the search itself is demonstrated in Section 10.5.

10.3.3 Sequence of Segments Properties

As we mentioned above, each path on a lower level can be represented by some segment sequence on the upper level. Intuitively, some two different paths can be represented by one segment sequence. Some of those path are called *connecting paths* and are defined in Definition 4. The main property of a connecting path is that it starts with an common edge of first two segments and ends with a common edge of last two segments in the sequence of segments.

Definition 4. *Connecting path in a sequence of segments:*

Common edges CE_i for $(S_1 \ldots S_l)$: $1 \leq i \leq l-1$: $CE_i = EDGES\_OUT(S_i) \cap EDGES\_IN(S_{i+1})$

Connecting path $p = (e_1 e_2 \ldots e_n) \in (S_1 \ldots S_l)$: $e_1 \in CE_1 \wedge e_n \in CE_{l-1} \wedge \exists i_2, i_3, \ldots i_{l-1} : 1 < i_2 < i_3 < \ldots < i_{l-1} < n : \{e_2, \ldots e_{i_2}\} \subseteq E_{S_2} \wedge \{e_{i_2}, \ldots e_{i_3}\} \subseteq E_{S_3} \wedge \ldots \wedge \{e_{i_{l-2}}, \ldots e_{i_{l-1}}\} \subseteq E_{S_{l-1}}$

In general, each segment sequence can represent a huge amount of paths of different lengths. This is because each segment represents a subgraph in which paths with different lengths can be found. Important to us is knowledge of a length of the shortest path that the particular segment sequence represents. Obviously, the shortest path is one of the connecting paths. The length of the shortest path is then referred to as a *weight* of the sequence of segments. We have chosen weight instead of length because length of a segment sequence means the length of the sequence but the more important to us is the length of the shortest path it actually represents. Therefore, if we want to compute all the paths to the length l we have to store all the segment sequences having its weight less or equal to l. This parameter l then forms a path length limit that is to be indexed.

Seemingly, to compute the weight of the sequence of segments $(S_1 \ldots S_l)$ we would have to compute all its connecting paths to find out which of them is the shortest. But an enhanced algorithm does not compute all the connecting paths but only one shortest connecting path for each combination of common edges picked from all CE_is, see Definition 4. Thus we have an upper bound on a number of connecting paths to be computed for each sequence of segments.

Due to the fact that the weight of a segment sequence represents the length of a shortest path it represents, it also represents some of the paths that are longer then its minimal weight. Therefore, using ρ-index we can compute surely all the paths to the length l but also some of the paths that are actually longer than the specified l. As we will show in the evaluation in Section 10.5 the amounts of paths longer then l is not insignificant, yet we realize that this fact highly depends on the nature of the data on which the ρ-index is being used.

10.4 The Search Algorithm

In this section we describe the algorithm for discovery of all paths to a certain length between two vertices in the indexed graph using ρ-index. Firstly, the algorithm looks up the segments the start and end vertex are assigned to. If we have more than two levels in the ρ-index it looks up to which segments on the upper level are assigned the segments acquired in the previous step. This continues until we reach the top level of the ρ-index or we get one common segment for both vertices. This process goes from the bottom of the structure to the top. From the definition of the graph segmentation each vertex or segment belongs to one segment on the upper level. Therefore, for each vertex in the original graph only one segment exists at each level that contains it.

10.4.1 The Transcription Graph

A special graph structure is used to represent the result throughout the algorithm computation. It is a transcription graph where the vertices and edges are replaced by subgraphs retrieved from the ρ-index. The vertices in the transcription graph are either the segments of the ρ-index or the vertices of the indexed graph. Those are considered to form the lowest level of the ρ-index. The transcription graph contains four special kinds of edges:

transitionTo denotes an existence of a transition (edge) between vertices at the particular level.

existsPathTo indicates an edge that can be replaced by a subgraph from ρ-index consisting of vertices at the same level and transitions between them, representing all sequences of segments lying between these two vertices. This edge may be only between two vertices that are assigned to one common segment on a higher level.

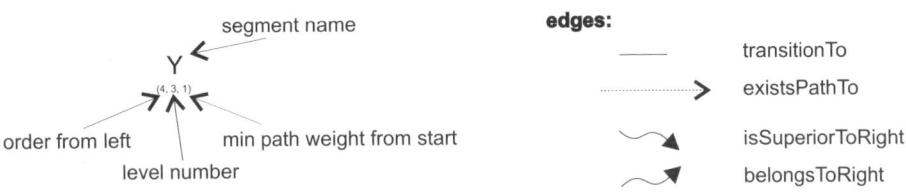

Fig. 10.4. Example of an initial transcription graph

belongsToRight represents the relationship of containment, a vertex from a lower level belongs to a vertex on a higher level.

isSuperiorToRight is an opposite of the previous relationship, it means that the vertex at a higher level contains the vertex on a lower level.

Figure 10.4 demonstrates an initial state of the transcription graph for a search of all paths between vertices 1 and 10 in ρ-index having four levels. The vertices are assigned to respective segments on upper levels and on the topmost level an existence of a path is supposed between the segments at the highest level.

The concept of the transcription process is to take the initial transcription graph and transform it to a graph which comprises of only vertices at the lowest level and all edges are of the *transitionTo* type. To achieve this, all the segments and edges at the higher levels need to be processed – transcribed – into entities at lower levels until we achieve the stop condition of the algorithm. Firstly, it replaces the *existsPathTo* edge by a respective subgraph of sequences of segments lying between the two vertices where all the edges are transitions. Secondly, all of the transitions concerning the particular segment, that is to be transformed into entities on the lower level, each transition originated or terminated at this segment is replaced by a subgraph of segments at a lower level connected to this segment by the type of edges connecting segments on different levels. This transformation is demonstrated in Figure 10.5 as the first step in the process. The transition between the segments X and Z is transformed into a transition between segments L and K, but on a lower level. This fact indicates, that there exists a border edge between segments X and Y which is originated in K and terminated in L, where K belongs to segment X and L is assigned to segment Y. If there existed any other border edges they would also appear in the transcription graph at this point.

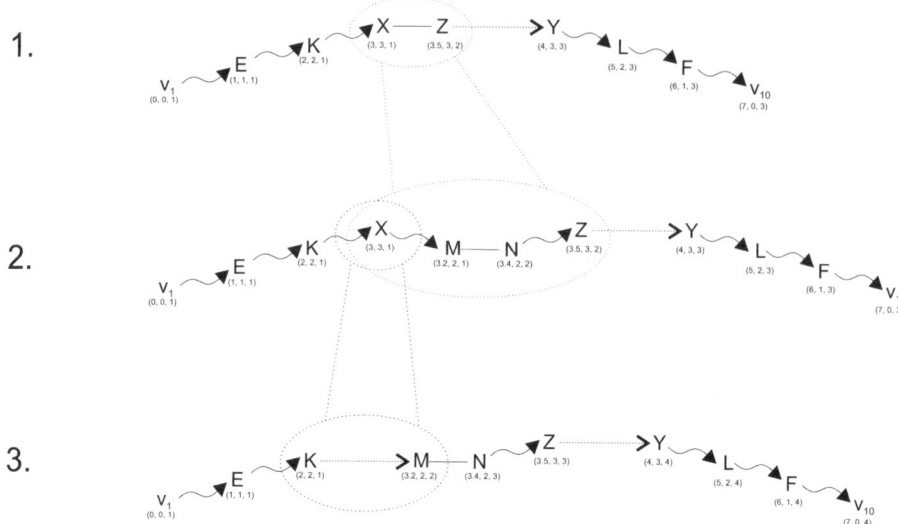

Fig. 10.5. Transcription of a transition to a lower level

Once the segment has only edges connecting it to other segments on a lower level it is transformed into lower level entities by connecting each entity on the left side with each entity on the right side with an *existPathTo* type of an edge going from left to right. This is demonstrated in Figure 10.5 by a step number 2 and 3. The transformed segment and all its connecting edges to lower levels are removed from the graph.

As for the transcription strategy during the transcription process, each vertex in the transcription graph is assigned two important numbers which are kept updated through the whole computation. The first number is the vertex's *order from left* and the other one is a length of a shortest path between the start vertex and this particular vertex. The left order number makes possible to have the vertices sorted by their position in the transcription graph as the algorithm processes its vertices strictly from left to right. Since the left order number is a floating point number, every time the process needs to insert a vertex between other two vertices there is always a gap between their left orders. Therefore, the transcription graph forms a special type of a directed graph referred to as *a network* which is also a DAG. Since, vertices can be ordered by its left order number and it is true that there is no edge pointing from a vertex with higher left order to a vertex with a smaller left order.

The length of a shortest path from the start vertex is used to limit the weight of segment sequences that are retrieved from the index to replace the path edges in the transcription graph. It considers the length of an already computed piece of path from the start vertex to the particular vertex. The segment sequences of a maximal weight of a difference of the already computed piece of the result and the maximal length of a desired path, our l, are retrieved and placed into the transcription graph. This fact assures that the algorithm will actually stop for any input because if it is not possible to reach the end vertex from a segment by a sequence of segments with a weight less then l considering the already minimal length of a path, the whole branch is removed from the transcription graph.

When the process finishes the resulting transcription graph represents either a network of all paths initiated in the start vertex and terminated in the end vertex with a length lower or equal to the predefined l and some paths longer than l due to the nature of the graph segmentation. All that with respect to the paths that are in the indexed graph. If there are no paths shorter than l between the start and end vertex the resulting transcription graph will have only two vertices and no edges.

The more detailed insight into the transcription graph principles and its transcription strategies, that is beyond the scope of this chapter can be found in [6].

10.5 Experimental Evaluation

In this section we present and discuss the results gained by the indexing structure and its search algorithm introduced in this chapter. As a testing data we have used generated synthetic data which's properties are described later in this section.

As follows, the set of experiments performed took as a testing data generated graphs having sizes growing from 5,000 to 30,000 vertices and from 10,000 to 60,000 edges. The graphs were generated iteratively using a small core graph in the first step. In each iteration a smaller graph was enlarged by randomly adding edges between newly added vertices or between a new vertex and an old vertex with a random direction. The probability of where the edge was placed was equal to the proportion of the number of vertices in the smaller graph to the number of vertices in the newly built graph. In the rest of this chapter we will refer to these graphs as G5000, G10000, G20000 and G30000 with respect to the number of vertices contained in the testing graph. The vertex degree distribution of the testing graphs is illustrated in Figures 10.6, 10.7, 10.8 and 10.9.

This way we gained graphs with different sizes and having the property that the smaller graph is always a subgraph of any of the larger graphs. This property is very important when we evaluate the experiments that compare the search

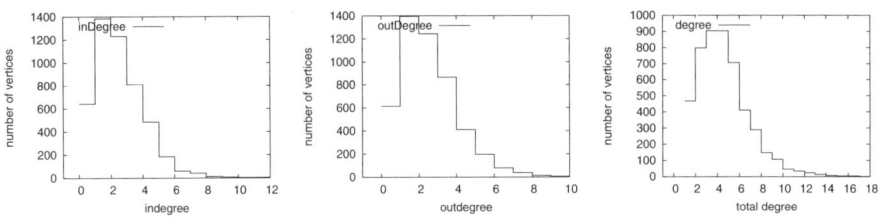

Fig. 10.6. Vertex degree distribution in the synthetic graph G5000

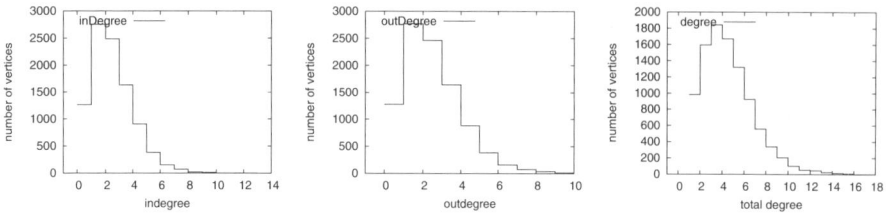

Fig. 10.7. Vertex degree distribution in the synthetic graph G10000

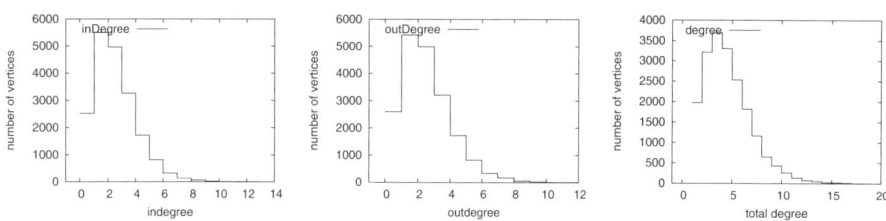

Fig. 10.8. Vertex degree distribution in the synthetic graph G20000

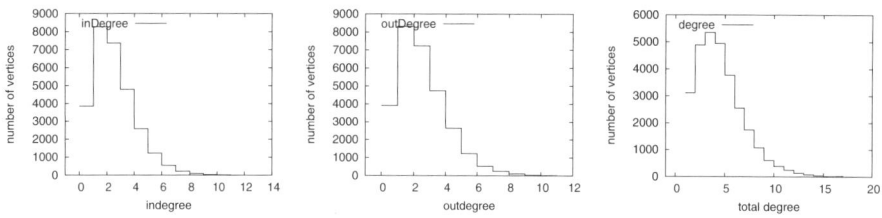

Fig. 10.9. Vertex degree distribution in the synthetic graph G30000

results in graphs with different sizes, because the result of a search performed on a smaller graph is also a subset of a search result of the same search performed on any larger graph. So its true that G5000 ⊆ G10000 ⊆ G20000 ⊆ G30000.

As we performed all the experiments described in the following sections we stated the maximal indexing length l to be 10. The ρ-index then was built to index all the paths up to this length and the search then returns all the paths to this length and some paths longer. Due to the space limitations of this chapter we will not present a detailed insight into what means *some* in exact numbers. Yet we briefly tackle this issue in Section 10.5.3.

As for the machinery on which we executed all our experiments concerning the ρ-index, the computer is a dual double core Athlon Opteron 2.4 GHz with a 12 GB of RAM. During the time the tests were run the computer was not dedicated to only that task so all the experiments were ran multiple times and the results depicted are averages of the results thus gained.

10.5.1 ρ-index Creation Time

First of all we present how much time the creation of the ρ-index consumes for certain sizes of the testing graphs and particular parameter setting. Figure 10.10 represents the experiments performed on our four testing graphs. The ρ-index created for each graph had 4 levels. The maximal size of a segment on the lowest level is represented by the values on the x-axis. The other parameters were chosen to be 10 at the second level, 5 on the third and 2 at the top level. Just to remind the parameters are the maximal sizes of the segment at the particular level.

The results of this evaluation showed that the ρ-index is sensitive to underfill of the structure. This can be observed for the case of the smallest graph when even the ideal parameter setting which is around the value 8 for the max segment size for the lowest level lead into a creation time which was greater than the best time of a graph twice as large. We assume that this is caused by a inadequate ρ-index setting. The ρ-index for this testing graph should have been created using only three levels or smaller maximal segment sizes at the second and third level.

The creation times of the remaining three graphs indicate that the ρ-index is highly dependent on the parameter setting. We can observe that the creation times form a curve of a parabolic shape for all graphs and the size of the testing graph determine the shift of the values on the y-axis. This implies that the optimum parameter setting can be easily predicted for graphs at this particular graph size category upon these experimental results. As a category we consider

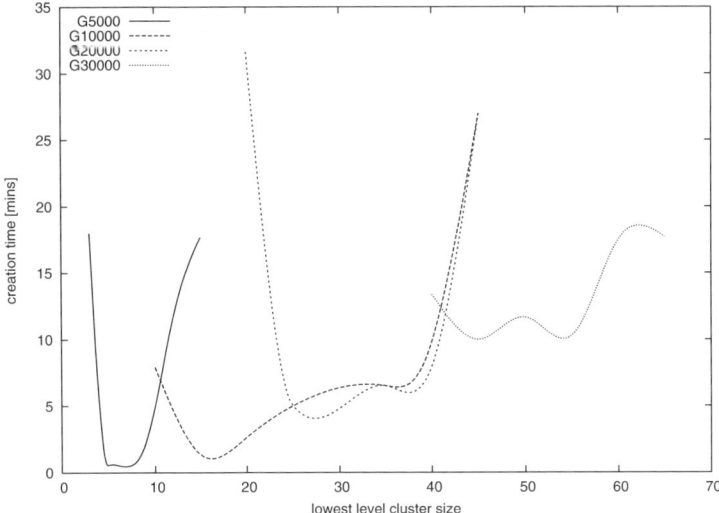

Fig. 10.10. A ρ-index creation time

a graphs of a similar size and connectivity. In this case the category is formed by graphs having from 10,000 to 30,000 vertices and 20,000 to 60,000 edges respectively.

10.5.2   Search Complexity

This group of experiments performed describe the complexity of the search algorithm using the ρ-index to search all paths to a certain length in respect to the size of the graph on which the search was performed. Figure 10.11 demonstrates the experiments where the parameter settings were fixed and the size of the graph grew. As we mentioned earlier in this section, the result of the search of the larger graph contains all the search results of the smaller graphs, thus they are comparable.

Both parts of Figure 10.11 refer to the same results of the same experiments. They only differ in the y-axis scale. The left part depicts the results in the whole scale, the right part depicts them ranging from 0 to 80,000 of processed vertices.

The particular curves represent the algorithms used to perform the search of all paths lying between two vertices. The lines labeled with the prefix *seq* represent a sequential algorithm. This algorithm represents an upper bound of a way to solve the problem of searching all path between two vertices to a certain length. It is a depth first search that tries to recursively build a path of a some maximum length. The number in the label states the maximal length of a searched path.

In Figure 10.11 are present two results for a sequential algorithm *seq*. This is caused by the nature of the ρ-index and its search algorithm which results in a fact that all paths to a specified length l to which the ρ-index was built are returned and some of the paths longer than l are also returned by the search algorithm. That implies for the result of the search using ρ-index that is true: $seq(l) \subseteq \rho\text{-index} \subseteq seq(l+k)$ for a particular k, where the set inclusion is meant on the results of the search algorithms. For that reason we also present the complexity of the algorithm $seq(12)$ which represents the sequential scan for all

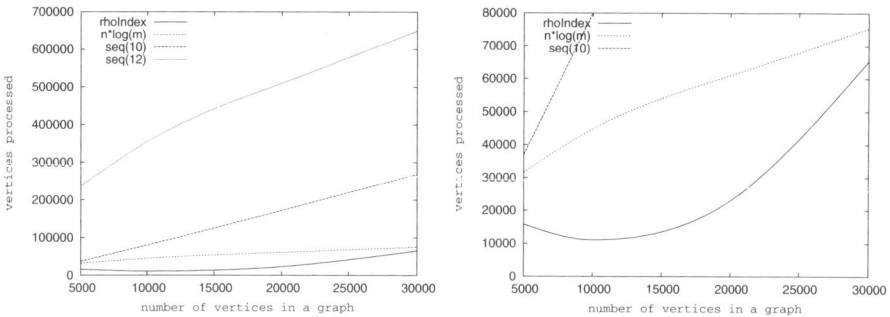

Fig. 10.11. A ρ-index search complexity with respect to the graph size

paths to the length 12. The rough comparison of the complexity measured for the sequential algorithm with the length set to 10 and 12 we can observe that the growth is exponential.

Another approach to the problem of searching all paths lying between two vertices in a directed graph is a direct computation using the Tarjan's algorithms described in [18] and [19]. The algorithm works in a time complexity $n * log(m)$ where n represents the number of edges in the graph and m the number of vertices in the graph. The algorithm takes a flow graph on the input and a start vertex and returns the path expressions (regular expressions where the letters are edges of the flow graph) representing all paths to all vertices in a graph on the output. A flow graph is a special type of a directed graph which allows only one source vertex in the graph and no cycles. There exist a non-trivial transformation of an arbitrary directed graph into a flow graph. This computational overhead of the graph transformation is not included in the complexity of the Tarjan's algorithm.

In the progress of the search computational complexity of our designed index structure and algorithm a decrease of the complexity can be observed for the graphs G5000 and G10000. As we mentioned in the previous subsection, this is due to the underfill of the search structure. The parameter settings used to built the ρ-indexfor each of the testing graphs were the optimal ones for each particular graph. Again only the max size of the segment on the lowest level varied and the rest of the parameter settings remained same for all testing graphs.

10.5.3   Search Complexity of Queries with Limited Maximal Length

To this point we always considered the maximal length of the searched path by the search algorithm to be the same as the maximal length that was used to create the ρ-index. In this subsection we explore the behavior of the search algorithm when the maximal length of the searched path is its parameter.

As we refer to the maximal length l of the indexed path, we refer to the maximal length of a searched path as $softL$. Setting this parameter does not limit the search to return paths longer than $softL$ but again it must not necessarily find all of them.

Thus Figure 10.12 represents searches executed on the graph G10000 and with the parameters set to 30, 10, 5 and 2. The ρ-index was computed with l equal to 10. The x-axis then represents the values of the $softL$ parameter and the curves represent the respective algorithms used to compute the result.

To make the Tarjan's approach comparable with ours and the sequential algorithm we approximated the computational complexity by limiting the input graph to only those vertices and edges that are reachable within $softL$ steps.

As for the number of the found paths, the sequential algorithm finds all paths to the length of $softL$, our algorithm finds all the paths to the length of $softL$ and some of the paths that are longer than that. Figure 10.13 represents the percentage of paths not found that have length greater than $softL$ for each particular length. Although we ran the experiment for the $softL$ value of 3 the curve representing returned results is not present here since it returns no paths

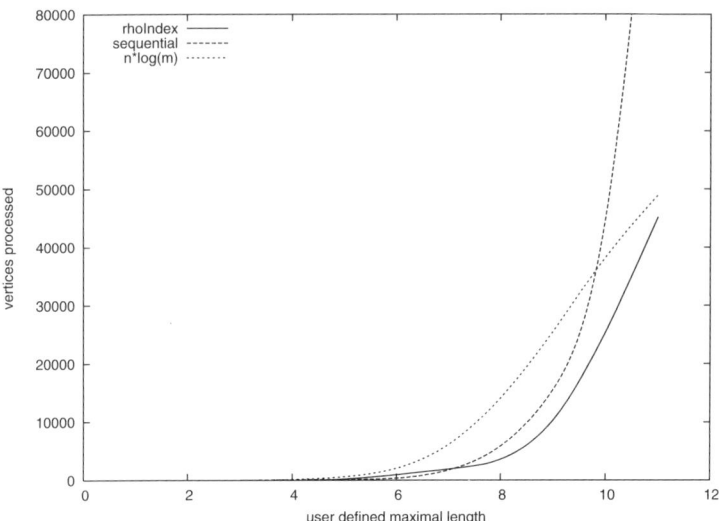

Fig. 10.12. A ρ-index search complexity of queries with different maximal search length

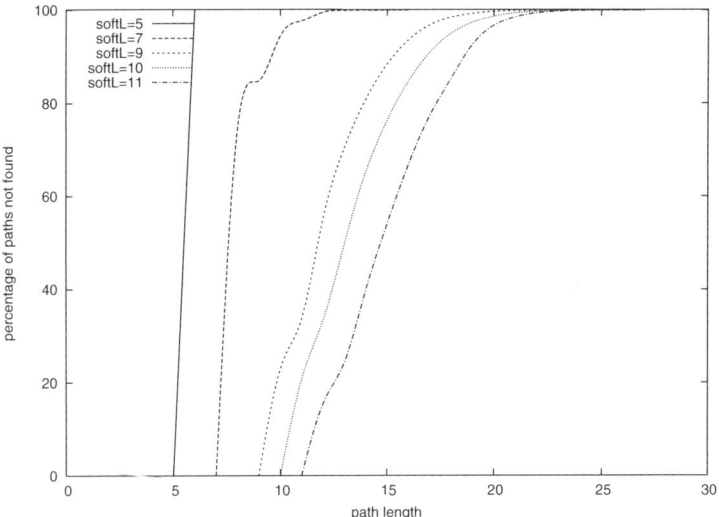

Fig. 10.13. A ρ-index percentage of paths longer than $softL$ not found

for this $softL$ value. For $softL$ value 5 it finds no path longer than 5 so the curve reaches immediately 100 percent at length 6.

At this point we have to point out that the amount of paths increases in exponential manner, what means that the amount of paths of length 12 between the testing start and end vertex actually present in G1000 is 1821 and the amount of paths of length 14 between the same two vertices is 12644. So even if we find really low percentage of the paths present in the indexed graph, their amount can easily reach tens of thousands. For illustration, for the $softL = l = 10$ and a path length of 24 the amount of found paths is 72,000 and the longest found path has a length of 42.

10.5.4 Search Complexity Affected by the Parameter Settings

Since the ρ-index can be created for one particular graph using different parameter settings and as we could see from this section, also having different properties, we explore the correlation between certain parameter setting and the complexity of the searches performed on the respective indexing structures built upon one particular testing graph.

Again we have chosen the testing graph G10000. The parameter settings differed in the maximal size of a segment on the lowest level, the upper level settings remained the same for all tests. Consequently, Figure 10.14 depicts the relation between the parameter settings and the average search complexity for thus created ρ-index. This curve is falling with the increase of the cluster size. The dashed curve in Figure 10.14 reflects the creation time of the ρ-index for that particular parameter setting. The time is in minutes multiplied by 1000 to make the curve visible in this scale. On the contrary, the progress of this curve is rising as the ρ-index structure is becoming underfilled. We have already seen this behavior in Figure 10.10 at all graph sizes at the rising part of the parabolas.

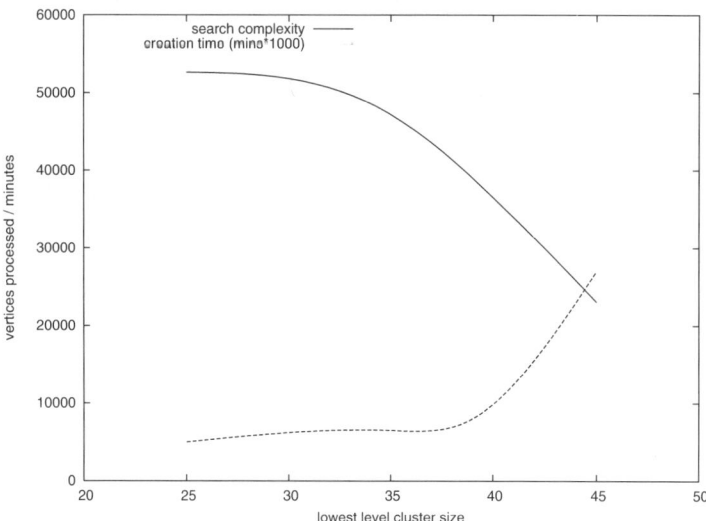

Fig. 10.14. A ρ-index search complexity related to the parameter setting

These facts represent a creation and search tradeoff. We gain better creation time results for certain parameter settings but on the other hand we get worse search complexity results. This tradeoff has even one more dimension which is the amount of paths returned that are longer then l. Due to the space limitations we are not able to discuss this dependence more in detail.

10.5.5   Deploying the ρ-index to the Real Life Data

The other set od data represents a piece of the CiteSeer [12] database of scientific publications and citations among them. Our data set was created taking one publication and deploying the breadth first search for all weakly accessible publications from our starting one until we got a certain amount of vertices in the built data set. Weak accessibility ignores the orientation of the edge between two vertices. The amount of publications in our data set we set to be 30, 000. The amount of edges acquired among the vertices in the data set was 63, 584. The distribution of the degrees of the vertices in this citation graph is demonstrated in Figure 10.15. The x-axis represents the particular vertex degree – respectively the amount of edges initiated, terminated and a total number of both in the particular vertex – and the y-axis then represents an amount of vertices having this degree. The x-axis is drawn using logarithmic scale to make a clearer view of the curve's progress. Also to the values on the y-axis the logarithm was applied to achieve better readability of the demonstrated distribution.

The indegree represents the number of citations of each particular publication. Notice that this distribution follows the power law that states that in the testing citation graph is a small number of vertices that have large indegree and a large number of vertices which's indegree is very small. This exactly conforms with the reality where most of the publications receive a small number of citations and was also presented in [4]. To the contrary, the outdegree represents the number of references that the particular publication refers to. This number is not always accurate since CiteSeer does not contain all the references for each publication in its database.

From the ρ-index evaluation demonstrated on the synthetic data in the previous sections we know that the structure is usable to search for paths between a pair of vertices. But now, using the CiteSeer's citation graph we can give those paths semantics, some meaning. Since each vertex in the citation graph

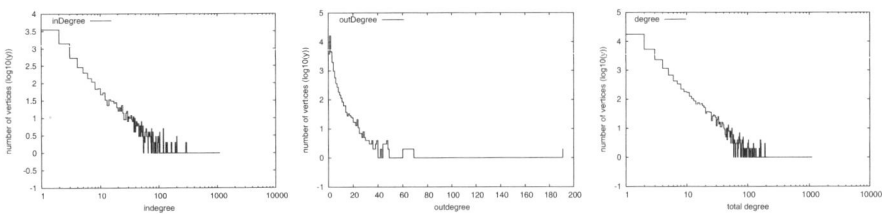

Fig. 10.15. Vertex degree distribution in the citation graph

Table 10.1. A summary of paths found between the reference and core publication

Path length	Amount of paths	Distinct vertices
4	2	5
5	7	13
6	17	32
7	27	51
8	33	58
9	48	59
10	62	60
11	46	61
12	29	61
13	22	61
14	11	61
15	2	61

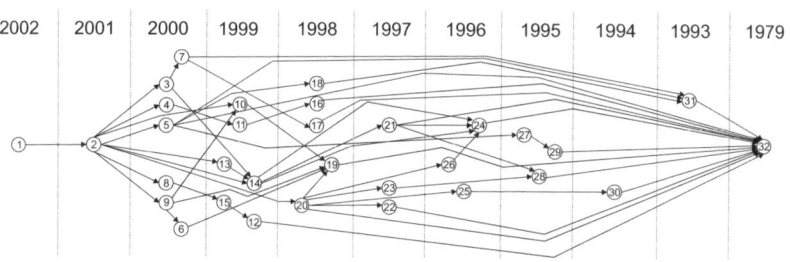

1 Predicting Query Performance
2 Relevance-Based Language Models
3 Relating the New Language Models of Information Retrieval to the Traditional Retrieval Models
4 Bridging the Lexical Chasm: Statistical Approaches to Answer-Finding
5 Improving the Effectiveness of Informational Retrieval with Local Context Analysis
6 Statistical Models for Tracking and Detection
7 The Mirror DBMS at TREC-8
8 OCELOT: A system for summarizing web pages
9 Dragon's Tracking and Detection Systems for the TDT2000 Evaluation
10 Topic Tracking in a News Stream
11 Probabilistic Latent Semantic Indexing
12 Automated Text Summarization in SUMMARIST
13 A General Language Model for Information Retrieval
14 A Hidden Markov Model Information Retrieval System
15 Summarizing Text Documents: Sentence Selection and Evaluation Metrics
16 Unsupervised Learning from Dyadic Data

17 MiRRor: Multimedia Query Processing in Extensible Databases
18 Effective Retrieval with Distributed Collections
19 Topic Detection and Tracking Pilot Study Final Report
20 On-line New Event Detection and Tracking
21 Boosting and Rocchio Applied to Text Filtering
22 Study on Retrospective and On-Line Event Detection
23 The TREC 5 Filtering Track
24 Context-Sensitive Learning Methods for Text Categorization
25 Text-Based Information Retrieval Using Exponentiated Gradient Descent
26 Training Algorithms for Linear Text Classifiers
27 Providing Government Information on the Internet: Experiences with THOMAS
28 Evaluating and Optimizing Autonomous Text Classification Systems
29 Corpus-Specific Stemming using Word Form Co-occurrence
30 Optimizing Ranking Functions: A Connectionist Approach to Adaptive Information Retrieval
31 Using Statistical Testing in the Evaluation of Retrieval Experiments
32 Information Retrieval

Fig. 10.16. Publication search result visualization

represents a publication and the edges incorporating such publication its citation relation, a path between a pair of vertices having length greater then two
represents an indirect citation relation. We would like to study this indirect citation relation between certain – important publications in this citation graph
to identify important publications in the particular context.

Our testing citation graph was built around Van Rijsbergen's *Information Retrieval* [21] which was identified as a very important publication in the IR field for
it's high number of citations by other publications. We followed an idea that if we
come across some newer publication that we consider interesting to our research
that falls into the same scientific field then there is a high probability that there
exists either direct or indirect citation of our core book. If there exists an indirect

citation then there is also a possibility that more then one indirect citation paths can be found. In this case we would like to study all paths to certain length lying between our recent – reference – publication and the core book. The vertices on these paths form a set of publications that deserve a further study of their importance by the user.

For our experiment we have chosen [11] as the reference publication. We deployed the ρ-index on our testing citation graph and searched for all paths to length 10. From the nature of the ρ-index that we discussed earlier in this chapter we got all the paths to the length of 10 and some longer. Table 10.1 summarizes the amounts of paths found according to their length and a total number of distinct vertices of all paths up to that length. Figure 10.16 demonstrates the network of the paths up to length 6 – as a length of a path we mean the number of vertices in a path. In that figure, the vertices represent publications that are placed on the background of a timeline to make the result more readable. Although, the ρ-index was created to index all the paths up to the length of 10 and as Table 10.1 shows ρ-index does index also some more, Figure 10.16 demonstrates only the paths to the length 6 since it would get very hard to follow when it have contained all the paths got from the ρ-index.

As can be seen in Figure 10.16, the result of the search is a network with one source which is our reference publication and one sink which represents our core publication. The result presents a chosen set of publications from the citation graph which relate to the reference publication because the reference publication indirectly cites them and they relate to the field of information retrieval since they indirectly cite our core publication. The resulting publications are ordered according to the year of their publication. Yet another ranking technique could be used to study the relevancy between the reference publication and the publication found using ρ-index but that is beyond the scope of this chapter.

The approached proposed used one core publication and as we seen we got for this one a fair amount of publications from the citation graph because the core publication is well known. If we used as a core publication not so well-known publication the system would not be able to retrieve a reasonably big set of publications. For this reason the approach could be improved to carry out the search with only one core publication but with a set of core publications. Consequently, ρ-index would find all the paths to each particular publication from the core set and put the result together. This improvement brings another interesting issue since the retrieved networks can overlap and that information can be also used for further recommendation process.

10.6 Concluding Remarks and Future Work

Our goal was to design an indexing structure that would make possible an effective discovery of paths having special properties in a large graph. The first objective was to find all paths to certain length l between any two vertices. Also we still get some amount of paths longer than the specified l as an approximation.

For brevity, in this chapter we did not presented all the experiments we have conducted in respect to explore the behavior of the designed ρ-index. For example we tackled the issue of the parameter settings modification also at the upper levels of the indexing structure and their impact on the particular ρ-index creation time and consequent search complexity.

Hence, to this relates also our next future work. Firstly we want to carry out more tests to be able to precisely predict the ρ-index properties under certain parameter settings and hence to be able to find optimal settings for the testing data. Afterwards, we would like to carry out tests on more testing data in order to investigate the scalability of the ρ-index in respect to the number of vertices and to the number of edges in the indexed data. In our near future work we would also like to implement the algorithm for discovering all connections between two vertices as the ρ-index allows such utilization according to [7].

As the experiment on the real life data shown it is interesting to study complex relationships between publications in the citation graph in order to help the user identify important publications in the particular scientific field. We presented a search for all paths between two publications where the result was presented as network ordered according to the date of publication. But also another information retrieval techniques could be used to rank the search result. We also identified possible new approaches using our indexing structure like a search for a relation between one reference publication and a set of core publications to improve the accuracy of the search.

The future work concerning the information retrieval part of our work is to confirm the contribution of a complex relation search in the citation graph. In the first place, we would like to verify using standard information retrieval techniques the relevance of the publications recommended by the system. Secondly, we want to study the relevance to this problem of using a set of core publications instead of just one single publication in the search. Last but not least we also want to incorporate and evaluate the search for a connection between two publications into the system because it provides a notable alternative to the indirect citation relation.

References

1. Agrawal, R., Borgida, A., Jagadish, H.V.: Efficient management of transitive relationships in large data and knowledge bases. In: Proceedings of the 1989 ACM SIGMOD international conference on Management of data, pp. 253–262. ACM Press, New York (1989)
2. Agrawal, R., Dar, S., Jagadish, H.V.: Direct transitive closure algorithms: design and performance evaluation. ACM Transactions on Database Systems 15(3), 427–458 (1990)
3. Agrawal, R., Jagadish, H.V.: Hybrid transitive closure algorithms. In: McLeod, D., Sacks-Davis, R., Schek, H.-J. (eds.) Proceedings of 16th International Conference on Very Large Data Bases, Brisbane, Queensland, Australia, August 13-16, 1990, pp. 326–334. Morgan Kaufmann, San Francisco (1990)
4. An, Y., Janssen, J., Milios, E.E.: Characterizing and mining the citation graph of the computer science literature. In: Knowledge Information Systems, vol. 6, pp. 664–678. Springer, New York (2004)

5. Anyanwu, K., Sheth, A.: The ρ-operator: Enabling querying for semantic associations on the semantic web. In: Proceedings of the twelfth international conference on World Wide Web, pp. 690–699. ACM Press, New York (2003)

6. Barton, S.: Indexing Graph Structured Data. PhD thesis, Faculty of Informatics, Masaryk University, Brno (May 2007)

7. Bartoň, S.: Indexing structure for discovering relationships in RDF graph recursively applying tree transformation. In: Proceedings of the Semantic Web Workshop at 27th Annual International ACM SIGIR Conference, pp. 58–68 (2004)

8. Bartoň, S., Zezula, P.: Rho-index - an index for graph structured data. In: 8th International Workshop of the DELOS Network of Excellence on Digital Libraries, pp. 57–64 (2005)

9. Brickley, D., Guha, R.V.: Resource Description Framework Schema specification (2000)

10. Cohen, E., Halperin, E., Kaplan, H., Zwick, U.: Reachability and distance queries via 2-hop labels. In: Proceedings of the 13th ACM-SIAM SODA, pp. 937–946 (2002)

11. Cronen-Townsend, S., Zhou, Y., Croft, W.B.: Predicting query performance. In: Proceedings of the 25th Annual International ACM SIGIR conference on Research and Development in Information Retrieval (SIGIR 2002), pp. 299–306 (August 2002)

12. Giles, C.L., Bollacker, K., Lawrence, S.: CiteSeer: An automatic citation indexing system. In: Witten, I., Akscyn, R., Shipman III, F.M. (eds.) Digital Libraries 98 - The Third ACM Conference on Digital Libraries, Pittsburgh, PA, June 23–26, pp. 89–98. ACM Press, New York (1998)

13. Lassila, O., Swick, R.R.: Resource Description Framework: Model and Syntax specification (1999)

14. Manber, U., Myers, G.: Suffix arrays: a new method for on-line string searches. In: SODA 1990: Proceedings of the first annual ACM-SIAM symposium on Discrete algorithms. Society for Industrial and Applied Mathematics, Philadelphia, PA, USA, pp. 319–327 (1990)

15. Matono, A., Amagasa, T., Yoshikawa, M., Uemura, S.: An indexing scheme for RDF and RDF Schema based on suffix arrays. In: Proceedings of SWDB 2003, The first International Workshop on Semantic Web and Databases, Co-located with VLDB 2003 (2003)

16. Purdom, P.W.: A transitive closure algorithm. BIT 10, 76–94 (1970)

17. Tarjan, R.E.: Depth first search and linear graph algorithms. SIAM Journal on computing, 146–160 (1972)

18. Tarjan, R.E.: Fast algorithms for solving path problems. J. ACM 28(3), 594–614 (1981)

19. Tarjan, R.E.: A unified approach to path problems. J. ACM 28(3), 577–593 (1981)

20. Thacker, S., Sheth, A., Patel, S.: Complex relationships for the semantic web. In: Fensel, D., Hendler, J., Liebermann, H., Wahlster, W. (eds.) Spinning the Semantic Web. MIT Press, Cambridge (2002)

21. Van Rijsbergen, C.J.: Information Retrieval, 2nd edition. Dept. of Computer Science, University of Glasgow (1979)

22. Warshall, S.: A theorem on boolean matrices. Journal of the ACM 9(1), 11–12 (1962)

23. Yan, X., Yu, P.S., Han, J.: Graph indexing based on discriminative frequent structure analysis. ACM Transactions on Database Systems 30(4), 960–993 (2005)

Full Perfect Extension Pruning
for Frequent Subgraph Mining

Christian Borgelt[1] and Thorsten Meinl[2]

[1] European Center for Soft Computing
c/ Gonzalo Gutiérrez Quirós s/n, 33600 Mieres, Spain
`christian.borgelt@softcomputing.es`
[2] Nycomed Chair for Bioinformatics and Information Mining
Dept. of Computer and Information Science
University of Konstanz, Box M712, 78457 Konstanz, Germany
`thorsten.meinl@uni-konstanz.de`

Summary. Mining graph databases for frequent subgraphs has recently developed into an area of intensive research. Its main goals are to reduce the execution time of the existing basic algorithms and to enhance their capability to find meaningful graph fragments. Here we present a method to achieve the former, namely an improvement of what we called "perfect extension pruning" in an earlier paper [4]. With this method the number of generated fragments and visited search tree nodes can be reduced, often considerably, thus accelerating the search. We describe the method in detail and present experimental results that demonstrate its usefulness.

11.1 Introduction

In recent years the problem how to find common subgraphs in a database of (attributed) graphs, that is, subgraphs that appear with a user-specified minimum frequency, has gained intense and still growing attention. For this task—which has useful applications in, for example, biochemistry, web mining, and program flow analysis—several algorithms have been proposed. Some of them rely on principles from inductive logic programming and describe the graph structure by logical expressions [7]. However, the vast majority transfers techniques developed originally for frequent item set mining. Examples include MolFea [11], FSG [12], MoSS/MoFa [3] , gSpan [16], Closegraph [17], FFSM [9], and Gaston [14]. A related, but slightly different approach, which is strongly geared towards graph compression, is used in Subdue [5].

The basic idea of these approaches is to grow subgraphs into the graphs of the database, adding an edge and maybe a node in each step, counting the number of graphs containing each grown subgraph, and eliminating infrequent subgraphs. Unfortunately, with this method the same subgraph can be constructed in several ways, adding its nodes and edges in different orders. The predominant method to avoid the ensuing redundant search is to define a *canonical form* of a graph that uniquely identifies it up to automorphisms: together with a specific way of growing the subgraphs it enables us to determine whether a given subgraph

D.A. Zighed et al. (Eds.): Mining Complex Data, SCI 165, pp. 189–205.

can be pruned from the search tree (see, for example, [1] for a family of such canonical forms and details of the procedure). As the properties of canonical forms and code words are widely used throughout this chapter, we briefly review them in Section 11.4.

To further improve the algorithms one may restrict the search to so-called *closed graph fragments* (Section 11.2), which capture all information about the set of all frequent subgraphs, but lead to considerably smaller output (in terms of the number of reported fragments). This restriction also enables us to employ additional pruning techniques, one of which is *perfect extension pruning*, as we called it in [4], or *equivalent occurrence pruning*, as it is called in [17]. Unfortunately, neither of these approaches, in the form in which they were originally described in these papers, works correctly, as they can miss certain fragments. This flaw we fix in this paper (Section 11.3).

In addition, the approach in [4] avoided redundant search with the help of a repository of found fragments instead of using the more elegant approach of canonical form pruning. As a consequence, perfect extension pruning was easier to perform, since it was not necessary to pay attention to the canonical form. With canonical form pruning, part of perfect extension pruning is easy to achieve, namely pruning the search tree branches to the right of the perfect extension (Section 11.5). This was first shown in Closegraph [17]. In this paper we show how one may also prune the search tree branches to the left of the perfect extension by introducing a (strictly limited) code word reorganization (Section 11.6). We demonstrate the usefulness of the enhanced approach with experiments on two molecular data sets (Section 11.7).

11.2 Mining Closed Graph Fragments

The notion of a closed fragment is derived from the corresponding notion of a closed item set, which is defined as an item set no superset of which has the same support, i.e., is contained in the same number of transactions. Analogously, a *closed graph fragment* is a fragment no superstructure of which has the same support, i.e., is contained in the same number of database graphs.

As an example consider the molecules (no chemical meaning attached—they were constructed merely for demonstration purposes) shown in Figure 11.1 as the given database of attributed graphs. A corresponding search tree (starting from sulfur as a seed and with fragments being extended only if they appear in at least two molecules) is shown in Figure 11.2 (how the extensions of fragments are chosen and ordered is explained below). The numbers below or to the left/right of the fragments state their support, i.e., the number of molecules a fragment is contained in. Infrequent fragments (i.e. with a support less than two molecules)

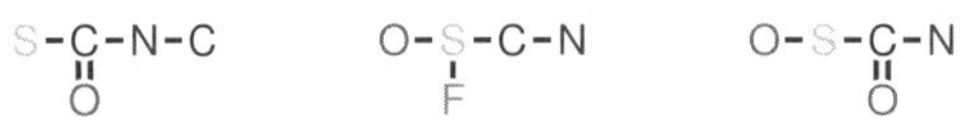

Fig. 11.1. Three simple example molecules

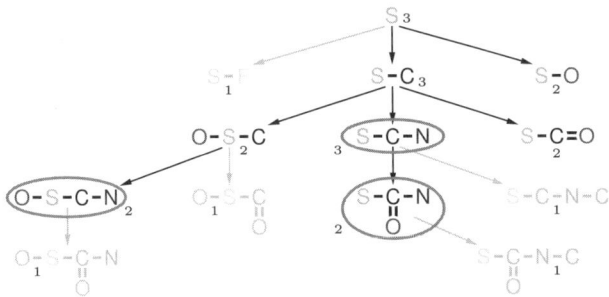

Fig. 11.2. Search tree for the three molecules in Fig. 11.1; infrequent fragments (contained in only one molecule) are drawn in gray/light colors, closed fragments are encircled

are drawn in gray/light colors. The encircled fragments are closed and thus constitute the output of the search (for a minimum support of two molecules). Note that, for example, the fragment O-S-C is not closed, since the fragment O-S-C-N, which contains O-S-C as a proper subgraph, has the same support (namely two molecules).

As for item sets, restricting the search for molecular fragments to closed fragments does not lose any information: all frequent fragments (drawn in black/dark color in Figure 11.2) can be constructed from the closed ones by simply forming all substructures of closed fragments that are not closed fragments themselves. Knowledge of the support of any non-closed frequent fragment is also preserved: its support is simply the maximum of the support values of those closed fragments of which it is a substructure. Consequently, restricting the search to closed fragments is a very convenient and lossless way to reduce the size of the output of a frequent subgraph mining algorithm.

11.3 Perfect Extensions

Perfect extension pruning is based on the observation that sometimes there is a fairly large common fragment in all currently considered database graphs (that is, in all graphs considered in a given branch of the search tree). From the definition of a closed fragment it is clear that in such a situation, if the current fragment is only a part of the common substructure, then any extension that does not grow the current fragment towards the maximal common one can be postponed until this maximal common fragment has been reached. That is, as long as the search has not grown a fragment to the maximal common one, it is not necessary to branch in the search tree. The reason is, obviously, that the maximal common fragment is part of all closed fragments that can be found in the currently considered set of molecules. Consequently, it suffices to follow only one path in the search tree that leads to this maximal common fragment and to start branching only from there.

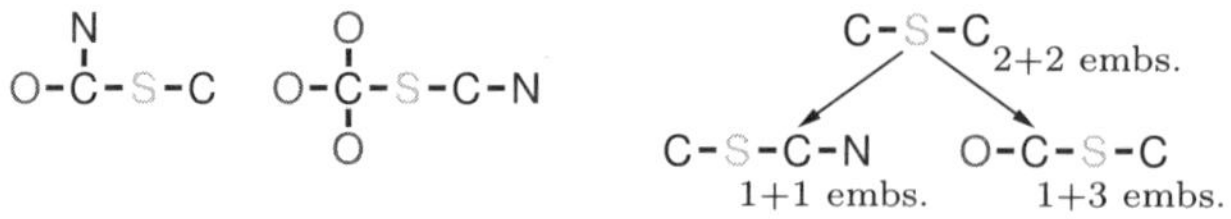

Fig. 11.3. Example of an imperfect extension

As an example consider again the simple set of molecules shown in Figure 11.1. If the search is seeded with a single sulfur atom, considering extensions by a single bond starting at the sulfur atom and leading to an oxygen atom can be postponed until the structure S-C-N common to all molecules has been grown (provided that the extensions of this maximal common fragment are not restricted in any way—a requirement we discuss in detail below).

Technically, the search tree pruning is based on the notion of a *perfect extension*. An extension of a fragment, consisting of an edge and possibly a node (if the edge does not close a ring), is called *perfect* if all of its embeddings (that is, occurrences of the fragment in the database graphs) can be extended in exactly the same way by this edge and node. (Note that there may be several ways of extending an embedding by this edge and node; then all embeddings of a fragment must be extendable in the same number of ways.) Obviously, if there is a perfect extension, all closed super-fragments of a given fragment can, in principle, be found by searching only the corresponding branch.

Note that one has to be careful when identifying perfect extensions. In the first place, it does not suffice to check whether the number of embeddings of the extended fragment is equal to or a multiple of the number of embeddings of the base fragment (as one may be tempted to think at first sight). This is only a necessary, but not a sufficient condition, as the example shown in Figure 11.3 demonstrates. Even though the total number of embeddings in the right branch is the same as for the root, the extension is not perfect, because the extension can be done only once in the left molecule, but three times in the right. The extension in the left branch is not perfect, because the number of extended embeddings, even though the same for each parent embedding, is reduced from the number of extensions of its parents. Such a reduction, which also occurs in the right branch for the left molecule, indicates that some symmetry has been destroyed by the extension, which therefore cannot be perfect. As a consequence, a test for perfect extension actually has to count and compare the number of embeddings per database graph.

A second problem (which was overlooked in both [17] as well as in [4]) is the behavior of rings (cycles) in the search, as we demonstrate with the example molecules shown in Figure 11.4. A search tree for these molecules (with only such fragments that are contained in both molecules) is shown in Figure 11.5. Here almost all extensions are perfect in the sense that they can be made in the same way in all molecules. However, the problem becomes clear when one considers adding a bond from the nitrogen atom to a carbon atom. This extension rules out certain ways of reaching the carbon atom via the oxygen atom and the

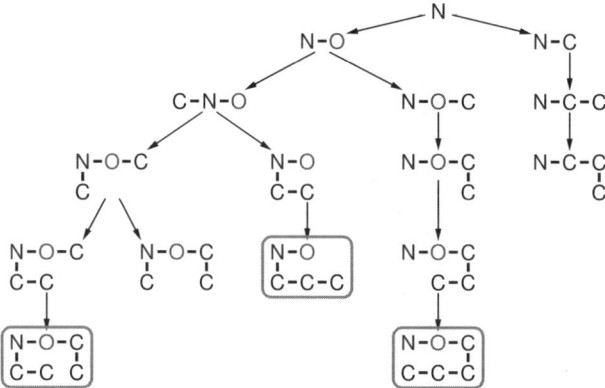

Fig. 11.4. Rings/cycles can cause problems

Fig. 11.5. Search tree for the two molecules in Figure 11.4; closed fragments are encircled

rest of the ring. Hence the bond leading to the carbon atom is only "locally" perfect, but not "globally", that is, when the ring structure is taken into account. As a consequence we cannot restrict the search to the corresponding branch, since we would lose fragments. This can be seen clearly from the location of the closed fragments in the search tree shown in Figure 11.5: there are three closed fragments (for a minimum support of 2, encircled in gray), but we cannot reach all of them if we see adding an edge from the nitrogen atom to a carbon atom as a perfect extension (even after the oxygen atom has been added, which could actually be seen as a perfect extension).

Obviously, the problem is that there are two ways of reaching the carbon atom that is directly connected to the nitrogen atom. Even though only one of them is possible in both molecules, both have to be considered, because part of the second possibility is the same in both molecules, thus leading to a relevant frequent fragment. Unfortunately there is no way to determine this locally, that is, by looking only at the grown fragment and its direct extension. In order to cope with this problem, we require that a perfect extension edge must be a bridge[1] (that is, the extension edge must be a bridge in all embeddings of the extended fragment). This is surely a safe (i.e. sufficient) condition as it rules out any possibility that the destination of the perfect extension edge can be reached in any other way, and thus fixes the flaw mentioned above. However, this

---

[1] An edge in an undirected graph is called a *bridge* if its removal increases the number of connected components of the graph.

requirement is not a necessary condition. As a closer inspection easily reveals, extensions closing a ring (that is, extensions by an edge leading to a node that is already in the base fragment) are also safe and thus can be allowed as candidates for perfect extensions: since the destination node is already in the fragment, there cannot be a problem with multiple ways of reaching it. Hence we can slightly relax the constraints.

Note, however, that these relaxed constraints are still only sufficient, but not necessary. There are other situations in which an extension can be considered perfect, even though it does not meet the abovementioned requirements. For example, if an edge leads to a new node and is part of rings of the same size and composition in all supporting graphs, it is harmless and thus can be considered a perfect extension. Even the extension by a bond from the nitrogen atom to the oxygen atom in Figure 11.5 can be considered a safe perfect extension, despite the fact that the rings have different size. Unfortunately, checking necessary and sufficient conditions for (safe) perfect extensions is complicated and costly and thus we confine ourselves to the rule that an extension edge must either be a bridge in all database graphs or must close a ring (cycle) in all database graphs in order to be considered perfect.

11.4   Canonical Codes for Graphs

As we already mentioned in the introduction, perfect extension pruning, as it was described in the previous section, is not a problem unless canonical forms are used to identify redundant fragments. However, since canonical forms are a lot more elegant than, for example, a repository of already processed fragments, need less memory and make it easier to parallelize the search, it is desirable to be able to use perfect extension pruning together with canonical form pruning. In this section we briefly review some fundamentals of canonical forms for (attributed) graphs, which are necessary to know in order to understand the code word reorganization we describe in Section 11.6.

The core idea of canonical forms of graphs is to describe an (attributed) graph by a *code word*, which uniquely identifies it up to automorphisms, and from which the graph can be reconstructed. The letters of such a code word describe the edges of the graph and which nodes they connect as well as the node and edge attributes (or labels). In order to capture the connection structure, the nodes are numbered (or, more generally, endowed with unique labels), since the node attributes are not enough to identify them uniquely: the same attribute may be assigned to several nodes in a graph. Of course, there are several possible ways of numbering the nodes, each of which gives rise to a different code word. In principle, all of these code words are taken into account and the lexicographically smallest (or greatest) code word is then defined to be the *canonical code word*. Note, however, that due to the way in which code words are used in the search (see below), the possible node numberings (and thus the possible code words) one has to consider can actually be restricted to those compatible with traversals of spanning trees of the considered graphs/fragments (see [1] for more extensive explanations).

Canonical code words are used in the search as follows: during the mining process fragments are grown by adding an edge in each step. This edge is characterized by the node from which it starts, by its attribute, and by the node it leads to. (Note that this does *not* imply directed edges; the "source" and "destination" node are solely defined by how the extension is done: the node that is extended is the "source" and the other node, the extension edge is incident to, is the "destination".) In addition, the nodes are numbered in the order in which they are added to the fragment. Hence the search process naturally constructs a code word for each grown fragment, namely by simply concatenating the descriptions of the edge extensions that led to it.

Of course, there are many possible ways of building a fragment by adding edges, each of which leads to a different code word. However, there is obviously only one way that leads to the canonical code word (since a code word fixes a specific order of the extensions). Hence we may choose to extend only those fragments further that have been built in such a way that their code word is canonical. Eliminating all other fragments is called *canonical form pruning*, which obviously eliminates all redundant search: each fragment is considered at most once. Note that the way in which the search process builds code words also explains why we can confine ourselves to node numberings (and thus code words) compatible with traversals of spanning trees (as mentioned above): no other code words can be constructed by the search.

For the rest of this paper we focus on code words resulting from node numberings that are obtained by breadth-first traversals of spanning trees, that is, the canonical form that is used in MoSS/MoFa [3]. Note, however, that the described approach is also applicable for code words resulting from node numberings that are obtained from depth-first traversals of spanning trees, that is, the canonical form that is used in gSpan [16] or Closegraph [17]. The necessary adaptations of the procedure are straightforward and thus not described in detail (they mainly concern the form of edge descriptions).

A breadth-first code word has the general form $a\ (i_s\ b\ a\ i_d)^m$, where a is node attribute, b an edge attribute, i_s the index (or number) of the source node of an edge, and i_d the index (or number) of the destination node of an edge (by definition, it is always $i_s < i_d$). The letter m denotes the number of edges of the fragment. Each parenthesized expression describes one edge.

As an example, consider the left molecule shown in Figure 11.1. If this molecule is built from left to right, that is, if we choose the left oxygen atom as the root of a spanning tree, a possible code is O 0-S1 1-C2 2=O3 2-N4. As can be seen from this code word, the bond added first is the one from the oxygen atom (index 0) to the sulfur atom (index 1), the bond added last is the one from the carbon atom (index 2) to the nitrogen atom (index 4). However, there is another possibility of building the same fragment, which leads to the code word O 0-S1 1-C2 2-N3 2=O4 (that is, the last two bonds are added in inverse order). If these two code words are compared lexicographically, the latter is smaller than the former (assuming that single bonds are "smaller" than double bonds, that is, - < =). Therefore we can conclude that the first code word is not the canonical code word, but neither is the second. The canonical code word for this molecule is actually S 0-C1 0-02 1-N3 1=04 (if we use the element order S < C < N < O and the bond order - < =, as we also do for all following examples in order to avoid confusion).

Note, however, that the canonical code word is C 0-N1 0-S2 0=03 2-04 if we use the order of the periodic table of elements (that is, C < N < O < S, together with - < =), showing that which code word is the canonical one also depends on the order we choose for the node and edge attributes. Empirical evidence suggests that it is recommendable to use an order that reflects the frequency of the attributes in the graph database to mine (less frequent attributes should precede more frequent ones), as this usually leads to fewer generated fragments and thus shorter search times.

Note also that (canonical) code words for graph fragments provide a natural way of ordering the fragments in the search tree: the children of a search tree node are listed from left to right in the order of lexicographically increasing code words. This makes precise what we mean by "to the left" and "to the right" of a search tree branch: "to the left" are fragments with smaller, "to the right" are fragments with greater (canonical) code words.

Checking whether a given code word is canonical usually requires testing all possible code words for a fragment (at least w.r.t. all possible node numberings resulting from traversals of spanning trees) and thus has essentially the same complexity as a graph isomorphism test. (Pseudo-code for such a canonical form check can be found, for example, in [1].) Nevertheless, canonical code words are very effective in pruning the search tree, because they use "global" information in contrast to only "local" rules, as they were used originally in [3]. These "local" or "simple" rules, however, can still be applied to support canonical form pruning, as they specify necessary (though not sufficient) prerequisites for code words to be canonical, which can be tested very efficiently and help to avoid a costly canonical form test in many cases.

For example, if we use a breadth-first (spanning tree traversal) canonical form (as it was described above), one may not extend a node that has an index smaller than another node in the fragment, which has already been extended (*maximum source extension*: only nodes with an index no less than the maximum source index may be extended). The reason is that an extension violating this rule necessarily leads to a non-canonical code word, as can easily be checked with a spanning tree rooted at the same node.

As an example consider the fragment S-C-N in the search tree in Figure 11.2: this fragment may not be extended by an edge from the sulfur atom (index 0) to an oxygen atom, because an atom with a higher index, namely the carbon atom (index 1), has already been extended (by attaching the nitrogen atom). Indeed, if we add such an edge, the code word of the resulting fragment is S 0-C1 1-N2 0-03, while the canonical code word for this fragment is S 0-C1 0-02 1-N3 (using again the element order S < C < N < O).

More details on canonical forms and the "local" or "simple" rules, which result from them and restrict the possible extensions of a fragment, can be found in [1]. Of course, there are "local" or "simple" rules not only for breadth-first, but also for depth-first (spanning tree traversal) code words, which specify a restricted set of extensions known as *rightmost path extensions*. However, for the discussion in this paper it suffices to know that, regardless of the canonical form used, the "local" or "simple" rules basically state that extensions that generated the sibling nodes to the left of a search tree branch may not be done in this search tree branch itself or in branches to the right of it.

11.5   Partial Perfect Extension Pruning

If one wants to combine perfect extension pruning with canonical form pruning as it was described above, the following problem has to be solved: growing the maximal common fragment can interfere with canonical form pruning and in particular with the extension restrictions resulting from it (note that this was no problem in [4] due to the use of a repository of found fragments to avoid redundant search). Obviously, perfect extensions should not lead to such a restriction, because otherwise search results may be lost. The fact that the code word of a fragment, as it is built by the search, is not canonical is no longer sufficient to prune it, since preferring perfect extensions may have changed the order of the extensions by which a fragment is built.

As an example consider again the search tree shown in Figure 11.2. If we simply confined the search to the sub-tree rooted at the fragment S-C-N, we would lose the fragment O-S-C-N in the leftmost branch. The reason is that the extension of S-C to S-C-N, due to canonical form restricted extensions, prevents an extension of the sulfur atom in this sub-tree (as described in the preceding section), because an atom with a higher number, namely the carbon atom, has already been extended in the preceding step.

Fortunately, this only affects search tree branches to the left of the perfect extension branch, since the corresponding extensions are ruled out by the perfect

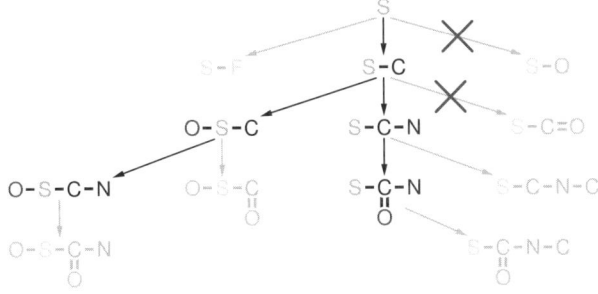

Fig. 11.6. Search tree for the three molecules in Figure 11.1 with partial perfect extension pruning (crossed out branches are pruned)

extension and the "local" or "simple" rules (see above). All extensions corresponding to branches to the right of the perfect extension are still possible for the fragment reached by the perfect extension. Therefore branches to the right can be pruned immediately without any loss: they cannot contain any closed fragment, because the perfect extension cannot be done in them without violating the canonical form, but has to be done in order to reach a closed fragment. This type of pruning we call *partial perfect extension pruning* (because it prunes only part of the branches aside from the perfect extension one). Note that Closegraph [17] uses only this form of pruning.

How partial perfect extension pruning changes the search tree for the molecules in Figure 11.1 is shown in Figure 11.6. Note that only non-closed fragments are removed from the search tree (compare to Figure 11.2, in which the closed fragments are highlighted). The gains consist in the fact that the two pruned fragments need not be processed: neither do they have to be checked for canonical form nor do we have to consider possible extensions of them.

11.6   Full Perfect Extension Pruning

Although partial perfect extension pruning is already highly effective, it is desirable to prune also the search tree branches to the left of the perfect extension, thus completing partial perfect extension pruning into *full perfect extension pruning*. In order to do so, we must not restrict the extensions of the fragment that resulted from a perfect extension as it would be required by canonical form pruning (with or without the "local" or "simple" rules). Otherwise we could lose (closed) frequent fragments, as we demonstrated above. In other words, we would like to have a search tree like the one shown in Figure 11.7 for the molecules shown in Figure 11.1.

The core problem with this is how we can avoid that the fragment O-S-C-N is pruned as non-canonical. The breadth-first search canonical code word for this fragment is S 0-C1 0-02 1-N3. However, with the search tree in Figure 11.7 it is assigned the code word S 0-C1 1-N2 0-03, because this reflects the order in

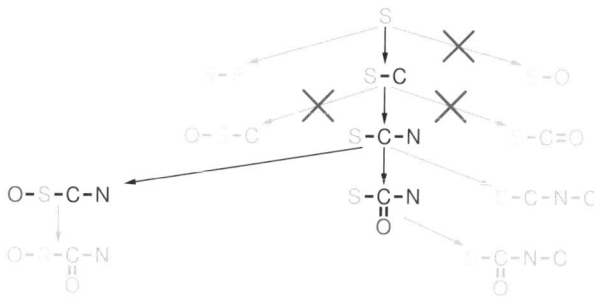

Fig. 11.7. Search tree for the three molecules in Figure 11.1 with full perfect extension pruning (crossed out branches are pruned)

which the bonds have been added. Since this code word is not canonical, the fragment would be pruned and neither extended nor reported.

In order to avoid this, we allow for a (strictly limited) reorganization of code words as they result from a search tree, which takes care of the fact that perfect extension edges may have been added earlier than required by the canonical form. Technically, we split the code word into two parts: the first, fixed part consists of the (possibly empty) prefix up to and including the last edge that was added by a non-perfect extension or by a perfect extension with no search tree branches to the left of it. The second, volatile part consists of the remaining suffix of the code word, which is made up only of perfect extensions edges, which had search tree branches to the left of it.

Note that we can check for the existence of branches to the left of a perfect extension branch after *minimum support pruning*, that is, after eliminating all fragments that occur in less than the user-specified minimum number of database graphs. The reason is that we can be sure that extensions leading to infrequent fragments in branches to the left will also lead to infrequent fragments in the perfect extension branch or in branches to the right of it and thus need not be considered in these branches.

The construction of the code word for an extended fragment is modified as follows: instead of always simply appending the description of the extension edge to the end of a code word, the description of the new edge may now be inserted anywhere in or even before the volatile part, but not in the fixed part. We may imagine this as first appending the new edge description and then shifting it to the left, as long as this makes the code word lexicographically smaller, but the new edge description does not enter the fixed part.

Note, however, that "shifting" an edge in the code word can make it necessary to renumber the nodes. For example, if in the fragment O-S-C-N the bond added last in the search (that is, the bond from the sulfur atom to the oxygen atom) is shifted left past the perfect extension bond (that is, the bond from the carbon atom to the nitrogen atom), the oxygen and the nitrogen atom get new indices. The reason is that the nodes must be numbered in the order in which they would be added if the edges were added in the order in which their descriptions are listed in the (reorganized) code word (see Figure 11.8).

Technically, we achieve this renumbering as follows: instead of actually shifting the extension edge from right to left, we rebuild the code word from left to right. First we traverse the fixed part, numbering all nodes in the order in which they are met. Then we continue with the volatile part until at least one of the two

1. Base fragment: S-C-N   canonical code: S 0-C1 1-N2
2. Extension to O-S-C-N   code: S 0-C1 1-N2 0-O3 (not canonical!)
3. Shift the non-perfect extension   code: S 0-C1 0-O3 1-N2
4. Renumber nodes   canonical code: S 0-C1 0-O2 1-N3

Fig. 11.8. Fixing a fragment's code word by shifting a non-canonical extension over perfect extensions (marked in gray) to the proper place and renumbering the nodes

nodes incident to the new edge is numbered. Note that this may already be the case before the first edge in the volatile part is considered. In this case no edge of the volatile part is processed in this step.

Finally we traverse the (remaining) volatile part edge by edge, each time comparing the next edge to the new edge. If the new edge (w.r.t. source and – possibly still to be assigned – destination index as well as edge attribute and destination node attribute) is lexicographically smaller, it is inserted at the current position in the volatile part and the rest of the volatile part is appended (renumbering nodes as needed). Otherwise any unnumbered node incident to the current volatile edge is numbered and the next volatile edge is considered. If all volatile edges have been traversed and the new edge has not been inserted, it is simply appended at the end of the code word.

To make the process clearer, we execute it step by step for the example shown in Figure 11.8. The root node (here the sulfur atom) is, of course, always in the fixed part. Hence it receives the initial node index, that is, 0. Since the next edge is already in the volatile part, this finishes processing the fixed part. Since by assigning the index 0 to the sulfur atom, one node incident to the new edge (sulfur to oxygen) is already numbered, we have to start immediately to compare edge descriptions. We compare two possibilities, namely appending the description of the new edge, which assigns the node index 1 to the oxygen atom, or appending the already present first perfect extension edge (sulfur to carbon), which assigns the node index 1 to the carbon atom. This yields two possible code word prefixes, namely S 0-O1 and S 0-C1. Since the latter is smaller (as C < O), it is fixed (that is, the new edge is not yet inserted) and we move to the next position. Here we compare the code word prefixes S 0-C1 0-O2 and S 0-C1 1-N2. Since the former is smaller (as O < N), the position of the new edge has been found and we fix the first prefix. In a final step, the remaining perfect extension edge is appended, assigning the node index 3 to the nitrogen atom. Note that the fixed part of the resulting code word now contains not only the root atom, but two bonds: the first perfect extension bond, which is rendered fixed by the fact that a non-perfect extension was inserted after it, and the new bond, which is fixed, simply because it is not a perfect extension. The volatile part contains only the second perfect extension (the bond from the carbon to the nitrogen atom).

Note that generally, provided the new edge is not a perfect extension itself, this edge is recorded for the restricted extensions as required by the "local" or "simple" rules of maximum source extensions (that is, extensions preceding this edge are ruled out). In other words, if the new edge is not a perfect extension, the place at which it is inserted is the new end of the fixed part of the code word (as described above). Note also that the resulting code word still has to be checked for canonical form. Since the reorganization is strictly limited, the resulting code word may not be canonical. For example, the new edge may actually have to be inserted into the fixed part in order to make the code word canonical. In this case the fragment must not be adapted, so that the code word becomes canonical, but has to be pruned.

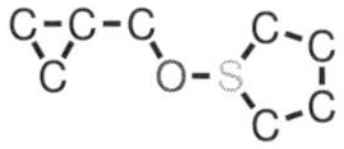

Fig. 11.9. Example molecule used to demonstrate full perfect extension pruning

To further illustrate the process, we study another complete example, which also shows the different cases that can occur. Consider the molecule shown in Figure 11.9. Our goal is to build this molecule using full perfect extension pruning.[2] As the order of the elements we use again S < C < O, which is in line with the order used in all preceding examples. As a consequence the search has to start at the sulfur atom, because all other starting points obviously lead to non-canonical code words (as even their first letter is greater).

Three extensions of this one-node fragment (code word: S) are possible: we may add one of the two ring bonds to carbon atoms (which lead to the same fragment S-C) or we may add the bond to the oxygen atom. Without perfect extension pruning, both child fragments (i.e. S-C and S-O) would have to be considered. However, the bond to the oxygen atom is a bridge, occurs in all molecules (only one in this example), and the number of embeddings of the extended fragment is the same as for the single sulfur atom. Hence adding this bond is a perfect extension, while the bond to a carbon atom is not eligible as a perfect extension, since it is a ring bond (and thus no bridge, see Section 11.3). This leads to the code word S 0-O1. The extension is marked as perfect, and the volatile part of the code word starts directly after the sulfur atom (as is indicated by a gray background).

Note that the other extension (leading to the fragment S-C) would have to be considered if we only used partial perfect extension pruning, since its code word, that is, S 0-C1, is smaller than S 0-O1. Only full perfect extension pruning allows us to eliminate this fragment from the search.

In the next step, all possible extensions are considered (no restriction by "local" or "simple" rules, because the preceding extension was perfect), which are the two ring bonds (again leading to the same fragment, now O-S-C) and the bond from the oxygen atom to the next carbon atom in the chain. The latter is a perfect extension and thus the other two extensions are pruned, resulting in the code word S 0-O1 1-C2. Since the new edge is a perfect extension, the volatile part grows to two edge descriptions (gray background).

In the third step, the two ring bonds incident to the sulfur atom are again eliminated due to the perfect extension to the next carbon atom in the chain, which is in the left ring: S 0-O1 1-C2 2-C3. Now there are no perfect extensions left, because all remaining bonds are part of rings (and thus no bridges).

It should be noted that the maximum source index is still 0 (sulfur atom), because all extensions made so far were perfect and thus their source indices are

---

[2] Mining only one molecule is, of course, not very useful in practice, but it keeps the example simple, and the process, at least w.r.t. pruning, is exactly the same as when mining a larger number of molecules.

not counted for the "local" or "simple" rules characterizing maximum source extensions. Without this special handling, we would not be allowed to add any of the bonds of the right ring. But since the sulfur atom is still extendable, we add, in the next step, one of the two ring bonds to it, which results in the code word S 0-01 1-C2 2-C3 0-C4. As one can immediately see from the source index 0 in the last bond, this code word is not canonical. Therefore we have to start the process of rebuilding the code word.

First, the sulfur atom is numbered 0 and this already determines three of the four parts of the description of the newly added bond from the sulfur to the carbon atom, namely 0-C? (the still to be assigned destination index is replaced by a question mark; alternatively it may be set to the next free node index, which is 1 in this case, as we did it before). This "incomplete" extension is compared to the first perfect extension in the volatile part. Since the incomplete description 0-0? of this extension is greater (as carbon precedes oxygen), the position of the new edge has been found and the description of this edge is appended. Therefore we have as a code word prefix S 0-C1, which forms the new fixed part of the code word. The three perfect extensions in the volatile part are renumbered accordingly (the indices of the destination nodes are increased by one) and their descriptions are appended, yielding the code word S 0-C1 0-02 2-C3 3-C4 (as before, the gray part is volatile).

The next extension adds the other (ring) bond from the sulfur atom to a carbon atom: we reorganize from S 0-C1 0-02 2-C3 3-C4 0-C5 to S 0-C1 0-C2 0-03 3-C4 4-C5. The sixth extension adds another ring bond, yielding the code word S 0-C1 0-C2 0-03 3-C4 4-C5 1-C6 (before reorganization). This time, the new edge is not inserted before all perfect extensions, but after the first, because its source node index is greater than that of the first perfect extension: S 0-C1 0-C2 0-03 1-C4 3-C5 5-C6. This has two effects: in the first place, the volatile part now consists of only the last two perfect extensions (as the insertion of a non-perfect extension edge after it renders the first perfect extension edge fixed). Secondly, the atom with the maximum source index (from which on extensions are still allowed) is now the one with index 1, namely the source atom of the added edge.

The next edge that is added is another ring bond and it is inserted before the volatile part, since its source index is smaller than the source index of the next perfect extension bond: S 0-C1 0-C2 0-03 1-C4 2-C5 3-C6 6-C7.

The next edge closes the right ring and it is inserted in the middle of the volatile part: S 0-C1 0-C2 0-03 1-C4 2-C5 3-C6 4-C5 6-C7 (since its source node index is larger than that of the first perfect extension bond in the volatile part, but smaller than that of the second perfect extension).

The last three edges, that is, the three bonds of the left ring (3 carbons), are added in the normal order (after the volatile part, or actually simply by appending to a fixed code word, since adding the first bond of the left ring renders the last perfect extension fixed). No code word reorganization is necessary in any of these steps. The final (canonical) code word is:

S 0-C1 0-C2 0-03 1-C4 2-C5 3-C6 4-C5 6-C7 7-C8 7-C9 8-C9.

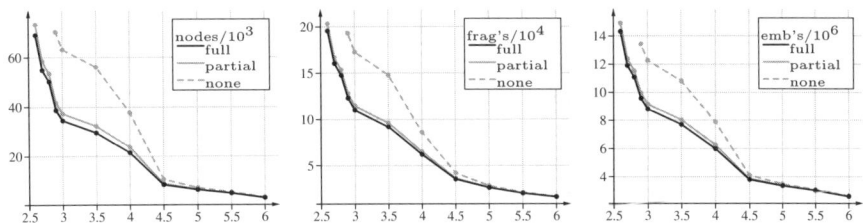

Fig. 11.10. Experimental results on the IC93 data without ring mining (extensions add only single bonds, both in rings and outside rings)

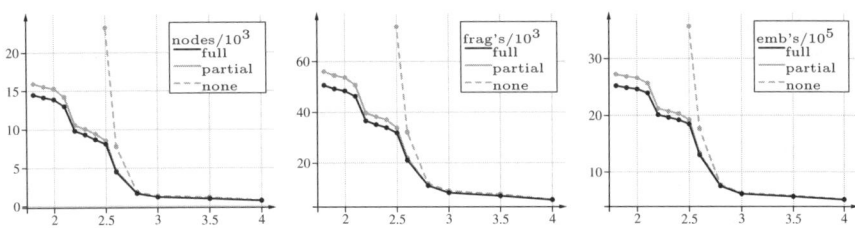

Fig. 11.11. Experimental results on the IC93 data with ring mining (extensions add single bonds that are not part of rings or complete rings)

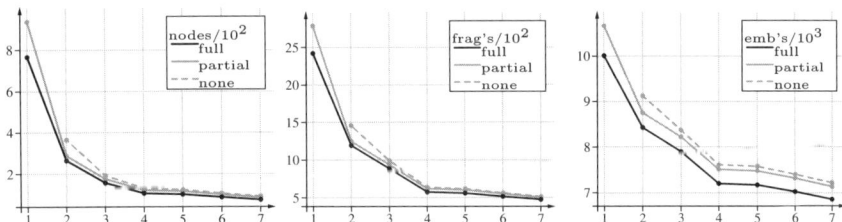

Fig. 11.12. Experimental results on the steroids data with ring mining (extensions add single bonds that are not part of rings or complete rings)

11.7 Experiments

In order to test full perfect extension pruning, we implemented it as an extension of the MoSS program[3], which is written in Java. As test datasets we used a well-known subset of the Index Chemicus 1993 [10] (IC93) and a small dataset of 17 steroids. The results on these datasets with different search modes are shown in Figures 11.10 to Figure 11.12, which display the number of search tree nodes (left), created fragments (middle), and created embeddings (right). The

---

[3] MoSS is available for free download under the GNU Lesser (Library) Public License at http://www.borgelt.net/moss.html.

horizontal axis shows the minimal support in percent (IC93) or as an absolute number (steroids). For the experiments of Figure 11.11 and Figure 11.12 we used ring mining, which means that rings in a user-defined size range (here: 5 to 6 bonds) are not built edge by edge, but added in one step. The technique underlying such ring mining was introduced in [8] for a repository of processed fragments to avoid redundant search, but later extended in [2] to work with canonical form pruning (using a code word reorganization technique that is similar to the one presented in this paper, but more complex).

In each diagram the dashed gray line refers to the basic algorithm without any perfect extension pruning, the gray solid line to partial perfect extension pruning and the black solid line to full perfect extension pruning. These results show that full perfect extension pruning indeed leads to some non-negligible gains (in the order of about 5 to 10%) over partial perfect extension pruning, even though the main gains clearly result from partial perfect extension pruning. Tests we ran during the development of the program indicated that relaxing the constraints for perfect extensions (that is, also edges closing rings/cycles are allowed as perfect extensions instead of only bridges) improved performance by up to an additional 3%.

11.8 Conclusions

In this paper we fixed the flaw of the original descriptions of perfect extension pruning by requiring that perfect extensions must be bridges, but still allowing edges that close rings/cycles apart from bridges. In addition, we introduced full perfect extension pruning, which consists in pruning not only the search tree branches to the right (partial perfect extension pruning as it is used in Closegraph [17]), but also those to the left of the perfect extension branch. To make this possible in combination with canonical form pruning, we allowed for a (strictly limited) reorganization of code words as they result from the search. The experimental results show that this method can actually further reduce the complexity of the search, although the main improvement comes from partial perfect extension pruning. Future work is directed at combining sibling perfect extensions into one extension, so that perfect extensions, once found, need not be rediscovered and reprocessed.

References

1. Borgelt, C.: On Canonical Forms for Frequent Graph Mining. In: Proc. 3rd Int. Workshop on Mining Graphs, Trees and Sequences, MGTS 2005, Porto, Portugal, 1–12. ECML/PKDD 2005 Organization Committee, Porto, Portugal (2005)
2. Borgelt, C.: Combining Ring Extensions and Canonical Form Pruning. In: Proceedings of the 4th International Workshop on Mining and Learning in Graphs (MLG 2006), ECML/PKDD 2006 Organization Committee, Berlin, pp. 109–116 (2006)

3. Borgelt, C., Berthold, M.R.: Mining Molecular Fragments: Finding Relevant Substructures of Molecules. In: Proc. IEEE Int. Conf. on Data Mining, ICDM 2002, Maebashi, Japan, pp. 51–58. IEEE Press, Piscataway (2002)
4. Borgelt, C., Meinl, T., Berthold, M.R.: Advanced Pruning Strategies to Speed Up Mining Closed Molecular Fragments. In: Proc. IEEE Conf. on Systems, Man and Cybernetics, SMC 2004, The Hague, Netherlands, IEEE Press, Piscataway (2004)
5. Cook, D.J., Holder, L.B.: Graph-Based Data Mining. IEEE Trans.on Intelligent Systems 15(2), 32–41 (2000)
6. Di Fatta, G., Berthold, M.R.: Distributed Mining of Molecular Fragments. In: Workshop on Data Mining and the Grid, IEEE Int. Conf. on Data Mining, pp. 1–9. IEEE Press, Piscataway (2004)
7. Finn, P.W., Muggleton, S., Page, D., Srinivasan, A.: Pharmacore Discovery Using the Inductive Logic Programming System PROGOL. Machine Learning 30(2-3), 241–270 (1998)
8. Hofer, H., Borgelt, C., Berthold, M.R.: Large Scale Mining of Molecular Fragments with Wildcards. Intelligent Data Analysis 8, 495–504 (2004)
9. Huan, J., Wang, W., Prins, J.: Efficient Mining of Frequent Subgraphs in the Presence of Isomorphism. In: Proc. 3rd IEEE Int. Conf. on Data Mining, ICDM 2003, Melbourne, FL, pp. 549–552. IEEE Press, Piscataway (2003)
10. Index Chemicus — Subset from 1993. Institute of Scientific Information, Inc (ISI). Thomson Scientific, Philadelphia, PA, USA (1993),
    http://www.thomsonscientific.com/products/indexchemicus/
11. Kramer, S., de Raedt, L., Helma, C.: Molecular Feature Mining in HIV Data. In: Proc. 7th ACM SIGKDD Int. Conf. on Knowledge Discovery and Data Mining, KDD 2001, San Francisco, CA, pp. 136–143. ACM Press, New York (2001)
12. Kuramochi, M., Karypis, G.: Frequent Subgraph Discovery. In: Proc. 1st IEEE Int. Conf. on Data Mining, ICDM 2001, San Jose, CA, pp. 313–320. IEEE Press, Piscataway (2001)
13. DTP AIDS Antiviral Screen (HIV Data Set) — Subset from 2001. Developmental Therapeutics Program (DTP), National Cancer Institute, USA (2001),
    http://dtp.nci.nih.gov/docs/aids/aids_data.html
14. Nijssen, S., Kok, J.N.: A Quickstart in Frequent Structure Mining Can Make a Difference. In: Proc. 10th ACM SIGKDD Int. Conf. on Knowledge Discovery and Data Mining, KDD 2004, Seattle, WA, pp. 647–652. ACM Press, New York (2004)
15. Washio, T., Motoda, H.: State of the Art of Graph-Based Data Mining. SIGKDD Explorations Newsletter 5(1), 59–68 (2003)
16. Yan, X., Han, J.: gSpan: Graph-Based Substructure Pattern Mining. In: Proc. 2nd IEEE Int. Conf. on Data Mining, ICDM 2003, Maebashi, Japan, pp. 721–724. IEEE Press, Piscataway (2002)
17. Yan, X., Han, J.: Closegraph: Mining Closed Frequent Graph Patterns. In: Proc. 9th ACM SIGKDD Int. Conf. on Knowledge Discovery and Data Mining, KDD 2003, Washington, DC, pp. 286–295. ACM Press, New York (2003)

12

Parallel Algorithm for Enumerating Maximal Cliques in Complex Network

Nan Du, Bin Wu, Liutong Xu, Bai Wang, and Pci Xin

Beijing Key Laboratory of Intelligent Telecommunications Software and Multimedia, Beijing University of Posts and Telecommunications, Beijing 100876, China
{dunan, wubin, xliutong, wangbai}@bupt.edu.cn, peixin@tseg.org

12.1 Abstract

Efficient enumeration of all maximal cliques in a given graph has many applications in the filed of Graph Theory, Social Network Analysis, Bioinformatics and etc. Recent researches indicate that many networks in our world are complex networks involving massive data. Being as the complete sub-graph, a maximal clique can represent a group of friends who all hang around together. It can also be used to find common sub-topologies in a set of protein structures. However, the large scale of real networks and the exponentially increasing computation time of the clique enumeration problem make most of the existing algorithms unsuitable in the real-world scenarios. Therefore, we present a parallel algorithm *Peamc* (Parallel Enumeration of All Maximal Cliques) which exploits several new and effective techniques to enumerate all maximal cliques in large-scale complex networks. Experimental results on true-life networks with up to 20 million vertices and 50 million edges show that *Peamc* can find all the maximal cliques with high efficiency and scalability.

12.2 Introduction

Many recent researches indicate that a large body of diverse systems in nature and society can be described in terms of complex networks[1] which often present a large intricate web of connections among the massive entities they are made of. For example, the Internet is a complex network of routers and computers linked by countless interwoven physical links; the WWW (World Wide Web) is a virtual network consisting of numerous web pages connected by hyperlinks; fads, rumors, epidemics and innovations spread on the social network, whose node corresponds to each of us in the human society and whose edges represent various social actions and relationships. Meanwhile, the computerization of data acquisition and storage leads to the emergence of large databases on the topology of various real networks, and the increased computing power has allowed us to investigate networks containing millions of nodes. Therefore, researchers from physics, computer science and sociology are now concerned with inferring the

D.A. Zighed et al. (Eds.): Mining Complex Data, SCI 165, pp. 207–221.
springerlink.com   © Springer-Verlag Berlin Heidelberg 2009

underlying linkage patterns or building blocks[2] to have a better understanding of the network's structural and functional properties.

Among the basic building blocks, maximal clique is a well-known sub-structure and has applications in many different domains. In graph theory, an algorithm for listing all maximal cliques in a graph can be used as a subroutine for solving many NP-complete graph problems. If a given subgraph is the maximal clique in some graph G, it is also the maximal independent set[3] in the complement graph of G. Apparently, the solutions to the maximum independent set problem, the maximum clique problem, and the minimum independent dominating problem must all be maximal cliques, and can be found by an algorithm that lists all maximal cliques and retains the ones with the largest or smallest size. *Lawler* (1976) observed that listing maximal independent sets can also be used to find 3-colorings of graphs: a graph can be 3-colored if and only if the complement of one of its maximal independent sets is bipartite. In social network analysis[4], a maximal clique represents a group of closely interrelated people, which is also called the community structure. For social network researchers, individuals belonging to the same community are probable to have properties in common. The communities in the blogs often correspond to topics of interest. Monitoring the aggregate trends and opinions revealed by these communities provides valuable insight to a number of business strategies and decisions. In bioinformatics, maximal cliques are used to search for the common sub-topologies in a set of protein structures[5]. Other more complex problems can also be modeled as finding a clique or independent set of a specific type. All these applications motivate many algorithms to enumerate all maximal cliques or equivalently, all maximal independent sets efficiently.

However, it is easy to see that enumerating all maximal cliques is also a NP-problem[6], where a graph with n vertices can have as many as $3^{n/3}$ maximal cliques, of which some are of the maximum size. Meanwhile, many real world networks often consist of millions of nodes and edges making the problem even more challenging. Given graph G with n vertices and m edges, in the worst case, *Peamc* runs with $O(\frac{\Delta \times M_C \times Tri^2}{P})$ time delay and in $O(m+n)$ using P processing elements, where Δ is the maximum degree of G, M_C represents the size of the maximum clique and Tri denotes the number of all triangles in G respectively.

12.3   Related Work

Traditionally, the study of networks has been the territory of graph theory. While graph theory initially focused on regular graphs, since the 1950s large-scale networks with no apparent design principles have been described as random graphs, which were first studied by the famous mathematicians *Paul Erdős* and *Alfréd Rényi*[1] . According to the *Erdős–Rényi* model, we start with n nodes and connect every pair of nodes with probability p. The obtained graph has expected $\frac{pn(n-1)}{2}$ edges distributed randomly. This model has guided our thinking about real world networks for many decades since its introduction. Based on this model, the maximal clique problem, being as one of the well-known sub-structures in a given graph, has been widely investigated since 1973. The basic maximal clique enumeration algorithm called Base BK[7] algorithm proposed by *Bron* and *Kerbosch* was first published in 1973. Since then, many algorithms have been developed to solve this classic problem from different perspectives.

In general, these algorithms could be classified into two groups. In terms of the first group, people use the depth-first searching method and adopt specific pruning policies to improve the efficiency. All possible combinations of vertices in the network constitute a search tree with each node corresponding to a candidate maximal clique. The basic idea is to treat the process of enumerating all the maximal cliques as a depth-first traversal of this tree. The algorithms of *Bron*, *Tsukyuama*[8] and *Makino*[9] are typical ones of this kind.

Bron's BK algorithm maintains three dynamically changing sets: $COMPSUB$, a global set containing the current growing clique; $CANDIDATES$, a local set holding all vertices that will eventually be added to the current $COMPSUB$; NOT, a local set containing all vertices that have been previously added to $COMPSUB$. Each candidate maximal clique corresponds to a node in the search tree and a function called $EXTEND$ traverses this search tree by the depth-first order recursively.

Makino's algorithm[9] develops a search tree by defining a child-parent relationship between two maximal cliques and traverses this search tree to list every maximal clique. In addition, given a lower bound f, this algorithm divides graph G into two sub-graphs: V_1 and V_2. V_1 consists of the vertices whose degree is larger than f and the left vertices constitute V_2. It first finds all maximal cliques in V_1 and and stores them in set Q. Then it finds all maximal cliques in V_2 while eliminates those in Q that are already contained by some maximal cliques in V_2. This strategy will relatively improve the efficiency of the algorithm in sparse graphs. However, there exist two major drawbacks. First, this algorithm involves too many expensive set operations. If the average size of the maximal cliques is large, the efficiency of the algorithm will decrease. Second, the selection of f depends on each specific problem and thus puts a great impact on the corresponding efficiency as well.

With respect to the second group, people have borrowed the join-and-pruning strategy. *Kose*'s algorithm[10] belongs to this class. It takes advantage of the fact that every clique of size $k(k \geq 2)$, is comprised of two cliques of size $k - 1$ that share $k - 2$ vertices. The basic principle is to build all possible 3-cliques by joining every two 2-cliques. Any 2-clique that cannot become a component of a 3-clique is declared maximal and output. This procedure is repeated according to the increasing order of the clique size until it is no longer possible to build a larger one. Nevertheless, this algorithm also has several drawbacks. On the one hand, building the candidate cliques in this join-and-pruning manner requires us to store both the $(k-1)$- and k-cliques, which will consume too much space when the problem size is great. On the other hand, every time a k-clique is formed, all $(k - 1)$-cliques contained within the new clique must be marked as being used to eliminate the duplicates. This process will also occupy too much space by the order of 2^n in the worst case.

All these current algorithms are successful approaches to listing all maximal cliques in different ways. However, just as mentioned before, the growing interests in complex systems have prompted many scientists to reconsider *Erdős–Rényi* modeling paradigm and ask a simple question: are the real networks behind such diverse complex systems as the Internet fundamentally random? Fortunately, in the past few years, several breakthroughs in the study of real world networks have brought about many new concepts and ideas. More and more grown evidence could show that real networks in our world can be modeled as complex networks. These networks are often large sparse graphs with the properties of short average path-length, power-law degree distribution and high clustering coefficient. Compared with the classic random network, which has short average path-length, Poisson degree distribution and law clustering coefficient, it is apparent that complex network is essentially a different kind of network model. Since most existing algorithms are based on the classic random network, it is necessary for us to propose a new method which is optimized on the complex network model and can be efficient to enumerate all the maximal cliques in practical scenarios.

12.4   Algorithm *Peamc*

For most complex networks, they are often sparse in global yet dense in local. Because triangle structure is the basic building block of the clique, our basic idea is to search for every possible triangle to form larger clique recursively until the finally obtained clique becomes the maximal one. During this process, we use several pruning strategies to improve the efficiency. Finally, by extending this basic idea, we provide a parallel model of the algorithm to make *Peamc* efficient and scalable enough when dealing with large-scale networks.

12.4.1   Notations and Definitions

In this paper, we consider simple graphs only, i.e., the graphs without self-loops or multi-edges. For graph G, $V(G)$ and $E(G)$ denote the sets of vertices and edges of G respectively.

Definition 1. $S \subseteq V(G)$, $\forall u, v \in S, u \neq v$ such that $(u, v) \in E(G)$, then S is a clique in G and n-clique denotes a clique of size n. If any other S' is a clique and $S' \subseteq S$ iff $S' = S$, then S is a maximal clique of G. In addition, the maximum clique is a maximal clique with the largest size in all.

Definition 2. Given vertex $v_i \in V(G)$, let $N(v_i) = \{v_j \in V(G) | (v_i, v_j) \in E(G)\}$, $N(v_i)$ is called the **neighbor** set of v_i and the **degree** of v_i is thus $|N(v_i)|$. Let $M(v_i) = \{v_j | i < j, v_j \in N(v_i)\}$, $T(v_i) = \{v_k | v_k, v_j \in M(v_i), i < j, j < k, (v_i, v_j) \in E(G), (v_j, v_k) \in E(G)\}$, $T(v_i)$ is called the **triangle neighbor** set of v_i.

Definition 3. *For a vertex $v_i \in V(G)$, if the neighbors of v_i are part of a clique, then there will be $\frac{|N(v_i)|(|N(v_i)|-1)}{2}$ total edges among them. The ratio of the number E of the edges that actually exist between the vertices of $N(v_i$ to the total number of edges is the* **clustering coefficient** *of v_i denoted as $C(v_i) = \frac{2E}{|N(v_i)| \times (|N(v_i)|-1)}$.*

12.4.2   Basic Idea and Method

To efficiently enumerate all maximal cliques, we must address the following two issues. The first one is how to provide a systematic way to enumerate all the maximal cliques without duplicates. The second one is how to design smart strategies to prune futile search subspaces to speed up the searching process.

In terms of the first issue, it is actually concerned with the correctness and completeness of the algorithm. From the priori discussion, we see that the average clustering coefficient value of random networks is rather different from that of complex networks. In a random graph, since the edges are distributed randomly, the *clustering coefficient* is the connection probability p. However, in most real networks the clustering coefficient is typically much larger than it is in a comparable random network which has the same number of nodes and edges as the real network. From definition 3, we see that the *clustering coefficient* is a direct indication to the existence of triangle structures. Because triangle structure or 3-clique is the basic unit of any clique whose size is larger than 3, we could take advantage of this fact to develop our traversal strategy to search for the vertices that could build triangles by depth-first order recursively. For each step of this traversal process, the large *clustering coefficient* property of the complex networks will be an important clue to find the maximal cliques quickly.

All vertices of G will be accessed by the ascending order of their index. For $v_i \in G$, a search tree rooted with v_i where each node corresponds to a candidate clique will be built and traversed. In the beginning, the vertices of $M(v_i)$ will be chosen by the ascending order of their index. Suppose the current vertex with the smallest index chosen from $M(v_i)$ is v_j. From v_i and v_j we build the *triangle neighbor* set $T(v_i)$. $\{v_i, v_j\}$ becomes the direct child of $\{v_i\}$ in the search tree. Again, a vertex v_k with the smallest index from set $T(v_i)$ will be chosen. Next, based on v_j and v_k, we build the *triangle neighbor* set $T(v_j)$. If $T(v_j) \neq \emptyset$, $\{v_i, v_j, v_k\}$ forms the child node of $\{v_i, v_j\}$. This process will repeat recursively until we reach the leaf node of the search tree where we could not build any triangle neighbor set. Apparently the traversal process of the search tree is actually the process of building triangles. Because every candidate vertex is accessed by the ascending order of its index, there will be no duplicate triangles and candidate cliques being generated. Since the search tree rooted with every vertex in G will be traversed, which means all the candidate cliques will be identified, the completeness of *Peamc* is thus ensured. To identify whether the candidate clique is a maximal clique, we present the following theorem.

Theorem 1. *Let clique* $S \subseteq V(G)$, *S is a maximal clique of G, iff $\forall v_i \in S$ and $v_j \in N(v_i) - S$ we have $S \nsubseteq N(v_j)$.*

Proof: One direction of the "if and only if" condition states that clique S is a maximal clique. If $\forall v_i \in S, \exists v_j \in N(v_i) - S$ such that $S \subset N(v_j)$, then in the subgraph $\{v_j\} \cup S$, $C(v_j) = 1$, and set $\{v_j\} \cup S$ is also a clique. However, we had S a maximal clique earlier, meaning that there is no superset of S being a clique, so a contradiction occurs. The other direction of "if and only if" condition states that $\forall v_i \in S$ and $v_j \in N(v_i) - S$, we have $S \nsubseteq N(v_j)$, which means there exists at least one vertex $v_k \in S$ such that $(v_j, v_k) \notin E(G)$, so for subgraph $\{v_j\} \cup S$, $v_j \in N(v_i) - S$, $C(v_j) \neq 1$. Consequently, S is a maximal clique.

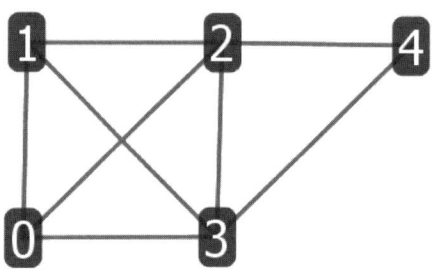

Fig. 12.1. 4-clique and 3-clique

To make things more concrete, an illustrated example is given as follows on the network shown in Fig.12.1, and the whole procedure is given in algorithm 1. In Fig.12.1, $C(v_0) = C(v_1) = \frac{2 \times 3}{3 \times (3-1)} = 1$, $C(v_2) = C(v_3) = \frac{2 \times 4}{4 \times (4-1)} = \frac{2}{3}$, and $C(v_4) = \frac{2 \times 1}{2 \times (2-1)} = 1$. We use set $CLIQUE$ to store the candidate maximal clique. Starting from v_0, we choose v_1 from $M(v_0) = \{v_1, v_2, v_3\}$. Based on v_0 and v_1, $T(v_0) = M(v_0) \cap M(v_1) = \{v_2, v_3\}$. Because $T(v_0) \neq \emptyset$, we thus choose v_2 from $T(v_0)$ and $CLIQUE = \{v_0, v_1\}$. From v_1 and v_2, $T(v_1) = T(v_0) \cap M(v_2) = \{v_3\}$. $T(v_1) \neq \emptyset$, v_3 is chosen, and $CLIQUE = CLIQUE \cup \{v_2\} = \{v_0, v_1, v_2\}$. From v_2 and v_3, $T(v_2) = T(v_1) \cap M(v_3) \neq \emptyset$, and $CLIQUE = CLIQUE \cup \{v_3\} = \{v_0, v_1, v_2, v_3\}$. At this moment, $\{v_0, v_1, v_2, v_3\}$ is a $4 - clique$. Since that $\{v_0, v_1, v_2, v_3\} \nsubseteq N(v_4)$, according to theorem 1, $\{v_0, v_1, v_2, v_3\}$ is a maximal clique. Similarly, starting from v_1, we can obtain the candidate clique $\{v_1, v_2, v_3\}$. However, since that $\{v_1, v_2, v_3\} \subset N(v_0)$, $\{v_1, v_2, v_3\}$ is not a maximal clique.

12.4.3 Pruning by Prediction

To address the concern on efficiency, at each step of the depth-first search, *Peamc* employs a pruning technique by predicting every possible maximal clique in advance. Fig.12.2 shows each of the search trees rooted with $\{v_0, v_1, v_2\}$ in Fig. 12.1.

After we have obtained the maximal 4-clique $\{v_0, v_1, v_2, v_3\}$, we have to backtrack to the node $\{v_0, v_1\}$. However, when we reach the leaf node $\{v_0, v_1, v_3\}$,

---

Algorithm 1. *FindClique($V(G), E(G)$)*

---

1. Read graph G
2. Generate set $M(v_i)$ for $\forall v_i \in V(G)$
3. **for** $\forall v_i \in V(G)$ **do**
4.   call `clique`($\{v_i\}, M(v_i), v_i$)
5. **end for**
6. **Function:** `clique`($CLIQUE, T, v_x$)
7. **for** $v_j \in T$ by the ascending order **do**
8.   build ***triangle neighbor*** set $T(v_x)$ from v_x and v_j
9.   **if** $T(v_x) \neq \emptyset$ **then**
10. $CLIQUE \leftarrow CLIQUE \cup v_x$
11.   choose v_k with the lowest index from $T(v_x)$
12.   call `clique`($CLIQUE, T(v_x), v_k$)
13.   **else**
14.   according to theorem 1
15.   **end if**
16. **end for**

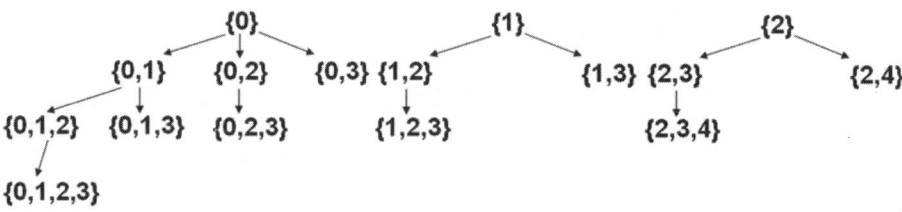

Fig. 12.2. Example of Search Trees

we finally find that it is not a maximal clique for that it is just contained in $\{v_0, v_1, v_2, v_3\}$. Similarly, for the branch $\{v_0\} \rightarrow \{v_0, v_2, v_3\}$, and even for the whole tree rooted with $\{v_1\}$, all the candidate cliques $\{v_0, v_2, v_3\}$ and $\{v_1, v_2, v_3\}$ are contained in $\{v_0, v_1, v_2, v_3\}$. As a result, these traversing processes contribute nothing at all. If these candidate maximal cliques could be predicted in advance, the corresponding tedious searching processes will be prevented from the beginning. Therefore, we come up with a pruning strategy as follows. We first give each obtained maximal clique a clique number. For example, there are 2 maximal cliques in Fig.12.1. $\{v_0, v_1, v_2, v_3\}$ is given 0, and $\{v_2, v_3, v_4\}$ is given 1. Then this clique number is attached to every vertex in the same maximal clique already obtained. Consequently, 0 is assigned to v_0, v_1, v_2 and v_3. 1 is assigned to v_2, v_3, v_4 respectively. As a result, both v_0 and v_1 have the clique number $\{0\}$; v_2 and v_3 have the clique number $\{0, 1\}$; v_4 has the clique number $\{1\}$. For the next time when we have obtained the *triangle neighbor* set $T(v_i)$ with given vertex v_j we can compare the clique numbers of all the vertices in $T(v_i)$ with those of v_i and v_j in advance. If they all share the same clique number, we can stop the further searching process, because all these vertices have ever been in some maximal clique already obtained before. For instance, when we come to the node $\{v_0, v_2\}$

with $T(v_0) = \{v_3\}$, since that v_0, v_2, and v_3 share the same clique number 0, the branch from $\{v_0\} \rightarrow \{v_0, v_2, v_3\}$ can now be pruned. In our experiments, this pruning strategy often improves the efficiency by a factor of 25% on average.

12.4.4   Parallel Model of *Peamc*

From the above discussion of the basic algorithm, we can find that the enumeration of all maximal cliques requires that the search tree rooted with every vertex of G must be traversed. Because every vertex is accessed by the ascending order of its index, there will be no duplicate triangles and candidate cliques being generated. In Fig.12.2, we see that from vertex v_0 we have candidate clique $\{v_0, v_1, v_2, v_3\}$. When we start from vertex v_1, only $\{v_1, v_2, v_3\}$ is built since vertex v_0 has lower index than vertex v_1. Thus $\{v_0, v_1, v_2\}$ is a duplicate triangle and will not be considered. This fact indicates that if $v_i \neq v_j$, the traversing of the search tree rooted with v_i, is independent of that rooted with v_j shown in Fig.12.2. Therefore, we have the following theorem.

Theorem 2. *The enumeration of all maximal cliques in graph G with n vertices can be partitioned into n independent enumerations of the maximal cliques from the search tree rooted with the each vertex.*

Practically, the number of processing elements on the existing parallel platform P is often smaller than n, so some mapping techniques are required. A naive mapping scheme is to assign n tasks to P processing elements sequentially. Let $I_p \in \{0, 1, .., P - 1\}$ denote the index of the processing element. Each partition of the n tasks assigned to each processing element is presented by $[I_p \times (\lfloor \frac{n}{P} \rfloor + 1]), (I_p + 1) \times (\lfloor \frac{n}{P} \rfloor + 1)]$. However, this scheme sometimes suffers from load unbalancing, simply because for a search tree rooted with vertex v_i, $|M(v_i)|$ may be significantly large. When $n \gg P$, it is highly possible that many tasks with vertices having large $|M(v_i)|$ may be assigned to the same partition, which will lead to a high load on a single processing element. To address such problem, we first sort the vertices of G by the descending order of $|M(v_i)|$ and then define the partition β on a single processing element I_p as follows

$$\beta = \{v_i | i = 0, ..., \lfloor \frac{n}{P} \rfloor + 1, v_i = u_j, u_j = \begin{cases} i \times P + I_p & \frac{i}{2} = 0 \\ (i+1) \times P - I_p - 1 & \frac{i}{2} \neq 0 \end{cases} \} \quad (12.1)$$

Let $\alpha = \frac{\sum_{i=0}^{\lfloor \frac{n}{P} \rfloor + 1} |M(v_i)|}{\lfloor \frac{n}{P} \rfloor + 1}$. This mapping scheme enables α of most partitions approximate with each other, which brings a better load balancingwhich will improve the efficiency by a factor of 30% on average.

12.4.5   Analysis of Peamc

Based on the experimental results of our research, we have found that the distribution of the maximal cliques whose size varies from 3 to M_C also holds a power-law feature, where M_C represents the size of the maximum clique. Let x denote the size of a maximal clique and y denote the number of the maximal cliques with size x, such that $y = x^{-\alpha}$. For the traversing of the search tree rooted with vertex v_i, the calculation of the *triangle neighbor* set $T(v_i)$ costs $a_0 \times (k \times 3^{-\alpha}) \times \Delta$, where $0 \le a_0 < 1$, Δ denotes the maximum degree of G. If $T(v_i) \ne \emptyset$, the next round calculation of $T(v_j)$ will cost $a_1 \times (k \times 4^{-\alpha}) \times \Delta$. If $T(v_j) \ne \emptyset$, the calculation of $T(v_k)$ will cost $a_2 \times (k \times 5^{-\alpha}) \times \Delta$. This process continues recursively until we reach the leaves of the search tree. Since that M_C denotes the length of the deepest path from the root to the leaf, the time to enumerate all the maximal cliques by traversing a single search tree is represented as :

$$T_0 = \Delta \times \{(a_0^{(T_0)} \times k \times 3^{-\alpha})^2 + ... + (a_{M_C-3}^{(T_0)} \times k \times M_C^{-\alpha})^2\}, a_i^{(T_0)} \in [0,1) \quad (12.2)$$

Because *Peamc* requires the traversing of the search tree rooted with every vertex of G, the total runtime is thus represented as

$$T = \sum_{i=0}^{n-1} T_i = \Delta \times \{(k \times 3^{-\alpha})^2[(a_0^{(T_0)})^2 + ... + (a_0^{(T_{n-1})})^2] + ...$$

$$+ (k \times M_C^{-\alpha})^2[(a_{M_C-3}^{(T_0)})^2 + ... + (a_{M_C-3}^{(T_{n-1})})^2]\}, a_i^{(T_j)} \in [0,1), \sum_{j=0}^{n-1} a_i^{(T_j)} = 1 \quad (12.3)$$

Since that $[(a_0^{(T_0)})^2 + ... + (a_0^{(T_{n-1})})^2] < ((a_0^{(T_0)}) + .. + (a_0^{(T_{n-1})}))^2 = 1$, we have $T < \Delta \times [(k \times 3^{-\alpha})^2 + ... + (k \times M_C^{-\alpha})^2] < \Delta \times M_C \times (3^{-\alpha})^2$. Moreover, because all vertices of G are traversed by the ascending order of their index, which eliminates many duplicate results, and the pruning strategy with prediction also reduces the search space greatly when the maximal cliques grow large, the actual runtime is much less than T. Thus, to enumerate all the maximal cliques in G, it will cost $O(\Delta \times M_C \times Tri^2)$ in the worst case, where $Tri - k \times 3^{\alpha}$, and require $O(\Delta \times n)$ time as a preprocessing to calculate set $N(v_i)$ of every vertex. To load the whole graph G with n vertices and m edges into memory will consume $O(m+n)$ space. If we have n processing elements, according to theorem 2, every processing element will traverse one search tree rooted with each vertex. Therefore, the parallel runtime is the time that elapses from the moment a parallel computation starts to the moment the last processing element ends, which will cost $O(\frac{\Delta \times M_C \times Tri^2}{n})$ theoretically. In practice, the number of the processing elements P is far less than n, so the practical time delay is $O(\frac{\Delta \times M_C \times Tri^2}{P})$ on average.

12.5 Experimental Evaluation

To evaluate the performance of Peamc, we have also implemented *Bron's BK* algorithm and *Kazuhisa's* algorithm. Our experiments are done on a *DAWN Cluster* (84 3.2GHz processors with 2Gbytes of main memory on each node, *Linux AS3*). The implementations of *Kazuhisa's* algorithm and Improved *BK* are strictly based on *Kazuhisa's* and *Bron's* papers. The code is also optimized by our best efforts with $C++$ *STL*. These algorithms have been first examined with graphs generated randomly. Then they are challenged with the complex networks and *Peamc* is also evaluated on large sparse call graphs built by real data taken from one telecom carrier for months in a city. The runtime in tables is expressed in seconds and we give the following notations for short: *M.D.* stands for *Max Degree*, *M.C.* for the number of *Maximal Cliques*, M_C is the size of the maximum clique, Ka for *Kazuhisa*'s Algorithm, BK for Improved BK, P_n and P'_n represent *Peamc* on n processing elements with and without pruning respectively and s stands for speedup in the end.

12.5.1   Random Networks

Our random graphs are generated as follows. For given r and n, we build a graph with n vertices such that v_i and v_j are adjacent with probability 0.5 if $i + n - j(mod\ n) \leq r$ or $j + n - i(mod\ n) \leq r$. We examine the case of $r = 10$ and $r = 30$ with $n = 1000, 2000, 4000, 8000, 16000, 32000$ respectively.

By comparing the results of $r = 10$ and $r = 30$ in Table 12.1 and 12.2, we see that Improved BK's performance keeps stable regardless of $|E(G)|$. However, in the case of $r = 30$, when $|V(G)|$ grows large, the clustering coefficient of every vertex is also increasing, thus *Peamc* outperforms Improved BK gradually. Since our implementation of *Kazuhisa*'s algorithm is optimized for the sparse graph according to the paper with $O(\Delta^4)$ time delay, *Kazuhisa*'s algorithm performs better in Table 12.3 than in Table 12.1 and 12.2.

Table 12.1. Results on random networks with $r = 10$

$|V(G)|$	$|E(G)|$	M.D.	M.C.	M_C	Ka	BK	P_1	P_{30}	s
1000	4534	16	2239	6	7	0.5	0.2	0.01	19.8
2000	8938	16	4396	6	27	4	0.4	0.02	19.7
4000	17987	24	8894	6	116	29	0.9	0.04	20
8000	36223	28	18112	6	519	243	1.8	0.09	18.5
16000	71904	35	35897	6	n/a	1934	3.6	0.19	18.8
32000	144350	44	68794	6	n/a	15389	7.3	0.38	19

Table 12.2. Results on random networks with $r = 30$

$|V(G)|$	$|E(G)|$	M.D.	M.C.	M_C	Ka	BK	P_1	P_{30}	s
1000	14432	64	19269	8	275	0.7	29	1.69	17.6
2000	28709	82	38060	8	1006	4	56	3.27	17.4
4000	58063	92	78054	8	12464	31	124	6.49	19
8000	116276	112	157278	8	50371	244	248	13.4	18.5
16000	231622	128	311572	9	n/a	1937	486	26	18.8
32000	464069	154	626207	9	n/a	15331	984	47.8	20.6

12.5.2   Complex Networks

For the complex network, the distribution with the number of the maximal cliques is confined by the power-law property, but the growth of the maximum degree of G does not conform to this limitation. Based on the fact that most complex networks in the real world are large sparse graphs, our researches and experiments show that the growth of the number of the triangles is slower than the growth of the maximum degree. Moreover, the size of the maximum clique is also less than the maximum degree. As a result, *Peamc* with $O(\Delta \times M_C \times Tri^2)$ outperforms *Kazuhisa*'s algorithm shown in Table 12.3.

Peamc is also evaluated on large sparse telecommunication call networks built upon the real datasets in a city for 10 months for a Telecom Operator in China. We regard each subscriber as a single vertex and two vertices will share an edge if the subscribers have once contacted with each other by their mobile phones. Results on the 10 large graphs are shown in Table 12.4.

Moreover, *Peamc* is further challenged with another cellphone call network of a month in a province with up to 2,423,807 vertices and 5,317,183 edges. We find 801,381 maximal cliques using 70 processors in 21,219 seconds and the size of the maximum clique is 26. Figure 12.3 gives the statistical results on the distribution

Table 12.3. Results on complex networks

$|V(G)|$	$|E(G)|$	M.D.	M.C.	M_C	Ka	BK	P_1
579	500	43	4	3	0.027	0.112	0.001
1161	1000	44	11	4	0.052	0.740	0.001
2301	2000	45	26	5	0.105	6.032	0.004
4557	4000	57	61	5	0.208	42.220	0.008
8760	8000	81	183	6	0.403	313.514	0.022
17376	16000	128	403	7	1.000	2449.000	0.067

Table 12.4. Results on the 10 cellphone call networks

G	$|V(G)|$	$|E(G)|$	M.D.	M.C.	M_C	P_1'	P_1	P_{30}
1	512,024	1,021,861	673	153,362	14	64	58	4
2	503,275	900,329	731	118,353	13	28	23	1
3	540,342	1,030,489	980	143,259	14	44	43	6
4	539,299	1,014,800	2,396	139,040	15	53	55	18
5	543,856	1,034,291	1,355	145,569	17	185	76	18
6	531,444	1,020,716	913	148,570	16	455	380	169
7	529,280	1,012,299	1,291	145,809	18	174	128	43
8	531,861	1,007,487	1,134	144,434	19	246	223	64
9	562,244	1,060,121	1,922	151,039	17	187	130	24
10	594,186	1,152,470	1,310	168,285	21	498	403	98

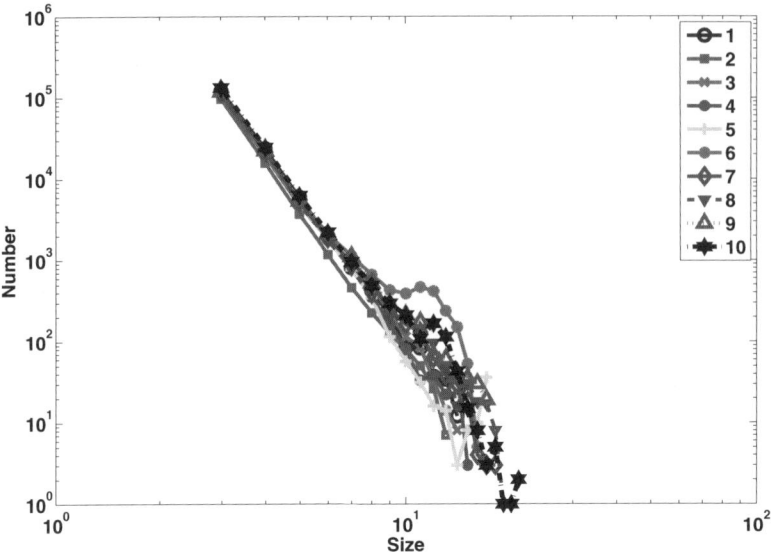

Fig. 12.3. Distribution of the maximal cliques

Table 12.5. Comparison between different schedules

G	Naive Schedule on P_{30}	Random schedule on P_{30}	Improvement
1	16	4	75%
2	4	1	76%
3	10	6	40%
4	21	18	14.28%
5	59	18	69.49%
6	171	169	1.16%
7	65	43	33.85%
8	132	64	51.52%
9	51	24	52.94%
10	198	98	50.50%

of all the maximal cliques obtained from the 10 graphs in Table 12.4 whose size varies from 3 to 21 respectively.

In Fig.12.4, X axis is the size of the maximal clique. Y axis is the corresponding number. We see that the call graph also has a power-law distribution of its maximal cliques. We have compared our schedule mechanism with the naive sequential schedule in Table 12.5. These experimental results are based on the cellphone call networks in Table 12.4 as well. Here we see that our schedule improves the whole efficiency by a factor of 46.37% on average.

12.6   Application

Many complex networks capture intricate web of connections among the units or sub-structures they are made of. An interest topic is to interpret the global organization of the whole network to identify the basic building blocks, which is crucial to the understanding of the network's structural and functional properties. In our researches, we see the maximum clique across 10 months is that of size 21 in 10th month, which indicates that the 21 customers have close relationship with each other. Since these customers could also appear in the previous 9 months to form smaller maximal cliques, by joining this maximum 21-clique with those maximal cliques enumerated in the ten months respectively, we can find the evolution of the maximum 21-clique in the 10 months, shown in Fig.12.4.

In Fig.12.4, we first focus on the customer denoted by ▲. The gray filled circles represent the customers of the maximum 21-clique and those dubbed by the small white blank circles denote their neighbors which also form maximal cliques in the 10 months respectively. We see that this ▲ customer is first introduced into the network in Jan and forms a triangle with other two persons in the quasi 19-clique

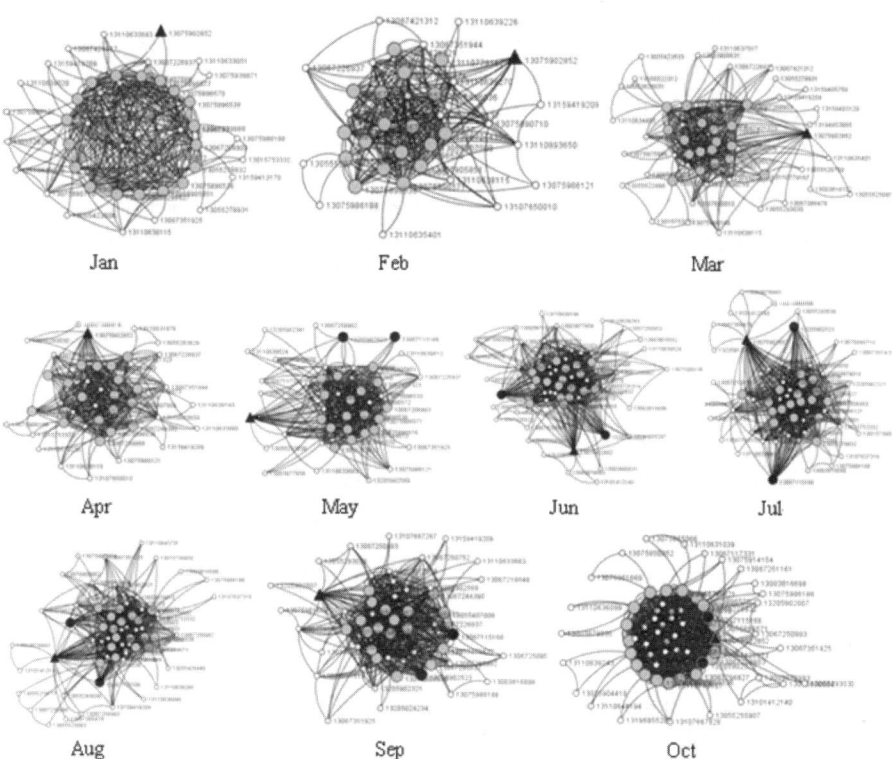

Fig. 12.4. Maximum 21-clique evolution

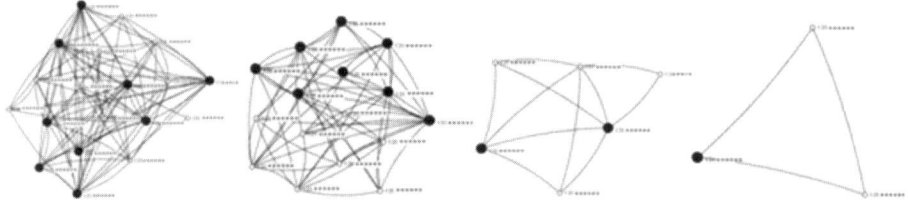

Fig. 12.5. Maximal 9-clique shrinking

(most pairs of its nodes are connected directly) which is part of the maximum 21-clique. This customer has more connections in the following months indicating he or she gradually joins in the social circle of the 21 customers and becomes a new VIP customer to the telecom career. Fig.12.4 also shows that only 19 persons of the maximum 21-clique have appeared in Jan and only after May, another two persons denoted by the black filled circles join in the network. The degree of these two persons is increased sharply in Jun and keeps stable in the later months, which means these two persons probably have known most of the 19 customers before and are recommended to use the same network with them, which is good to the telecom career's business. Moreover, we find that more and more other customers appearing as small white nodes inside the maximum 21-clique depend on this structure heavily and together they form a bigger quasi-clique.

By contrast, Fig.12.5 gives the shrinking process of a maximal 9-clique from Mar to Jun. The shrinking of the maximal clique among the call graphs has a close relationship with the specific time and the constitution of the structure. After one person in the 9-clique quit in Apr, the whole structure has shrank sharply in the next month. Consequently, we can infer that this person may hold an important position and have a heavy impact on the others in the community of the 9 persons, so if we can retain such kind of people in advance, more lost profits will be prevented, which is another piece of good news to the telecom operator.

12.7   Future Trends

In recent years, easy connections brought about by cheap devices, modular content, and shared computing resources are having a profound impact on our social structures. People now increasingly take their required information from one another rather than from institutional sources like corporations, media outlets, religions, and political bodies. As a result, people are implicitly involved in many social networks which are formed by our friend lists in the instant messaging softwares, by the bloggers who comment on a certain topic in your blogspace, or by the users who write collaboratively in a wiki site. In social network analysis[4], maximal clique is one of the typical cohesive sub-structures[11] in a given network, which represents a group of closely related friends who seem to dominate everything social. However, the clique definition puts too strict restriction on the structure of the sub-graph. Moreover, the detected maximal cliques are usually small in size, and people can get tremendous number of them. Therefore, people may need some approaches to identify the more general community structures where not every pair of vertices is required to connect with each other; yet, the vertices within the same sub-structure have higher density of edges while vertices between these structures have lower density of edges. Compared with clique, although this kind of community is not yet a complete graph, it is indeed dense relatively. Since that maximal clique is the densest sub-graph in a given network, perhaps we can use the overlapped maximal cliques as the clustering kernel and may use a k-means like method to carry out an agglomerative process. Finally, each of the obtained clusters may correspond to a community.

12.8 Conclusion

In this paper, a novel parallel algorithm *Peamc* is provided, which exploits several effective techniques to efficiently enumerating all maximal cliques in complex network. Since most networks in our real world conform to the complex network model, our algorithm enjoys more attractive advantages in practice. A comprehensive performance study to compare Peamc with the existing algorithms on the real data sets has illustrated that Peamc is more efficient and scalable. Moreover, a promising application is shown to present and model the sub-structures' evolution and shrinking among continuous sparse cellphone call networks in the end.

References

1. Albert, R., Barabási, A.L.: Statistical mechanics of complex networks. Rev. Mod. Phys. 74(1), 47–97 (2002)
2. Milo, R., Itzkovitz, S.: Network motifs: Simple building blocks of complex networks. Science 298, 824–827 (2002)
3. Johnson, D.: On generating all maximal independent sets. Information Processing Letter 27(3)
4. Scott, J.: Social Network Analysis: A Handbook. Sage Publications, London (2002)
5. Faisal, N., Abu-Khzam.: On the relative efficiency of maximal clique enumeration algorithms, with application to high-throughput computational biology. In: International Conference on Research Trends in Science and Technology, vol. 557
6. Tomita, E.: The worst-case time complexity for generating all maximal cliques. Theoretical Computer Science 363, 28–42 (2006)
7. Bron, C., Kerbosch, J.: Finding all cliques of an undirected graph. Communications of the ACM 16, 575–577 (1973)
8. Tsukiyama, S.: A new algorithm for generating all the maximal independent sets. SIAM Journal on Computing 6, 487–504 (1977)
9. Makino, K., Uno, T.: New algorithms for enumerating all maximal cliques. In: 9th Scandinavian Workshop on Algorithm Theory, pp. 260–272
10. Kose, F., Weckwerth, W., Linke, T., Fiehn, O.: Visualizing plant metabolomic correlation networks using clique-metabolite matrices. Bioinformatics 17, 1198–1208 (2001)
11. Zeng, Z., Wang, J., Karypis, G.: Coherent closed quasi-clique discovery from large dense graph databases. In: The 12th ACM SIGKDD, pp. 797–802 (2006)

13

Community Finding of Scale-Free Network: Algorithm and Evaluation Criterion

Sen Qin[1] and Guanzhong Dai[2]

[1] School of Science, Hangzhou Dianzi University, No.1, Street No.2, Xiasha Higher Education Zone, Hangzhou, Zhejiang, P.R. China, 310018
qinsen0425@gmail.com
[2] College of Automation, Northwestern Polytechnical University, No.127, Youyi West Road, Xi'an, Shaanxi, P.R. China, 710072
daigz@nwpu.edu.cn

Abstract. Recently, topology structures of many social, biological and technological networks have been discovered to display a scale-free property. For a network, a community is a natural division of network nodes into groups in which there are more links between nodes within the groups than to nodes outside of it. Many methods of community finding have been proposed to seek a fast, feasible and reasonable partition algorithm for the whole network nodes. In this chapter, we introduce the topology of the network to evaluate the feasibility and correctness of a community finding algorithm. A relationship between the rough number of communities and the magnitude of the number of hub nodes in the network is given in detail firstly. Then, an algorithm based on Laplace matrix spectral decomposition is proposed and its key technology, threshold selection of Euclidean distance between nodes, is discussed. Based on the scale-free topology of complex network, the evaluation criterion of community finding algorithm including three conditions is obtained. Numerical results show that the algorithm of community finding is an effective one and the evaluation criterion is feasible, fast and easy to operate.

Keywords: Complex network; Community finding; Scale-free network; Algorithm; Evaluation criterion.

13.1 Introduction

In the last several decades, although our capabilities of collecting and processing data have been increasing rapidly, we are deluged and baffled by various and massive data, such as scientific data, financial data, etc. These complex data are a great challenge to traditional methods of data mining [1]. However, we should not be pessimistic since there are many new approaches to find the useful information from these complex data. A new approach is to use complex network theories to seek the common characteristic of these data [2]. For example, in the WWW database, each webpage can be considered a node of network, and the hyperlinks between webpages are the links of the network. Then we can obtain the set of those very important nodes according to analysis of the network's

D.A. Zighed et al. (Eds.): Mining Complex Data, SCI 165, pp. 223–242.
springerlink.com   © Springer-Verlag Berlin Heidelberg 2009

topology. Network topology analysis can be considered to be a data mining method in a widespread sense, which is one of the methods of mining complex data sometimes.

Complex networks have drawn increasing interests in the community of mathematics, computer science and physics. Many networks in the real world, including the WWW and the Internet, metabolic and protein networks, social networks, etc., have been revealed that these topology structures exhibit a scale-free property [3, 4, 5]. That is to say, the probability that a node of these networks has k links obeys a power-law distribution $P(k) \sim k^\gamma$, with the exponent γ that ranges between 2 and 3. The scale-free property is proposed formally by Barabási and Albert [3], so a network model having a power-law distribution is also titled by a BA model. According to the evolving process of the BA model and the studies of Barabási, Albert and Jeong [6], there are two main mechanisms to produce a scale-free power-law distribution: growth and preferential attachment. Growth means that the numbers of nodes and links are increasing with the evolving time t, which accords with the evolving tendency of most networks. In the preferential attachment mechanism, the probability of a new node will be linked to node i depends on the degree k_i, such that $\Pi(k_i) = k_i / \sum_j k_j$. Therefore, the latter mechanism indicates that there is a higher probability to be linked to a node that already has a large number of links, which can effectively explain the phenomenon of "rich get richer" in many real world networks [7]. In fact, preferential attachment is a deep depiction of "Survival of the fittest" and "Matthew effect" [8]. In general, these two mechanisms can help to explain the existence of hub nodes in a network, which hub nodes have very large degrees and play a key role in influencing the topology of the whole network. An exceptional characteristic of scale-free network is that, it displays an amazing robustness against random attack, but when very few hub nodes are failed or removed its topology structure is very fragility.

General speaking, most networks contain some groups in which the nodes are more highly linked to each other within group than to the nodes outside of it. These groups are usually called communities, clusters or modules of the networks [9]. The links of the nodes in a community are dense, while they are very sparse between communities. An example of community structure is shown in Fig. 13.1. Let a well-defined community be an intrinsic community of the network. Since detecting the community structure of a network can provide a lot of help in understanding and visualizing the structure of networks, it is a very important issue that how to achieve a suitable community structure by means of an excellent algorithm of community finding. There are many successful algorithms to detect community structure and they can be classified to two categories roughly, traditional algorithms for some networks with small scale and some new algorithms for some complex networks. Some traditional methods of community finding, including spectra bisection, the Kernighan-Lin algorithm, hierarchical clustering, etc., can be used to discover community structure of those networks with small scale [10]. For instance, the karate club network Zachary can be divided roughly into two communities by these algorithms. However, none of these methods is ideal for the scale-free

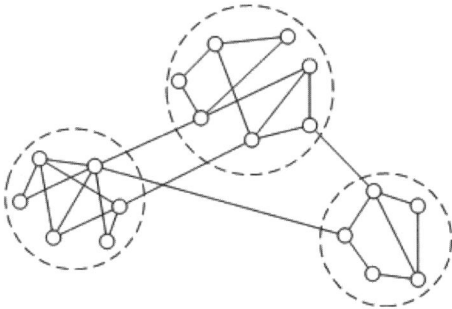

Fig. 13.1. An example of community structures [16]. There are three communities, which are partitioned by the dashed circles, and the links between communities are very sparser denoted by four blue lines.

complex networks, such as the Internet, the WWW, and biological networks. Some new methods, including Girvan and Newman algorithm [11], the algorithm based on the edge betweenness score [12], local community finding algorithm based on the greedy maximization of local modularity [13], the algorithm based on the maximum modularity [14], the algorithm based on information centrality [15], etc., are able to work well with these complex networks. For the two categories of these algorithms, the evaluation criterion is typically the algorithm's time complexity.

On the other hand, there is a very difficult issue that how to determine the optimal number of well-defined communities obtained by most algorithms of community finding. If the number of well-defined communities is not clearly known, some algorithms will not be able to run correctly. For the Kernighan-Lin algorithm, it is necessary to know the number of communities in advance [10]. If the number of nodes is very large, or the number of communities is unknown, the algorithm will be failure. The new algorithms have the same issue as the traditional community finding algorithms. For GN algorithm, it can not determine the appropriate iterative step when the number is not determined. Although in the Newman algorithm [17] the appropriate number of communities can be determine according to the local maximum of the modality indices, the number of the communities can not be obtained.

For each node, since the probabilities of connecting new edges to other nodes are different according to preferential attachment of the scale-free network, not all nodes have the same influences to distribute and transform the information of the network. The centrality of a network aims at the foundational question how to measure the importance of nodes. The importance of nodes in a complex network has attracted a lot of attention of researches since it concerns crucial subjects such as networks resilience to attacks [12]. In this chapter, it is believed that there is a close relation between the number of the important nodes and the number of well-defined communities. So some centrality indices of each node and the whole network are considered to design the evaluation criteria of community finding algorithms.

Therefore, how to evaluate the merits of the various community finding algorithms and give a reasonable evaluation criterion, is an urgent problem of community finding research. In addition, it is very important to obtain the prior knowledge of the number of community finding. These two aspects are concerned in this chapter. A new evaluate criterion of community finding is given according to the topology analysis of the scale-free network, and a method predicted the magnitude order of the number of communities for a scale-free network is obtained.

13.2 Basic Conceptions

In this section, we give some important conceptions: complex network, degree distribution, power law distribution, scale-free network, community structure, community finding algorithm, centrality index.

The *complex network* that we consider in this chapter is graph consisting of nodes connected by links, which has non-trivial topological structure [4]). Clearly, the meanings of nodes and edges in different real-world networks are different. The nodes of the WWW are the web pages and the edges are the hyperlinks that point from one web page to another. The topology of the Internet is considered at two different levels: router level and autonomous system level. And in the former level, the nodes are the routers and edges are the physical connections between them, while the nodes are domains in the latter level. In the movie actor network, the nodes are the actors, and two nodes have a common edge if the corresponding actors have acted in a movie together.

The *degree* of a node in a complex network, k, is the total number of links connected to the node. And the *degree distribution* of the network, $P(k)$, is the probability that a randomly selected node has k degree [18].

The popular distributions of real-world complex networks are various, which can be broadly divided into two categories: one is relatively even distribution; another is seriously uneven distribution. Poisson and Gauss distribution can approximately describe the former distribution, whose probability distributions have a ball-shaped curve [6]. And for seriously uneven distribution, it can be described by power-law distribution commonly:

$$P(k) = Ck^{-\gamma}, \quad \gamma > 0, \quad k = 1, 2, \ldots \tag{13.1}$$

The power-law distribution has an "L" shaped curve and on log-log scale the distribution appears as a straight line, which is shown as Fig. 13.2.

Traditionally, the degree distribution of the random graph network described by Erdös and Rányi [19] follows a binomial distribution, or a Poisson distribution in the limit of large number of nodes N, that is to say, $P(k) = e^{-\mu}\lambda^{\mu}/k!$, and the coefficient $\mu = p(N-1)$, where p is the probability that each random selected nodes are connected. However, the topology structures of many real-world networks don't exhibit the characteristics of random graph. Real world networks are nonrandom since some important evolving mechanisms are suggested to control their topology structures in the whole evolving processes. Recently the data of

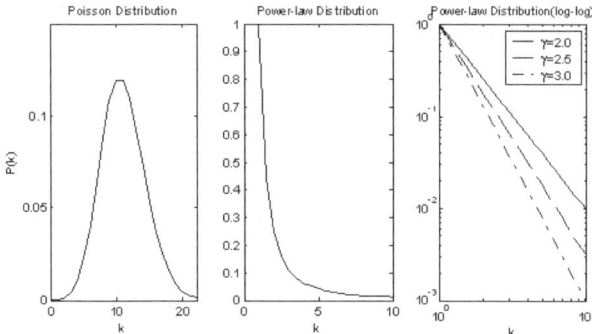

Fig. 13.2. Poisson distribution, power-law distribution and its log-log plot

complex networks obtained by the researchers of many fields show that: some nodes have a tremendous number of links to other nodes, whereas most nodes have just a handful links. And these important nodes are called hub nodes in the network, which have thousands or even millions of links. In the sense, Barabási et al. suggest the network appears to have no scale. Generally speaking, a complex network is a scale-free network if its degree distribution follows power-law distribution. From Equation (13.1), we can find that:

$$P(C'k) = C(C'k)^{-\gamma} = CC'^{-\gamma}k^{-\gamma} = C_1 k^{-\gamma}. \tag{13.2}$$

That is, the exponent of power-law distribution γ is independent on the variable k. In other words, if the multiplicative factor C' is taken as a scale of variable k, the functional form $P(k) = Ck^{-\gamma}$ remains unchanged when the scale k changes to $C'k$. Therefore, the exponent γ of power-law distribution of the network is scale-free in this sense.

Community structure of a network means that the nodes are often found to cluster into highly knit groups with a high density of within-group edges and a low density of between-group edges. Since the component of a node is the set of nodes that can be reached from it by paths running along edges of the graph, which is the connected subgraph in undirected graph, the community structure is not equivalent to the connected subgraphs or components.

Community finding algorithm is an algorithm to detect and partition the community structure of a network. All community finding algorithms fall into two rough classes depending on whether they focus on the addition or removal of edges to or from the network. The main idea of the first category of algorithms is agglomerative. That is, the network is divided into N communities and there are no edges between each pair of nodes at first, where the total number of nodes is N, the new edges are added to link two nodes according to a certain connected-edge rule of the algorithm. For example, the rule can be that the higher the similarity between pairs of nodes is, the more priority the link between the nodes is connected. The key technology of another category of algorithms is divisive. Divisive community finding algorithms try to find the lowest similarity of node pairs of the origin network, then to remove the edge of this node pairs. Repeat the process and the whole network to be partitioned to increasingly small parts gradually. The process can be terminated in any case and the whole network is partitioned some communities finally.

The researches of the centrality of complex networks, especially social networks, were beginning in the middle of the last century [20]. Three classical centrality indexes: degree, betweenness and closeness, were proposed by Freeman [21] to describe the characteristic of a node communicating or controlling the whole network. And he had proved that three centrality indices assign the star or wheel network the maximum values and the circle and the complete graph the minimum values. Degree index of a node describes its potential communication activity in the network. Betweenness index describes its potential of a node for control of communication. And closeness index describes its potential independence or efficiency of community. The measures of centrality for a network can be classified into two categories: node centrality and entire network centrality. Some new centrality indexes, such as group centrality, stress centrality, modified betweenness centrality, have been studied recently [22, 23].

Degree centrality of a node [21] is defined as

$$C_D(i) = k_i/(n-1), \tag{13.3}$$

where k_i is the degree of node v_i. Let $C_D(i^*)$ be the largest value of $C_D(i)$ for any node in the network and C_D be a general formula for determining the degree centrality index of the network. For the sake of simplicity, we adopt that $C_X(i^*)$ is the largest value of $C_X(i)$ for any node in a network, where X can be D (degree index), B (betweenness index), and C (closeness index). Then the expression of C_D can be given by

$$C_D = \frac{\sum_{i=1}^{n}[C_D(i^*) - C_D(i)]}{(n^2 - 3n + 2)/(n-1)}. \tag{13.4}$$

Let g_{ij} be the number of geodesics linking node v_i and v_j. And $g_{ij}(k)$ is number of geodesics links v_i and v_j that contained v_k. Then the probability that node v_k falls on a randomly selected geodesic linking v_i and v_j is

$$b_{ij}(k) = g_{ij}(k)/g_{ij}. \tag{13.5}$$

So the *betweenness centrality* [21] can be calculated by

$$C_B(k) = \frac{2\sum_{i<j}\sum b_{ij}(k)}{(n^2 - 3n + 2)}. \tag{13.6}$$

As far as the betweenness of the whole network be concerned, it is the average value of the betweenness index of each node, so it can be calculated by the following formula:

$$C_B = \frac{\sum_{i=1}^{n}[C_B(i^*) - C_B(i)]}{n-1}. \tag{13.7}$$

The independence or efficiency of a node is determined by its closeness to all other nodes in the network. Let $d(i, k)$ be the number of edges in the geodesic linking node v_i and v_k. Then *closeness centrality* index [21] is defined by

$$C_C(k) = \frac{n-1}{\sum\limits_{i=1}^{n} d(i, k)}. \tag{13.8}$$

Then the closeness index of the whole network is the average value of the corresponding index of each node, i.e., the expression of this formula is

$$C_C = \frac{\sum\limits_{i=1}^{n} [C_C(i^*) - C_C(i)]}{n-1}. \tag{13.9}$$

Note that the variation ranges of three indexes are from 0 to 1, we can select an appropriate percent as the single threshold of three indexes. If three indexes of one node are larger than the threshold, the node will be joined into the set of the central nodes. In fact, these centrality indices imply some competing methods of how centrality might affect group processes. Since the communities of a complex network are some separate groups in some sense, the centrality indices can be used to design the algorithms of community finding of complex network, such as GN algorithm.

13.3 Community Finding Algorithm Based on Laplace Matrix Spectral Decomposition

We design the community finding algorithm based on Laplace matrix spectral decomposition in the section. And the key threshold of Euclidean distance is discussed to obtain an optimal value of the network. The main merit of this algorithm is that the community structures of the network can be decomposed thought one-time since other many algorithms take progressive iteration. In addition, the appropriate number of community structures can be determined according to the characteristic of its Laplace matrix.

In many complex networks, there exist many isolated nodes or very small cliques. It is no sense to analyze these nodes or cliques for community finding or centrality. Hence only the largest connected component L of the network is considered and other connected components are ignored for community finding. For a complex network, let M be the Laplace matrix of the largest connected component, which is a $n \times n$ square matrix. The ith diagonal element of the Laplace matrix M is the degree of node v_i and the non-diagonal elements are defined as follow:

$$m_{ij} = \begin{cases} -1, & \text{if there is an edge between node } v_i \text{ and } v_j; \\ 0, & \text{otherwise.} \end{cases} \tag{13.10}$$

Hence, M is the symmetric matrix and the total of the elements of each line or column is always zero.

The set of the eigenvalues of a matrix is defined to its spectrum. It is easily proved that the spectrum of the Laplace matrix, M, has some properties as follows:

(i) All spectra of M are non-negative.
(ii) 0 is an element of spectrum set of M.
(iii) The algebraic multiplicity q of 0 is the number of well-defined communities of the network.

In fact, if the network is full-connected, the algebraic multiplicity of eigenvalue 0 is 1. For example, the spectra of the Laplace matrix of the network determined by Figure 13.1 are listed as Table 13.1.

Table 13.1. The spectral decomposition of the Laplace Matrix determined by Figure 13.1. It is obvious in a complex network that the number of nodes equals to the number of spectra of the corresponding Laplace matrix. From this table, we can find that the smallest eigenvalue is zero, and the fourth smallest eigenvalue (1.1286) obviously has a great gap with the previous three eigenvalues. It implies that there are three well-defined communities in the network, which accords with the result shown in Figure 13.1.

Order	Eigenvalues					
1-6	0.0	0.3553	0.4762	1.1286	1.8912	2.2505
7-12	2.7901	3.4170	3.6729	4.0925	4.5766	4.8204
13-18	5.0000	5.0648	5.2332	5.8186	6.6236	6.7886

As mentioned above, the number of eigenvalue 0 is 1 indeed when the network is full-connected. However, in a network composed by some large components adding to a few edges between these components, the Laplace matrix has also a certain number of eigenvalues very closing to zero, and the remaining eigenvalues have a gap away from zero remarkably. So we can believe that these near-zero eigenvalue is the dimension of zero eigenvalue.

Suppose that the number of communities is q if there are q eigenvalues very closing or equaling to zero. Let $\lambda_i (i = 1, 2, \ldots, n)$ be the ith smallest eigenvalue of the Laplace matrix and ξ_i be the corresponding eigenvector. Hence, we have

$$\lambda_1(= 0) \leq \lambda_1 \leq \cdots \leq \lambda_{q-1}\lambda_q \leq \lambda_{q+1} \leq \cdots \leq \lambda_n. \tag{13.11}$$

Since q approximately equals to the algebraic multiplicity of zero, it is very important that selecting an appropriate value of q according to the distributed characteristic of eigenvalues of the Laplace matrix. We select the largest value of q satisfying the following conditions as the appropriate value q^*:

$$\beta(\lambda_q - \lambda_{q-1}) < \lambda_{q+1} - \lambda_q, \quad \beta = 3 \quad \text{and} \quad \lambda_q < 1. \tag{13.12}$$

The former term of Equation (13.12) is used to measure the gap between the whole eigenvalues, and the latter is used to ensure that the selected eigenvalue closes to zero. And the value β measures the gap between all eigenvalues. A large value of β implies a more precise community structure of the network. If there is no one to satisfy Equation (13.12) for all eigenvalues, we will give a smaller value of β ($\beta = 2$ or 1) to judge all the eigenvalues to satisfy the conditions once again. From Table 13.1 we can find $\beta = 5$.

Since there are only q^* communities in a network, the Euclidean distance between each pair of nodes can be described by the distance of the former q^* components of their eigenvectors. That is, the Euclidean distances between node v_i and node v_j are calculated by

$$D_{ij} = \|\xi_i - \xi_j\| \triangleq \sqrt{\sum_{k=1}^{q^*} (\xi_{ik} - \xi_{jk})^2}. \qquad (13.13)$$

The algorithm of community finding based on Laplace matrix spectral decomposition are as follows:

(i) Construct the Laplace matrix M of the network, then calculate the eigenvalues and eigenvectors of M and sort in ascending order.

(ii) Determine the appropriate value of q^* according to the conditions of Equation (13.12), which is the number of network well-defined communities.

(iii) Select the former q^* components of each eigenvector and calculate the Euclidean distance of pairs of nodes according to Equation (13.13).

(iv) Give the threshold D^* of Euclidean distance and reconnect an edge of the original empty network if the distance of two nodes is less than the threshold. Redo the processes until there are not the distances more than the threshold.

(v) The original network is partitioned to q^* communities according to its subgraph of the reconstructed network.

In this algorithm, there is no suitable selection method to determine the threshold D^* of Euclidean distance currently. Numerical simulation results in the following shows that, if the given threshold makes that the percentages of the numbers of nodes and edges of the reconstructed network to the whole network exceed 80% and 95%, respectively, it is very appropriate to community finding.

13.4 Evaluating Criterion of Community Finding Algorithm

Since many real-world networks have scale-free characteristic, this unique topology can be used to detect their community structures. Meanwhile, it also provides a new thinking to find the evaluation criteria of community finding algorithm. In a scale-free network, its degree distribution displays power-law decay, so the exponent of power-law distribution is an important coefficient of its topology structure. On the other hand, the number of hub nodes of a scale-free network has a close relationship to the power-law exponent. The relationship between the range of the exponent and the number of hub nodes in a scale-free network has discussed by Wang and Dai [24]. They found that there exist lots of hub nodes whose degree distribution exponents are between 1 and 2. And when the exponents are between 2 and 3, there are a certain number of hub nodes, which is less than the number of hub nodes in the first case. Only a few hub nodes exist in the network when the exponent exceeds 3.

It is rational that each well-defined community of the network includes only several or one hub nodes, so the numbers of communities and hub nodes have the same order of magnitude. Therefore, the exponent of the network is prior information of the number of well-defined communities. According to the distribution of power-law exponent of real-world networks, four likelihoods of community number are considered as follows:

(i) Since the degree distribution exponents of real-world cannot be less than 1, it is only a theoretical possibility when the exponent is $0 < \lambda \leq 1$. In fact, the network closes to a random graph network and the community finding is no sense for this kind of network.

(ii) If the range of the power-law exponent is $1 < \lambda \leq 2$, it is proved that there are a lot of hub nodes in the network, so its community structure is very dispersed and the number of well-defined communities have the hundred order of magnitude at least.

(iii) If the exponent is between 2 and 3, which is a majority of the case for real-world networks, there are a moderate number of hub nodes, so the network has dozens of different communities.

(iv) When the exponent is $\lambda > 3$, a very small number of nodes have connected a very large number of edges according to the characteristic of power-low distribution. Hence, there are very few hub nodes in the network and the number of communities doesn't exceed ten.

After the community structure is obtained by any algorithm of community finding, we can reconstruct the network with the inverse process of the algorithm. The topology structures of the original network and the new network reconstructed by community finding algorithm should remain basically unchanged. From these likelihoods of power-law exponent we can find that, if the number of the communities detected by a certain community finding algorithm is inconsistent with the order of magnitude of the hub nodes, the algorithm is useless or non-optimal. In other words, if the degree distribution of the reconstructed network doesn't follow the power-law distribution any longer, or it does but the new exponent has a large deviation to the original exponent, the algorithm is not an effective one.

The power-law degree distribution of a network is one of the topology characteristics, specifically which considers the network connectivity features. However, network centrality indices mainly measure the activity, control or independence of communication. So network centrality indices of the original network and the new network will have no obvious deviation if the algorithm of community finding is effective.

Therefore, the criterion evaluating the correctness and efficiency of a certain community finding algorithm is that three fundamental conditions as follows must be satisfied:

(i) The number of the well-defined communities detected by the algorithm has the same order of magnitude to the hub nodes of the network according to the exponent of power-law distribution.

(ii) There is not an obvious deviation between the exponent of the original network and the corresponding exponent of the reconstructed network.

(iii) There is not conspicuous change between three network centrality indices, degree index, betweenness index and closeness index, of the original network and the reconstructed network.

13.5   Implementation

In this section, we firstly construct a new reply network of a BBS data and prove that its topology has scale-free characteristic. Then, the community finding algorithm based on Laplace matrix spectral decomposition is implemented for this network. Finally, though the calculations and comparisons of the power-law exponent of degree distribution and three network centrality indices of the network, the evaluation criterion of community finding algorithm is validated.

The data used in the chapter are downloaded from China Forum, Current Affairs Board, which listed in Table 13.2.

From the table we can find that, the number of nodes and edges of the largest connected component account for 99.9% and 98.8% of the effective network. Hence, the main information of the whole network is contained in the largest connected component, and it is enough to only analyze the corresponding characteristics, such as degree distribution, community structure and centrality indices, of the largest connected component.

By means of the community finding algorithm mentioned above, the numbers of well-defined communities of six months are shown as Figure 13.3.

In the community structure of a BBS network, there is a largest community which contains a lot of nodes. And the number of nodes in other communities is significantly less than that of the largest community, which is shown as Figure 13.3. It shows that there exists only one hot-topic talk which is concerned by many users in BBS forum each month. The hot-topic talk enables them to participate in the discussion and post some different perspective articles. Since other communities contain few number of nodes, the contents which be concerned by these communities are not hot-topic talk.

According to the data of the forum in April, with the different selection thresholds of Euclidean distance, the numbers of nodes and edges of the largest connected component of the reconstructed network are listed in Table 13.3.

It is shown from Table 13.3 that, when the Euclidean distance thresholds are increasing, the numbers of nodes and edges of the reconstructed network are increasing accordingly, and the same tendency in the largest connected component. Since the threshold increases, the exiguous distances between nodes are

Table 13.2. Statistical data of virtual user networks month by month. The URL of this webpage http://bbs.people.com.cn/bbs/. The total number of BBS reply articles is 745740, those articles posted between Feb. 1, 2003 and Jul. 31, 2003 to form the original network. Data preparation process is as follows. Firstly, the articles are downloaded with multi-thread from the website to storage to HTML files. Secondly, the important information of an article, such as posting time, poster, reply conditions, are extracted and deposited to the corresponding database. Lastly, the effective posters are summarized according to the constructed method of article reply network. These nodes are linked by their reply relationship to form the original network [25]. And in the table, the effective network is the network with no circle, no multi-edge and removing the nodes whose original posters are not in the networks.

Month	NO[a]	EO[b]	NE[c]	EE[d]	NL[e]	EF[f]
Feb.	3017	75142	1939	9983	1912	9968
Mar.	3936	116940	2401	13132	2375	13118
Apr.	4053	119146	2533	14092	2519	14085
May.	4445	146560	2750	16978	2715	16960
Jun.	5001	141378	2885	15872	2839	15848
Jul.	4497	146574	2625	16278	2572	16250

[a]The number of nodes for the original network.
[b]The number of edges for the original network.
[c]The number of nodes for the effective network.
[d]The number of edges for the effective network.
[e]The number of nodes for the largest connected component.
[f]The number of edges for the largest connected component.

Table 13.3. The reconstructed networks with different selection threshold

Threshold	NR[a]	ER[b]	NLR[c]	ELR[d]
0.0005	447	1752	155(34.7%)	1249(71.3%)
0.0006	710	5608	421(59.3%)	5356(95.5%)
0.0007	1009	18008	744(74.7%)	17836(99.0%)
0.0008	1297	50113	1039(81.2%)	49965(99.7%)
0.0009	1520	107279	1258(82.8%)	107108(99.8%)
0.0010	1725	192621	1647(95.5%)	192442(99.9%)

[a]The number of nodes for the reconstructed network.
[b]The number of edges for the reconstructed network.
[c]The number of nodes for the largest connected component of the reconstructed network.
[d]The number of edges for the largest connected component of the reconstructed network.

overlooked, which forms the more edges. So the nodes and edges of the whole network are all increasing. Of course, the nodes and edges of the largest connected component also increase.

When these thresholds arrange in sequence from small to large, the theoretical and actual values of the ratios of the number of nodes of the largest connected component with the former threshold to that of the latter threshold are shown in Figure 13.4. For example, when the selected threshold is 0.0005, the number of nodes the largest connected component is 155, while the latter threshold is

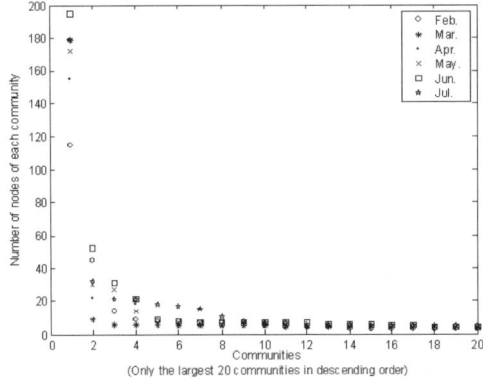

Fig. 13.3. The numbers of communities of the BBS poster networks from Feb. to Jul.

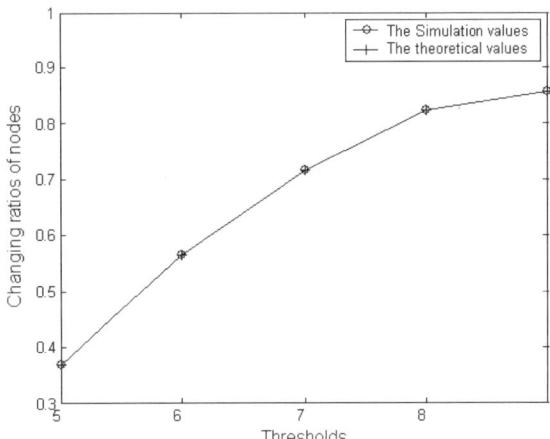

Fig. 13.4. The comparisons of the simulation values to the theoretical values of node numbers with different thresholds

0.0006, the node number is 421, the ratio of 155 to 421 is 0.368, which is denoted by the first point (drawn by "+"). This value is the theoretical value when the thresholds are changing. The first point drawn by "o" denotes the actual ratio of the number of nodes with the former threshold to the number of nodes with the latter threshold. Clearly, these two curves have any difference, which implies that the nodes with the former threshold must be contained in the set of nodes with the latter threshold when these thresholds are in ascending order.

Now we give the relationships between selection methods of the threshold and some nodes' degrees. Our main attention focuses on the change of the largest ten nodes with different threshold selections, which as shown in Figure 13.5. With the first subfigure as an example, when the basis threshold is 0.0005, the ten

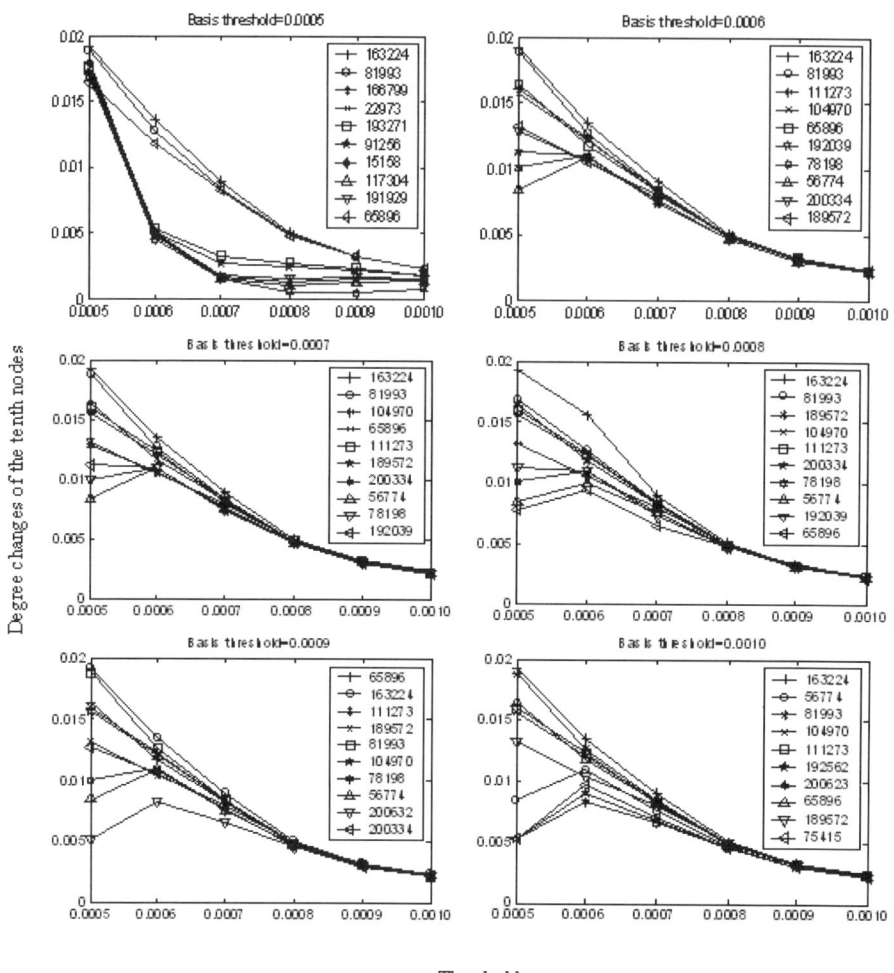

Fig. 13.5. Degree changes of the tenth nodes with different thresholds

points denotes the ratios of the degrees of the largest ten nodes to the sum of degrees of all nodes. And the label of this figure describes the identifications of posters in the BBS forum. The other points in one curve denote the ratio of the ten degrees to the sum of degrees with different thresholds.

When the basis threshold is 0.0005, the changes of the ratios of these node degrees are quite acute. For example, the degree ratio of the node that poster identification is 65896 only is the last one of ten ratios with threshold 0.0005, while the threshold increases, the ratio is always in the top three of these ratios. In addition, there is a significant difference between the ratios of ten nodes when the threshold increases, which is not same as the changes of other five

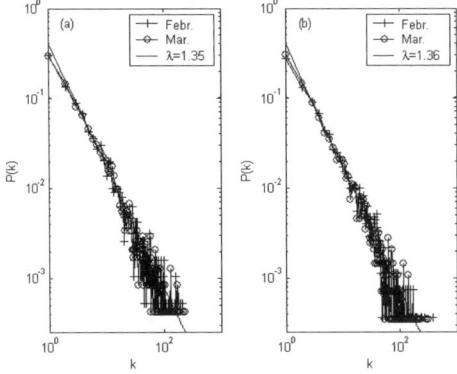

Fig. 13.6. The power-law degree distribution of virtual user network (a) and the reconstruct network (b)

Table 13.4. The Number of communities

Month	Feb.	Mar.	Apr.	May.	Jun.	Jun.
NC[a]	87	145	99	159	122	139
VN[b]	115	179	155	172	195	179

[a] The number of well-defined communities.
[b] The number of nodes of the largest connected component.

subfigures. It is shown that, the numbers of nodes and edges are very small so that the network characteristics are not comprehensive analyzed with a too small threshold selection.

In the case of the basis threshold 0.0010, since the numbers of nodes and edges of the whole network are all large, there are an obvious differences although the degrees of nodes is with great changes, which is shown as the last subfigure. For example, the degree ratio of the node that its identification is 200623 accounts for the seventh place, but with the decreasing threshold, the ratio value changes very small and it does not appear the set of the first ten nodes. This case about the node is the same as node 192562 and node 75415. So the numbers of nodes and edges have increased substantially due to the distance threshold is selected too large. Thereby, the changes of degrees of these nodes are hidden and the accurately community structure can not be obtained according to the topology of the reconstructed network.

The changes of the ratios of these node degrees are very little when the basis threshold is 0.0007 or 0.0008. From the Table 13.3, when the threshold is selected this two values, the proportions of the numbers of nodes and edges of largest connected component to that of the whole network are 80% and 99%, respectively. Since the threshold selection directly influences the topology of the reconstructed network, the appropriate distance threshold can be as an important coefficient of the Laplace matrix spectral decomposition algorithm of community finding.

Now we demonstrate the validity of the evaluation criterion of community finding according to the topology structure of the BBS poster networks.

According to Equation (13.1), if the topology of a complex network obeys the power-law degree distribution, it is approximately a slope line in the log-log plot. As shown in Figure 13.6(a), the degree distribution of the BBS poster network has a power-law distribution approximately and its exponent $\lambda = 1.35$. Since the power-law exponent has the range of $1 < \lambda < 2$, the number of well-defined communities have the hundred order of magnitude according to the evaluation criterion. The number of well-defined communities of the largest connected components each month and the number of nodes of the largest community are listed in Table 13.4. The number of communities obtained by community finding algorithm has the same magnitude order as the number indicated by the power-law exponents of the networks each month.

In addition, although there are lots of well-defined communities in the BBS networks, comparing the data of Table 13.4 with Table 13.2, the number of nodes contained in the largest connected component is very large. It shows that there exists a small quantity of hot-topic talks each month on the BBS forum.

The topology of the reconstructed network has exhibited scale-free property and its power-law exponent is $\lambda = 1.36$, which is shown in Figure 13.6(b). This exponent is quietly closed to the original exponent 1.35, which verifies the second condition of the evaluation criterion. We calculate the three network centrality indices of the original network and the reconstructed network, which are shown in Figure 13.7. We can find that the two curves have no much difference, and there is an incremental trend of these indexes from Feb. to Jul. Since the network

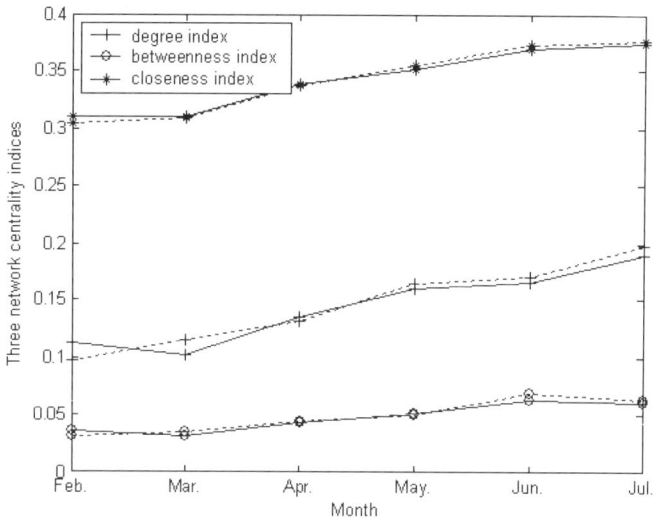

Fig. 13.7. The Three network centrality indices comparisons of the original networks (solid lines) to the reconstructed networks (dotted lines)

centrality indices have no significant changes before and after community finding of the network, the algorithm basis on Laplace matrix spectral decomposition satisfies the third condition of evaluation criterion. In addition, the trend implies that some central posters have a higher popularity than other general posters and their articles attract lots of replied articles month by month.

In a word, for the algorithm basis Laplace matrix spectral decomposition, the three conditions of evaluation criterion of community finding algorithm are satisfied well according to the data of the BBS poster networks. Hence, the evaluation criterion of community finding algorithm is validated.

13.6 Future Studies

Further study mainly focuses on two aspects: the community finding algorithm based on scale-free topology of complex networks and the evaluation criterion of community finding. Although lots of real-world complex networks have a scale-free topology structure, the inherent evolving mechanisms forming this special nature in these networks are to be further thinking and excavated. If the evolution of complex network is considered as a reverse process of community finding, the mechanisms of evolution accords with the key rules of community finding algorithm in the sense. On the other hand, the seeking of evolving mechanisms is the basis of the analysis of network topologies.

Degree distribution is one of important topology characteristics of complex network, but there are many other characteristics, such as average shortest path length, clustering coefficient, etc., that be introduced to design some new community finding algorithms of complex networks. And the centrality indices, such as betweenness, degree, have introduced to the algorithm of community finding and some valuable results have obtained by many researchers. So we can expect that, there are many more centrality indices that can be used to design community finding algorithms.

In the community finding algorithm based on Laplace matrix spectral decomposition, we focus on the method of threshold selection of Euclidean distance. And the method relies mainly on the ratios of nodes and edges of the largest connected component to those of the whole network. Since the number of communities is obtained before the Euclidean distances between nodes are calculated, and it is also an important coefficient of distance calculation formula, the relationship between the number of communities and the method of selection threshold is worth studying further.

For the evaluation criterion of community finding algorithm, lots of existed results have focused on the complexity of runtime to measure the advantages and disadvantages of an algorithm. And there are some other simple evaluation criterions. For example, if an algorithm can be validated by some real-world networks, such as Zachary network, whose community structures are known in advance or those topologies are very simple, it is an effective one. In the past literature, there is no use of the network topology to evaluate the algorithm of community finding. We have mainly adopted the relationship between the power-law exponent of scale-free complex networks and the number of communities, and the changes of network centrality indices of the original and reconstructed network, to design the evaluation criterion in the chapter. Some issues have emerged about the evaluation criterion immediately. Firstly, although the algorithm based on the Laplace matrix spectral decomposition satisfies the evaluation criterion and it is proved an effective algorithm, the comparison of different community finding algorithms according to the evaluation criterion is still interesting. Secondly, three network indices are selected as one of three conditions of the criterion to measure community finding algorithm. So it is worth discussing whether many other important network centrality indices can be competent for the task or not. Lastly, as mentioned above, there are some important topology characteristics of complex networks. Average shortest path length describes the distance from arbitrary node to the other node in average, and can be as the measurement of speed of information dissemination in the sense. If the value of average shortest path length scales logarithmically or slower with the network size for a given mean degree, the network shows the small-world effect [26]. And clustering coefficient reflects the cliquishness of the mean closest neighborhood of a network node, iconically in the friendship network, the probability that your friend's friend is also your friend directly. However, we only introduce the degree distribution of complex network to give some conditions in the criterion. Therefore, in the evaluation criterion of community finding algorithm, the two characteristics maybe serve as its relevant conditions through further analysis.

13.7   Conclusions

In this chapter, we have mainly presented some basis conceptions, such as complex network, degree distribution, community structure, community finding algorithm and network centrality index at first. Then, a new algorithm of community finding based on Laplace matrix spectral decomposition is proposed. And the key technology of the algorithm, the method of threshold selection of Euclidean distances between nodes has been discussion. The evaluation criterion of algorithms including three conditions is given according to the power-law exponent of scale-free network and three network centrality indices. Finally, the algorithm is implemented by the poster networks of a BBS forum in a period of time. And three conditions are satisfied for the community finding algorithm, which proves that it is an effective algorithm.

There are three major innovations in the chapter. One is combining scale-free topology characteristic of complex network and network centrality indices to analyze the algorithm of community finding. And the evaluation criterion of community finding algorithm is also basis on the topology analysis of complex network. It is the most important idea which is different from the previous method distinctly. The Euclidean distance threshold in the algorithm of Laplace matrix spectral decomposition is given by means of the analysis of ratio of nodes and edges of the original network to that of the reconstructed network, which is the second innovation. The third one is that, the relationship between the power-law exponent of scale-free network and the number of its communities is expounded in detail, which can be considered as the deep embody about the hub nodes of a network.

Acknowledgments

This work is supported by the Scientific Research Starting Foundation of Hangzhou Dianzi University (No. KYS091507073) and partly the National High Technology Research and Development Program of China (No. 2005AA147030).

References

1. Han, J., Kamber, M.: Data Mining: Concepts and Techniques, 2nd edn. Morgan Kaufmann Publishers, New York (2006)
2. Newman, M.E.J.: The structure and function of complex networks. SIAM Review 45(2), 167–256 (2003)
3. Barabási, A.L., Albert, R.: Emergence of scaling in random networks. Science 286(5439), 509–512 (1999)
4. Albert, R., Barabási, A.L.: Statistical mechanics of complex networks. Rev. Mod. Phys. 71(1), 47–97 (2002)
5. Dorogovtsev, S.N., Mendes, J.F.F.: Evolution of Network: From biological nets to the Internet and WWW, 1st edn. Oxford University Press, London (2003)
6. Barabási, A.L., Albert, R., Jeong, H.: Mean-field theory for scale-free random networks. Physica A 272, 173–187 (1999)
7. Wang, X.F., Chen, G.R.: Complex networks: small-world, scale-free and beyond. IEEE Circuits and Systems Magazine 3(1), 6–20 (2003)
8. Dai, G.Z., Wang, L., Qin, S.: Scale-free property of complex systems and its macro-control. Science and Technology Review 24(5), 11–16 (2006) (in Chinese)
9. Palla, G., Derenyi, I., Farkas, I., Vicsek, T.: Uncovering the overlapping community structure of complex networks in nature and society. Nature 435, 814–818 (2005)
10. Newman, M.E.J.: Detecting community structure in networks. The European Physics Journal B 38, 321–330 (2004)
11. Girvan, M., Newman, M.E.J.: Community structure in social and biological networks. Proc. Natl. Acad. Sci. 99(12), 7821–8726 (2001)
12. Barthélemy, M.: Betweenness centrality in large complex networks. The European Physical Journal B 38, 163–168 (2004)
13. Clauset, A.: Finding local community structure in networks. Phys. Rev. E 72, 026132 (2005)
14. Newman, M.E.J., Girvan, M.: Finding and evaluating community structure in networks. Phys. Rev. E 69, 026113 (2004)
15. Capocci, A., Servedio, V.D.P., Caldarelli, G., Colaiori, F.: Detecting community in large networks. Physica A 352, 669–676 (2005)
16. Qin, S., Dai, G.Z., Li, Y.L.: Mining system for community finding and centrality of virtual user network on the internet. In: Sixth IEEE Conference on Data Mining-Workshops (ICDMW 2006), pp. 284–289. IEEE Press, New York (2006)
17. Newman, M.E.J.: Fast algorithm for detecting community structure in networks. Phys. Rev. E 69, 066113 (2004)

18. Dorogovtsev, S.N., Mendes, J.F.F.: Evolution of networks. Advances in Physics 51, 1079–1187 (2002)
19. Erdös, P., Rényi, A.: On the evolution of random graphs. Publ. Math. Inst. Hung. Acad. Sci. 5, 17–61 (1960)
20. Newman, M.E.J., Park, J.: Why social networks are different from other types of networks. Phys. Rev. E 68(3), 036122 (2003)
21. Freeman, L.C.: Centrality in social networks conceptual clarification. Social Networks 1(3), 215–239 (1979)
22. Flake, G., Lawrence, S., Giles, C.: Efficient identification of Web communities. In: Sixth ACM SIGKDD International Conference on Knowledge Discovery and Data Mining, pp. 150–160. ACM, Boston (2000)
23. Lehmann, K.A., Kaufmann, M.: Decentralized algorithms for evaluating centrality in complex networks. Technical report of the Wilhelm- Schickard- Institut. WSI 2003-10, ISSN 0946-3852 (2003)
24. Wang, L., Dai, G.Z.: On degree distribution of complex network. Journal of Northwestern Polytechnical University 24(4), 405–409 (2006) (in Chinese)
25. Kuo, Z.B., Zhang, C.S.: Reply networks on a bulletin board system. Phys. Rev. E. 67(3), 036117 (2003)
26. Watts, D.J., Strogatz, S.H.: Collective dynamics of 'small-world' networks. Nature 393, 440–442 (1998)

The k-Dense Method to Extract Communities from Complex Networks

Kazumi Saito[1,*], Takeshi Yamada[2], and Kazuhiro Kazama[3]

[1] School of Administration and Informatics, University of Shizuoka
   52-1 Yada, Suruga, Shizuoka, Shizuoka 422-8526 Japan
   `k-saito@u-shizuoka-ken.ac.jp`
[2] NTT Communication Science Laboratories, NTT Corporation
   2-4 Hikaridai, Seika, Soraku, Kyoto 619-0237 Japan
   `yamada@cslab.kecl.ntt.co.jp`
[3] NTT Network Innovation Laboratories, NTT Corporation
   3-9-11 Midori-cho, Musashino, Tokyo 180-8585 Japan
   `kazama@ingrid.org`

Abstract. To understand the structural and functional properties of large-scale complex networks, it is crucial to efficiently extract a set of cohesive subnetworks as communities. There have been proposed several such community extraction methods in the literature, including the classical k-core decomposition method and, more recently, the k-clique based community extraction method. The k-core method, although computationally efficient, is often not powerful enough for uncovering a detailed community structure and it only discovers coarse-grained and loosely connected communities. The k-clique method, on the other hand, can extract fine-grained and tightly connected communities but requires a substantial amount of computational load for large-scale complex networks. In this paper, we present a new notion of a subnetwork called k-dense, and propose an efficient algorithm for extracting k-dense communities. We applied our method to the three different types of networks assembled from real data, namely, from blog trackbacks, word associations and Wikipedia references, and demonstrated that the k-dense method could extract communities almost as efficiently as the k-core method, while the qualities of the extracted communities are comparable to those obtained by the k-clique method.

14.1 Introduction

In many scientific and engineering domains, complicated relational data structures are frequently represented by networks or, equivalently, graphs. For example, WWW (World Wide Web) sites are often represented by *hyperlink networks*, with pages as nodes and hyperlinks between pages as edges, the interactions between genes, proteins, metabolites and other small molecules in an organism are represented by *gene regulatory networks*, and the relationships between people and other social entities are characterized by *social networks*. Extracting a set of

---

[*] This work has been done while the first author was with NTT Communication Science Laboratories.

D.A. Zighed et al. (Eds.): Mining Complex Data, SCI 165, pp. 243–257.
springerlink.com   © Springer-Verlag Berlin Heidelberg 2009

cohesive subnetworks as communities from those large-scale complex networks plays an important role to understand their basic structures and functions [12], and it hopefully inspires researchers to discover new knowledge and insights underlying those complex networks.

For this task, two types of approaches have been studied extensively by many researchers [5, 7, 9, 10, 14, 15]. One approach is based on *clustering* techniques that divide all nodes of a network into several communities [5, 7, 15]. Another approach is based on *core decomposition* techniques that extract only certain cohesive portions as communities from a given network; thus the union of extracted communities is not necessarily equal to the original network [9, 10, 14]. The former approach is suitable for understanding the entire structure of a relatively small network by regarding all nodes as equally important, while the latter can be used for finding the major building blocks of a relatively large network.

In this paper, we focus on the core decomposition approach that extracts certain cohesive portions as communities from a large-scale network. For this task, we can employ the classical *k-core decomposition* method, or simply, the *k-core method* [14], or the recently proposed method called CFinder [10] that extracts a set of k-clique communities (therefore for simplicity, we call it the *k-clique method* in this paper). However, the k-core method is usually not powerful enough for uncovering the detailed community structure although it is computationally quite efficient, while the k-clique method often requires a substantial amount of computational load for large-scale networks.

In this paper, we present a new concept of subnetwork called k-dense, and propose an efficient algorithm for extracting k-dense communities. In Section 14.2, we present the notion of k-dense and k-dense communities as well as we review the notions of k-core and k-clique communities. We also describe an efficient algorithm for extracting k-dense communities. In Section 14.3, we first explain the three types of networks used in our experiments, and then we describe criteria for evaluating extracted communities. We then report our experimental results. In Section 14.4, we discuss some related work and future directions.

14.2   Extracting k-Dense Communities

In this section, we present a new notion of a subnetwork called k-dense as well as we review the notions of k-core and k-clique communities. We also describe an efficient algorithm for extracting k-dense communities.

14.2.1   The k-Core Community

For a given network (or equivalently graph) $G = (V_G, E_G)$, let $V_G = \{1, \cdots, N\}$ be a set of nodes (or vertices) and $E_G = \{e_1, \cdots, e_M\}$ a set of links (or edges), where $e_m = \{i, j\} \subset V_G$ and $i \neq j$, meaning we focus on undirected networks without self-links.

Now for a given node i in the network G, we denote $F_G(i)$ as a set of *adjacent nodes* of i as follows:

$$F_G(i) = \{j : \{i, j\} \in E_G\}. \tag{14.1}$$

A subnetwork $C(k)$ of G is called k-*core* if each node in $C(k)$ has more than or equal to $(k - 1)$ adjacent nodes in $C(k)$[1]. More specifically, for a given order k, the k-core is a subnetwork $C(k) = (V_{C(k)}, E_{C(k)})$ consisting of the following node set $V_{C(k)} \subset V_G$ and link set $V_{C(k)} \subset V_G$:

$$V_{C(k)} = \{i : |F_{C(k)}(i)| \geq k - 1\}, \tag{14.2}$$

$$E_{C(k)} = \{e_m : e_m \subset V_{C(k)}\}. \tag{14.3}$$

Here $|A|$ denotes the number of elements in the set A. Hereafter we focus on the subnetwork of maximum size with this property as $C(k)$, and its connected components $C^s(k)$ $(1 \leq s \leq S_{C(k)})$, each of which is referred to as a k-*core community*. Here $S_{C(k)}$ denotes the number of communities (or connected components) in $C(k)$.

Figure 14.1 shows an example of k-core communities, where the subnetwork $C^1(3)$ in the outer box is a 3-core community in which each node has at least two adjacent nodes in $C^1(3)$, and $C^1(4)$ and $C^2(4)$ in the inner boxes are both 4-core communities.

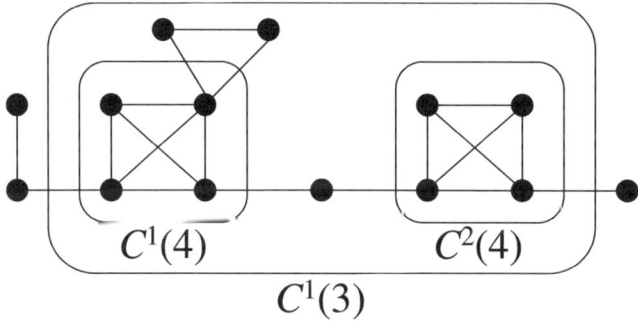

Fig. 14.1. An example of k-core communities

14.2.2 The k-Dense Community

For a given set of nodes $V \subset V_G$, we denote a set of *common adjacent nodes* $F_G(V)$ of V that is a natural extension of $F_G(i)$ in Equation 14.1 as follows:

$$F_G(V) = \bigcap_{i \in V} F_G(i). \tag{14.4}$$

---

[1] Our definition of k-core is, in fact, $(k-1)$-core in the conventional definition. We use this definition for compatibility with the other core concepts.

We propose that a subnetwork $D(k)$ of G is called k-*dense* if each pair of adjacent nodes in $D(k)$ has more than or equal to $(k-2)$ common adjacent nodes in $D(k)$. More specifically, for a given order k, the k-dense is a subnetwork $D(k) = (V_{D(k)}, E_{D(k)})$ consisting of the following node set $V_{D(k)} \subset V_G$ and link set $E_{D(k)} \subset E_G$:

$$V_{D(k)} = \bigcup_{e_m \in E_{D(k)}} e_m, \tag{14.5}$$

$$E_{D(k)} = \{e_m : |F_{D(k)}(e_m)| \geq k - 2\}. \tag{14.6}$$

Note that in this definition, the link e_m and the set of nodes $\{i, j\}$ connected by e_m are identified.

The rationale behind this definition is that when we group two nodes connected with a link together into a same community, we need some solid evidence or witness to support a strong positive relation between them: the fact that they are just connected by a single link may not strong enough. The existence of more common adjacent nodes in the same community suggests stronger positive relation.

We can easily see that k-dense implies k-core, i.e., $D(k) \subset C(k)$. This is because $|F_{D(k)}(e_m)| \geq k - 2$ implies $e_m \subset V_{C(k)}$. In fact, if $e_m = \{i, j\}$ and $|F_{D(k)}(e_m)| \geq k - 2$, then the node i needs to have more than or equal to $(k - 2)$ adjacent nodes other than the node j. Since the nodes i and j are adjacent, we can confirm that $|F_{D(k)}(i)| \geq k - 1$. Again we focus on the subnetwork of maximum size with this property as $D(k)$, and its connected components $D^s(k)$ $(1 \leq s \leq S_{D(k)})$, each of which is referred to as a k-*dense community*.

Figure 14.2 shows an example of k-dense communities, where the subnetwork $D^1(3)$ and $D^2(3)$ are both 3-dense communities in which each link has at least one common adjacent node in $D^1(3)$ and $D^2(3)$ respectively. Please note that the $D^2(3)$, which was first introduced as 4-core $C^2(4)$ in Figure 14.1, is in fact 4-dense, or more accurately, 4-clique. The definition of k-clique is described in the next subsection.

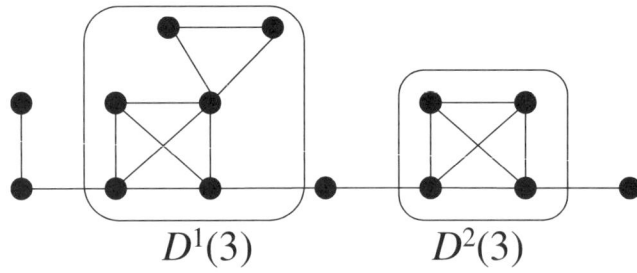

Fig. 14.2. An example of k-dense communities

14.2.3   The k-Clique Community

A subnetwork $Q(k) = (V_{Q(k)}, E_{Q(k)})$ of G consisting of k nodes is called k-*clique* if each node pair in $Q(k)$ is mutually adjacent. Since a k-clique is a complete subnetwork of size k, we can easily see that a k-clique implies a k-dense and k-core, i.e., $Q(k) \subset D(k) \subset C(k)$. In contrast, a k-core or k-dense of size k is a k-clique, i.e., $|V_{C(k)}| = k$ implies $Q(k) = D(k) = C(k)$.

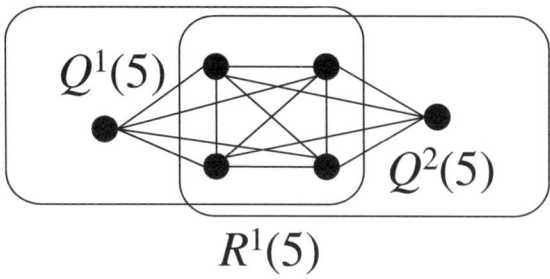

Fig. 14.3. An example of k-clique communities

Two k-cliques are called adjacent if they share $(k - 1)$ nodes. The k-*clique community* proposed by Palla et al. [10] is defined as a union of all the k-cliques that can be reached from each other thorough a series of adjacent k-cliques. Hereafter we denote each k-clique community as $R^s(k) = (V_{R^s(k)}, E_{R^s(k)})$. Note that the k-core $C(k)$ (or the k-dense $D(k)$) is the union of all k-core communities (or k-dense communities), each of which is also k-core (or k-dense) itself. On the other hand, the union of all k-clique communities is not itself a k-clique or a k-clique community. In this paper, we introduce a new notion called a k-clique union $R(k)$ in order to directly compare it with $C(k)$ and $D(k)$. Namely, a k-clique union is defined as a subnetwork $R(k) = (V_{R(k)}, E_{R(k)})$ constructed by the following operation:

$$R(k) = (\bigcup_{s=1}^{S_{R(k)}} V_{R^s(k)}, \bigcup_{s=1}^{S_{R(k)}} E_{R^s(k)}). \tag{14.7}$$

Figure 14.3 shows an example of a k-clique community, where the subnetwork $Q^1(5)$ and $Q^2(5)$ are both 5-cliques, which share 4 nodes in common, resulting in their union $R(5)$ being a 5-clique community.

14.2.4   Higher-Level k-Dense Community

Because a link e_m can be identified as a 2-clique $Q(2)$, we can reformulate the definition of the k-dense subnetwork $D(k) = (V_{D(k)}, E_{D(k)})$ given in Equation (14.6) as follows:

$$\forall Q(2) \subset D(k) \Rightarrow |F_{D(k)}(V_{Q(2)})| \geq k - 2. \tag{14.8}$$

This implies that any pair of adjacent nodes in $Q(2)$ needs to have more than or equal to $(k - 2)$ common adjacent nodes. By extending Equation (14.8) for any h and h-clique $Q(h)$ such that $1 \leq h < k$, we can define the h-level k-dense subnetwork $D(k; h) = (V_{D(k;h)}, E_{D(k;h)})$ as follows:

$$\forall Q(h) \subset D(k; h) \Rightarrow |F_{D(k;h)}(V_{Q(h)})| \geq k - h \qquad (14.9)$$

By identifying $Q(1)$ with a single node, $D(k; 1)$ is identical to the k-core $C(k)$, i.e., $D(k; 1) = C(k)$. On the other hand, by setting $h = k - 1$ we can easily see that any node i in $D(k; k - 1)$ belongs to at least one k-clique. This implies that $(k - 1)$-level k-dense $D(k; k - 1)$ is identical to the k-clique union $R(k)$, i.e., $D(k; k - 1) = R(k)$. Note that a $(k - 1)$-level k-dense community corresponds to a union of k-clique communities that share at least one node. To summarize, the k-dense together with its higher level extension can be regarded as a general concept that naturally interpolates between the k-core and the k-clique union.

A higher-level k-dense community is a subnetwork of corresponding lower-level k-dense community. Our preliminary experiments show that for many networks, most of the higher-level k-dense communities with $h > 2$ are more or less approximated by the corresponding 2-level k-dense communities, while we need a substantial amount of commutation load for extracting higher level k-dense communities. Therefore, from a viewpoint of the applicability to large-scale networks, we should only focus on the 2-level k-dense and we refer it simply as k-dense.

14.2.5   Other Notions of Subnetworks

Distance-based cliques that generalize the notion of a clique, such as an n-step clique[2], or an n-club are widely used in several settings of social network analysis theory [3]. Here a subnetwork is called an *n-step clique* if and only if the geodesic distance of any node pair in the original network is less than or equal to n, while a subnetwork is called an *n-club* if and only if the diameter (the maximum geodesic distance among all node pairs) of the subnetwork is less than or equal to n. Clearly a 1-step clique or 1-club is nothing but a clique, and all nodes in an n-step clique or n-club need not be adjacent for $n \geq 2$. However, since many real-world networks have small diameters, it is widely recognized that the distance is a rather coarse measure to identify meaningful network structures.

A concept called *n-plex* generalizes the notion of a clique in another direction [3]. A subnetwork $P(n) = (V_{P(n)}, E_{P(n)})$ is called an *n-plex* if and only if it satisfies the following condition:

$$\min_{i \in V_{P(n)}} \{|F_{P(n)}(i)| \geq |V_{P(n)}| - n\}. \qquad (14.10)$$

---

[2]   In order to avoid confusion, we use the term "n-step clique" here, although the same term "n-clique" is often used for this generalized clique as well.

A clique is simply a 1-plex as a special case. Since finding maximum n-plexes is $\mathcal{NP}$-hard for all natural numbers $n > 0$, any approach based on the n-plex would have some intrinsic limitation on applicability to large-scale networks.

14.2.6 The k-Core Extraction Algorithm

For a given core order k, we describe a basic algorithm for extracting a k-core from a network $G = (V_G, E_G)$. The basic idea is to recursively eliminate the set of nodes that do not satisfy the k-core condition. The basic procedure for calculating $(V_{C(k)}, E_{C(k)}) = \mathcal{A}_k^{core}(V_G, E_G)$ is described as follows:

1. Initialize $V_{C(k)} = V_G$, $E_{C(k)} = E_G$;
2. Compute $T = \{i : |F_{C(k)}(i)| < k - 1\}$;
3. If $T = \emptyset$, output $(V_{C(k)}, E_{C(k)})$ and terminate;
4. Set $V_{C(k)} = V_{C(k)} - T$, $E_{C(k)} = E_{C(k)} - \{e_m : e_m \cap T \neq \emptyset\}$, and return to **2**.

Here T denotes a set of nodes for elimination under the k-core condition.

The algorithm above extracts k-core for a fixed k. However, it is not realistic to assume that we know an appropriate core order k in advance. Fortunately, the maximum core order is bounded to $\max_i\{F_G(i)\}$, and thus by making use of the property $C(k + 1) \subset C(k)$, we can iteratively compute k-core for every k quite efficiently as follows:

1. Initialize $V_{C(1)} = V_G$, $E_{C(1)} = E_G$, and set $k = 2$;
2. Compute $(V_{C(k)}, E_{D(k)}) = \mathcal{A}_k^{core}(V_{C(k-1)}, E_{C(k-1)})$;
3. If $V_{C(k)} = \emptyset$, then terminate;
4. Set $k = k + 1$, and return to **2**.

Note that extracting a k-core at $k = 2$ is equivalent to eliminating a set of isolated nodes. Hereafter this method is referred to as the k-core method.

14.2.7 The k-Dense Extraction Algorithm

As mentioned earlier, by using the fact that $D(k) \subset C(k)$, we can utilize the k-core method for extracting a k-dense. The basic procedure for calculating $(V_{D(k)}, E_{D(k)}) = \mathcal{A}_k^{dense}(V_G, E_G)$ is described as follows:

1. Initialize $V_{D(k)} = V_G$, $E_{D(k)} = E_G$;
2. Compute $(V_{D(k)}, E_{D(k)}) = \mathcal{A}_k^{core}(V_{D(k)}, E_{D(k)})$
3. Compute $L = \{e_m : |F_{D(k)}(e_m)| < k - 2\}$;
4. If $L = \emptyset$, output $(V_{D(k)}, E_{D(k)})$ and terminate;
5. Set $E_{D(k)} = E_{D(k)} - L$ and return to **2**.

Here L denotes a set of links for elimination under the k-dense condition.

By using the property $D(k + 1) \subset D(k)$, we can iteratively compute k-dense for every k as follows:

1. Initialize $V_{D(2)} = V_{C(3)}$, $E_{D(2)} = E_{C(3)}$, and set $k = 3$;
2. Compute $(V_{D(k)}, E_{D(k)}) = \mathcal{A}_k^{dense}(V_{D(k-1)}, E_{D(k-1)})$;
3. if $V_{D(k)} = \emptyset$, then terminate;
4. Set $k = k + 1$, and return to **2**.

In this procedure, the initial value for k is set to $k = 3$ because $D(2) = C(2)$. Hereafter this method is referred to as the k-dense method.

Please note that by using the property $R(k) \subset D(k)$, it may be possible to construct an efficient algorithm for extracting k-clique communities by utilizing the k-dense method. Although finding all the k-cliques is NP-hard and intractable, it may be computationally acceptable to first extract the set of k-dense communities, and then only for a particular subset of interest from the set, extract k-clique communities.

14.3   Evaluation by Experiments

In this section, we first describe three networks used in our experiments, and then we describe criteria for evaluating extracted communities. Finally, we report our experimental results.

14.3.1   Experimental Data and Evaluation Criteria

We have applied the three core extraction methods, the k-core method, the k-dense method and the k-clique method, to three different types of networks assembled from real data, namely, a trackback (TB) network obtained from a Japanese Blog (Weblog) space [13], a word association network that was retrieved and converted from [11] and also used by Palla et al in [10] to evaluate their k-clique method called CFinder. an English Wikipedia reference network [17]. Hereafter these networks are referred to as *Blog*, *Word* and *Wikipedia* respectively. Table 14.1 shows basic statistics of these networks.

As evaluation criteria, we have employed the total size of communities, the size of the maximum community, the number of communities, and the normalized entropy for measuring the distribution of each community size. Below we describe some details of each criterion.

Let $C(k)$, $D(k)$ and $R(k)$ be the set of communities extracted by the k-core method, the k-dense method and the k-clique method respectively. More specifically, for each order k we compare $C(k) = \{C^s(k) : 1 \leq s \leq S_{C(k)}\}$, $D(k) =$

Table 14.1. Basic statistics of networks

network	# nodes	# links	avg. # links
Blog	12,047	39,960	3.317
Word	7,207	31,784	4.410
Wikipedia	536,724	1,337,902	2.493

$\{D^s(k) : 1 \leq s \leq S_{D(k)}\}$, and $R(k) = \{R^s(k) : 1 \leq s \leq S_{R(k)}\}$. Here $S_{C(k)}$, $S_{D(k)}$ and $S_{R(k)}$ are the number of communities (or connected components) in $C(k)$, $D(k)$ and $R(k)$ respectively. Let $X(k)$ be either $C(k)$, $D(k)$ or $R(k)$, the total size of communities means the number of nodes that belong to a union of extracted communities, i.e., $|V_{X(k)}|$. Note that in the case of k-clique communities, some nodes may belong to multiple communities, but we avoid their duplicate counts. The size of the maximum community and the number of communities correspond to $\max\{|V_{X(k)}^s|\}$ and $S_{X(k)}$, respectively. The *normalized entropy* is a measure defined as follows:

$$\mathcal{E}(X(k)) = \frac{-1}{\log(S_{X(k)})} \sum_{s=1}^{S_{X(k)}} \frac{|V_{X(k)}^s|}{\sum_s |V_{X(k)}^s|} \log \frac{|V_{X(k)}^s|}{\sum_s |V_{X(k)}^s|}. \tag{14.11}$$

This measure is close to 1 when the variance in community sizes is small, i.e., community sizes $|V_{X(k)}^s|$ are almost the same, while it is close to 0 when one community becomes dominant i.e., $X(k)$ consists of one extremely large community and other much smaller ones. Note that we can define this measure $\mathcal{E}(X(k))$ only when there exist more than one communities, i.e., $S_{X(k)} \geq 2$.

14.3.2   Evaluation Using the Blog Trackback Network

The k-core, k-dense and k-clique methods are applied to the Blog trackback network labeled *Blog* in Table 14.1. Let k_{max} be the maximum core order, up to which non-empty communities can be extracted. The results of the k-dense method and the k-clique method both show that $k_{max} = 15$, while the result of the k-core method shows $k_{max} = 20$. Figure 14.4(a) shows the total size of the extracted communities, $|V_X(k)|$, for each order k, obtained by the k-core method (labeled as k-core), the k-dense method (labeled as k-dense) and the k-clique method (labeled as k-clique) respectively. We can see that for each k, the total sizes of communities obtained by the k-dense method and the k-clique method are almost the same, while the total size of communities obtained by the k-core method is substantially larger than both of them.

Figure 14.4(b) shows the size of the maximum community, $\max\{|V_{X(k)}^s|\}$, for each order k obtained by the k-core, k-dense and k-clique methods. When k is in the range of $7 \leq k \leq 13$, the maximum sizes of communities obtained by the k-dense and k-clique methods are almost the same, while the size obtained by the k-core method is substantially larger than them. These experimental results indicate that the results of the k-core method are not well balanced compared to the other two methods in the sense that the k-core method is likely to produce one giant community and other much smaller communities.

Figure 14.4(c) shows the number of extracted communities $S_{X(k)}$ for each order k obtained by the k-core, k-dense and k-clique methods. It can be seen that the number of communities obtained by the k-dense method is relatively smaller than that obtained by the k-clique method when $k \leq 6$, and as k becomes larger, the difference between the two becomes much smaller. When $k \leq 6$, the number of communities extracted by the k-clique method is larger than 100 and the size of

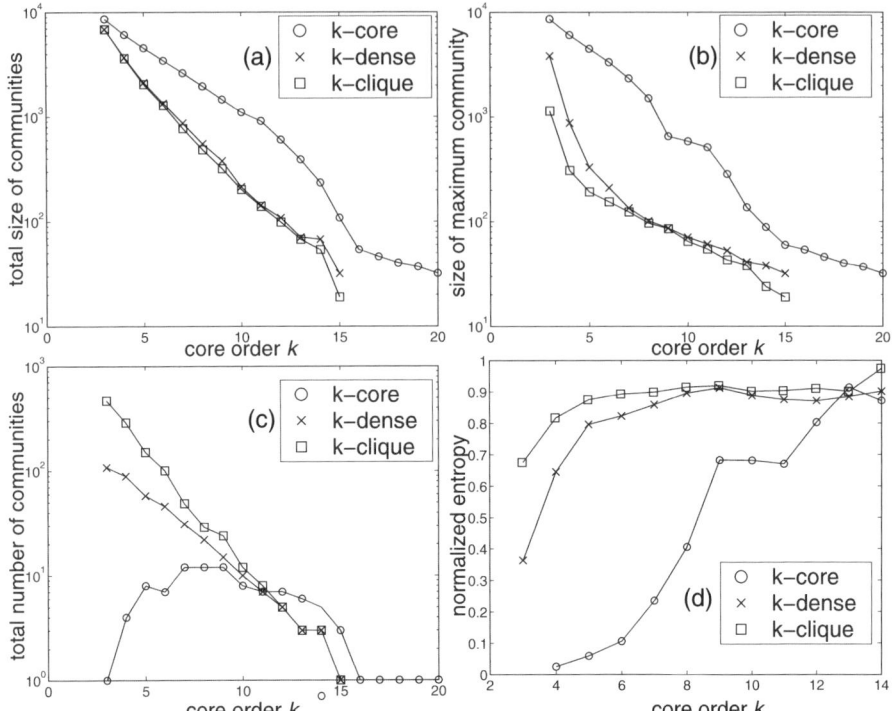

Fig. 14.4. Total size of communities (a), size of the maximum community (b), total number of communities (c), and normalized entropy (d) of the blog trackback network

each community is small, which suggests that the k-clique method inconveniently produces a large number of small communities. On the other hand, the number of communities extracted by the k-core method is at most 10, which suggests that it may be insufficient to analyze detailed community structure.

Figure 14.4(d) shows the normalized entropy $\mathcal{E}(X(k))$ for each order k obtained by the k-core, k-dense and k-clique methods respectively. Here only the results when k is in the range of $3 \leq k \leq 14$ are displayed because this measure is valid only when the number of communities is more than one. This figure indicates that the variance in community sizes is small when extracted by the k-dense method or the k-clique method, while the variance is large when extracted by the k-dense method, especially when k is small.

14.3.3   Evaluation Using the Word Association Network

The k-core, k-dense and k-clique methods are applied to the word association network labeled *Word* in Table 14.1. The results of the k-dense method and the k-clique method both show that $k_{max} = 7$, while the result of the k-core method shows that $k_{max} = 8$.

Figure 14.5(a), (b) and (c) respectively shows the total community size $|V_X(k)|$, the maximum community size $\max\{|V^s_{X(k)}|\}$ and the number of communities $S_{X(k)}$ for each order k for the three core extraction methods.

We can observe that although the statistics of these two networks shown in Table 14.1 are different, the experimental results of these two networks are quite similar. Namely, the total sizes of the communities extracted by the k-dense method and the k-clique method are the same for each k. When $4 \leq k \leq 5$, the maximum community size obtained by the k-dense method is somewhat larger than that obtained by the k-clique method, while the number of communities extracted by the k-clique method is larger than 100. Finally, the k-core method can extract only one community for each k, and the total size of the communities and the size of the maximum community are larger than a few thousands even at the maximum order $k = 8$.

Figure 14.5(d) shows the normalized entropy $\mathcal{E}(X(k))$ for each order k obtained by the k-dense and k-clique methods. In this figure, we can observe a remarkable difference between the two methods when $k = 4$ and the number of communities obtained by the k-clique method is almost one thousand.

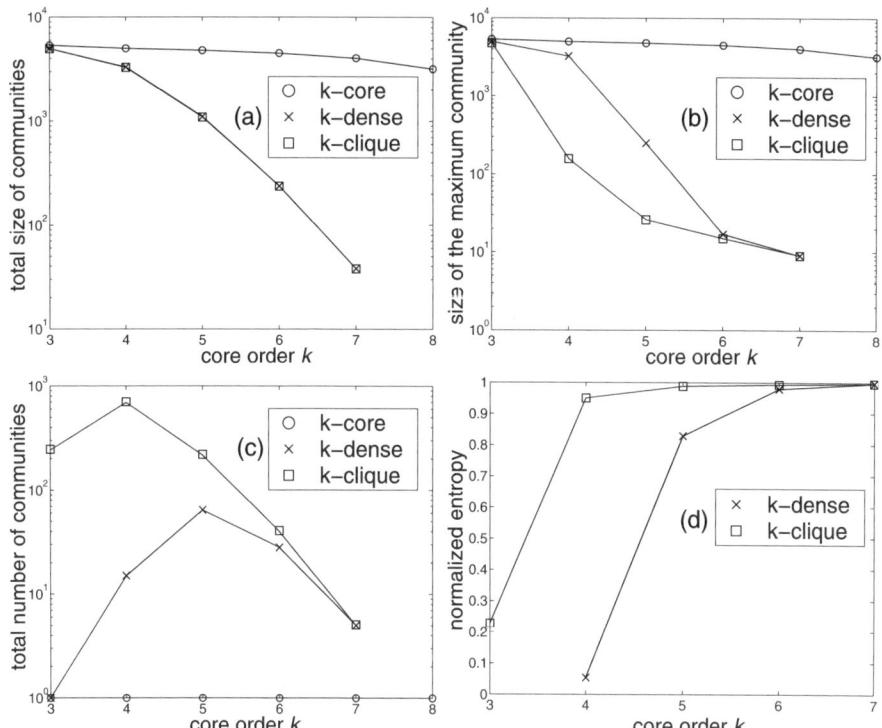

Fig. 14.5. Total size of communities (a), size of the maximum community (b), total number of communities (c), and normalized entropy (d) of the word association network

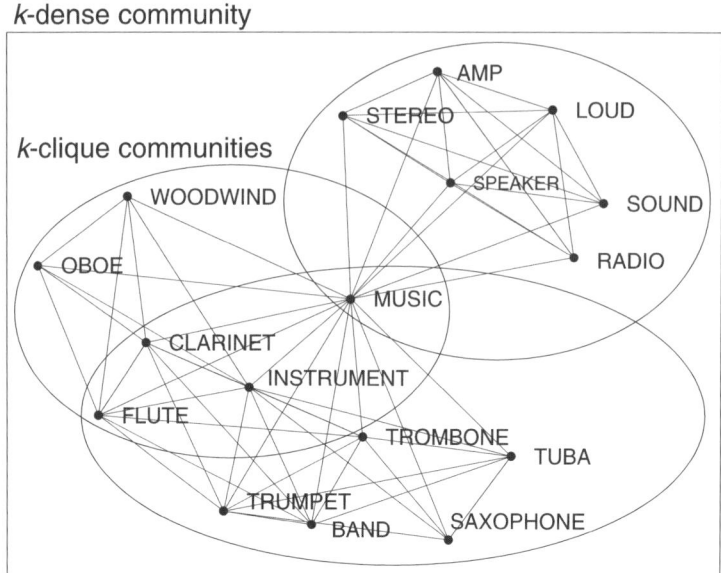

Fig. 14.6. An visualized example of k-dense and k-clique communities extracted from the word association network

Figure 14.6 shows visualization results of communities extracted by the k-dense method and the k-clique method at $k = 6$. Here the spring method proposed by Kamada and Kawai [8] is employed for visualization. As shown in this figure, the k-dense method extracts the subnetwork as only one community, while the k-clique method decomposes it into the three communities, allowing node overlaps, i.e., the node labeled "MUSIC", for example, belongs to all three 6-clique communities. In this example, we admit that it is arguable whether we should allow node overlapping or not and whether we should decompose it into three communities or treat it as one community. At least by directly visualizing the subnetwork extracted by the k-dense method, which is substantially easier than visualizing the whole original network, it would be possible for us to easily find such substructures without farther decompositions. In addition, this subnetwork is an example that any of the h-level k-dense community with $h \geq 2$ is the same.

14.3.4 Evaluation Using Wikipedia Reference Network

We have applied the k-dense and k-core methods to the Wikipedia reference network labeled *Wikipedia* in Table 14.1. It was not possible, at least for us, to apply the k-clique method by using the CFinder program [10] to this large Wikipedia reference network with 536,724 nodes and 1,337,902 links. The result of the k-dense method shows that $k_{max} = 36$, while the result of the k-core method shows that $k_{max} = 48$.

Figure 14.7(a), (b) and (c) respectively shows the total community size $|V_X(k)|$, the maximum community size $\max\{|V^s_{X(k)}|\}$ and the number of communities $S_{X(k)}$ for each order k for the k-core and k-dense methods. As shown in these figures, the results are similar to those of the previous two networks. Namely, in comparison to the k-dense method, the k-core method extracts communities with substantially larger sizes. As for the maximum community size shown in Figure 14.7(b), the k-dense method extracts a 36-clique at $k = 12$, and it is the maximum community up to $k = 36$. The k-core method extracts the maximum community whose size is more than or around one thousand even when k is large (up to around 30).

As shown in Figure 14.7(c), the number of communities extracted by the two methods are the same when $23 \leq k \leq 31$. In fact, in this range of k, the communities extracted by the k-core and the k-dense methods are more or less equivalent and they consist of a clique with a few additional nodes. An example of such cliques is a complete graph that consists of nodes starting from "List of authors by name A" to "List of authors by name Z".

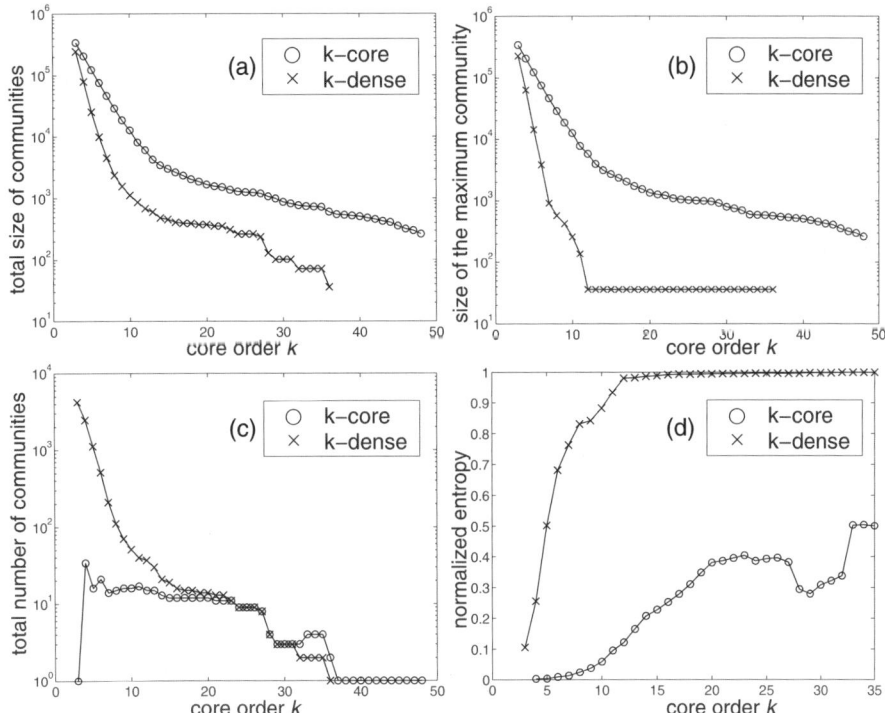

Fig. 14.7. Total size of communities (a), size of the maximum community (b), total number of communities (c), and normalized entropy (d) of the Wikipedia reference network

Figure 14.7(d) shows the normalized entropy $\mathcal{E}(X(k))$ for each order k for the two methods respectively. Again we can observe that for $k \geq 6$ the k-dense method extracts a set of communities with almost equal size, while the sizes of communities extracted by the k-core method are unbalanced.

14.4   Discussion and Related Work

The CFinder program provided by Palla et al. [10] as a part of their k-clique method, first enumerates maximal cliques from a given network and then constructs the clique-clique overlap matrix by counting the number of common nodes between each pair of cliques, and then it extracts k-clique communities for a given k. Here we can alternatively use Carraghan and Pardalos' algorithm [1] or Applegate and Johnson's dfmax program [4] for calculating the maximum clique, and several methods including an algorithm proposed by Tsukiyama et al. [16] for enumerating maximal cliques.

However, this type of problems is known to be $\mathcal{NP}$-hard [6]. Moreover, the number of maximal cliques becomes extremely large for certain types of large-scale networks. Actually in our own experiments using the Wikipedia reference network described above, the CFinder program could not proceed its calculation at the step of the clique-clique overlap matrix calculation. In contrast, the k-dense method proposed in this paper can produce all of the communities of all possible k's in less than one minute. Here all of our experiments were done by using a Dell PC with an Intel 3.4GHz Xeon processor with 2GB of memory.

The computational complexity of the k-dense method is closely related to that of the *clustering coefficients* calculation, which is widely used to characterize complex networks [12]. Because the k-dense condition requires that each link must be shared by at least $(k-2)$ different node-link triangles that are formed by the link and its common adjacent node, the k-dense calculation as well as the calculation of the clustering coefficients, involves in repeatedly counting the number of those triangles in a network. Although the number of triangles for each link changes and needs to be recalculated as nodes and links that do not satisfy the k-dense condition are eliminated, the recalculation can be done quite efficiently because only the links in the neighborhoods of the eliminated nodes and links are affected by the elimination.

In this paper, for simplicity we focused on only undirected networks without self-connections, i.e., we treated each link as a set of nodes. However, it is possible to extend our framework to coping with directed networks. As one such approach, we can follow the work by Batagelj and Zaversnik [2], in which they extend the notion of a k-core to directed networks. As another direction, we can consider a problem to extract bipartite cores or cliques. One pioneering work in this direction includes Web Trawling [9].

14.5   Conclusion

In this paper, we presented a new concept of a subnetwork called k-dense, and we then derived an efficient algorithm for extracting k-dense communities. In our experiments using three types of real networks assembled from blog trackbacks, word associations and Wikipedia references, we demonstrated that the k-dense method could extract communities almost as efficiently as the k-core method, while the qualities of the extracted communities are comparable to those obtained by the k-clique method. we plan to perform more extensive experiments using a wider variety of networks in order to clarify relative strength and weakness of the k-dense method.

References

1. Applegate, D., Johnson, D.S.: Clique-finding program dfmax.c, C program, ftp://dimacs.rutgers.edu/pub/challenge/graph/solvers/
2. Batagelj, V., Zaversnik, M.: Generalized Cores, arXiv:cs.DS/0202039 (2002)
3. Brandes, U., Erlebach, T. (eds.): Network Analysis. LNCS, vol. 3418, pp. 1–6. Springer, Heidelberg (2005)
4. Carragan, R., Pardalos, P.M.: An exact algorithm for the maximum clique problem. Oper. Res. Lett. 9, 375–382 (1990)
5. Flake, G.W., Lawrence, S., Giles, C.L.: Efficient identification of Web communities. In: Proceedings of the Sixth ACM SIGKDD International Conference on Knowledge Discovery and Data Mining, pp. 150–160 (2000)
6. Garey, M., Johnson, D.S.: Computers and intractability: a guide to the theory of NP-completeness. W.H. Freeman, San Francisco (1979)
7. Girvan, M., Newman, M.E.J.: Community structure in social and biological networks. In: Proceedings of the National Academy of Sciences of the United States of America, vol. 99, pp. 7821–7826 (2002)
8. Kamada, T., Kawai, S.: An algorithm for drawing general undirected graph. Information Processing Letters 32, 7–15 (1989)
9. Kumar, R., Raghavan, P., Rajagopalan, S., Tomkins, A.: Trawling the Web for emerging cyber-communities. In: Proceedings of the 8th International World Wide Web Conference (1999)
10. Palla, G., Derényi, I., Farkas, I., Vicsek, T.: Uncovering the overlapping community structure of complex networks in nature and society. Nature 435, 814–818 (2005)
11. Nelson, D.L., McEvoy, C.L., Schreiber, T.A.: The University of South Florida word association norms, http://w3.usf.edu/FreeAssociation
12. Newman, M.E.J.: The structure and function of complex network. SIAM Review 45(2), 167–256 (2003)
13. Kimura, M., Saito, K., Kazama, K., Sato, S.: Detecting Search Engine Spam from a Trackback Network in Blogspace. In: Proceedings of KES 2005, Vol. IV, pp. 723–729 (2005)
14. Seidman, S.B.: Network Structure and Minimum Degree. Social Networks 5, 269–287 (1983)
15. Shi, R., Malik, J.: Normalized cuts and image segmentation. IEEE Trans. PAMI 22(8), 888–905 (2000)
16. Tsukiyama, S., Ide, M., Ariyoshi, H., Shirakawa, I.: A new algorithm for generating all the maximal independent sets. SIAM J. Comput. 6, 505–517 (1977)
17. Yamada, T., Saito, K., Kazama, K.: Network analyses to understand the structure of Wikipedia. In: Proceedings of AISB 2006, vol. 3, pp. 195–198 (2006)

Part IV

Data Clustering

15

Efficient Clustering for Orders

Toshihiro Kamishima and Shotaro Akaho

National Institute of Advanced Industrial Science and Technology (AIST),
AIST Tsukuba Central 2, Umezono 1–1–1, Tsukuba, Ibaraki, 305–8568 Japan
mail@kamishima.net
http://www.kamishima.net/
s.akaho@aist.go.jp

Abstract. Lists of ordered objects are widely used as representational forms. Such ordered objects include Web search results or best-seller lists. Clustering is a useful data analysis technique for grouping mutually similar objects. To cluster orders, hierarchical clustering methods have been used together with dissimilarities defined between pairs of orders. However, hierarchical clustering methods cannot be applied to large-scale data due to their computational cost in terms of the number of orders. To avoid this problem, we developed an k-o'means algorithm. This algorithm successfully extracted grouping structures in orders, and was computationally efficient with respect to the number of orders. However, it was not efficient in cases where there are too many possible objects yet. We therefore propose a new method (k-o'means-EBC), grounded on a theory of order statistics. We further propose several techniques to analyze acquired clusters of orders.

15.1 Introduction

The term *order* indicates a sequence of objects sorted according to some property. Such orders are widely used as representational forms. For example, the responses from Web search engines are lists of pages sorted according to their relevance to queries. Best-seller lists, which are item-sequence sorted according to sales volume, are used on many E-commerce sites.

Orders have also been exploited for sensory test of human respondents' sensations, impressions, or preference. For such a kind of surveys, it is typical to adopt a scoring method. In this method, a respondents' sensation is measured using a scale on which extremes are represented by antonymous words. One example is a five-point-scale on which *1* and *5* indicate *don't prefer* and *prefer*, respectively. If one very much prefers an apple, he/she rates the apple as *5*. Though this scoring method is widely used, it is not the best way for all types of sensory test. For example, as pointed out in [1], a trained expert, e.g., a wine taster, can maintain a consistent mapping from his/her sensation level to rating score throughout a given session. However, users' mappings generally change for each response, especially if the intervals between responses are long. Hence, even if two respondents rate the same item at the same score, their true degrees of sensation may not be the same. When effects of such demerits cannot be ignored, a ranking method is used. In this method, respondents show their degree of sensation by orders, i.e., object sequences according to the degree of a target sensation.

D.A. Zighed et al. (Eds.): Mining Complex Data, SCI 165, pp. 261–279.
springerlink.com   © Springer-Verlag Berlin Heidelberg 2009

In this case, respondents' sensation patterns are represented by orders, and analysis techniques for orders are required.

Orders are also useful when the absolute level of observations cannot be calibrated. For example, when analyzing DNA microarray data, in order that the same fluoresce level represents the same level of gene expression, experimental conditions must be calibrated. However, DNA databases may consist of data sampled under various conditions. Even in such cases, the higher level of fluoresce surely corresponds to the higher level of gene expression. Therefore, by treating the values in the microarray data as ordinal values, non-calibrated data would be processed. Fujibuchi et al. adopted such use of orders in searching a gene expression database for similar cell types [2].

And clustering is the task of partitioning a sample set into clusters having the properties of internal cohesion and external isolation [3]. This method is a basic tool for exploratory data analysis. Clustering methods for orders are useful for revealing the group structure of data represented by orders such as those described above.

To cluster a set of orders, classical clustering has been mainly used [4, chapter 2]. In these studies, clustering methods were applied to ordinal data of a social survey, sensory test, etc. These data sets have been small in size; the number of objects to be sorted and the length of orders are at most ten, and the number of orders to be clustered are at most thousands. This is because an scoring method has been used to acquire responses for a large-scale survey. Responses can easily be collected by requesting for respondents to mark on rating scales that are printed on paper questionnaire forms. On the other hand, using printed questionnaire forms is not appropriate for ranking method, because respondents must rewrite entire response orders when they want to correct them. Therefore, in a ranking method, respondents generally reply by sorting real objects. For example, respondents are requested to sort glasses of wine according to their preference. However, it would be costly to prepare so many glasses. Due to this reason, ranking method has been used for a small-scale survey, even if its advantage to an scoring method is known as described above. But now, adoption of computer interface clear this obstacle in using a ranking method. Respondents can sort virtual objects instead of real objects. Further, methods to implicitly collect preference orders have proposed [5, 6]. These technical progress has made it easier to collect the large number of ordinal data.

We can now collect a large-scale data that consist of orders. However, current techniques for clustering orders are not fully scalable. For example, to cluster a set of orders, dissimilarities are first calculated for all pairs of orders, and agglomerative hierarchical clustering techniques are applied. This approach is computationally inefficient, because computational cost of agglomerative hierarchical clustering is $O(N^2 \log(N)))$ under non-Euclidean metric [7], where N is the number of orders to be clustered. To alleviate this inefficiency in terms of N, we proposed a k-means-type algorithm **k-o'means** in our previous work [8]. The computational complexity was reduced to $O(N)$ in terms of the number of orders. Though this method successfully extracted a grouping structure in a set of orders, it was not efficient yet, if the number of possible objects to be sorted was large. In this paper, to alleviate this inefficiency, we propose a new method, k-o'means-EBC. Note that EBC means **Expected Borda Count**, which is a classic method to find an order so as to be as concordant as possible with a given set of orders. And incompleteness in orders are processed based on a theory of order statistics. Additionally, we propose several methods for interpreting the clusters of orders.

We formalize this clustering task in Section 15.2. Our previous and new clustering methods are presented in Section 15.3. The experimental results are shown in Sections 15.4 and 15.5. Section 15.6 summarizes our conclusions.

15.2 Clustering Orders

In this section, we formalize the task of clustering orders. We start by defining our basic notations regarding orders. An object, entity, or substance to be sorted is denoted by x_j. The universal object set, X^*, consists of all possible objects, and L^* is defined as $|X^*|$. The order is denoted by $O = x_a \succ \cdots \succ x_j \succ \cdots \succ x_b$. Note that subscript j of x doesn't mean "The j-th object in this order," but that "The object is uniquely indexed by j in X^*." The order $x_1 \succ x_2$ represents "x_1 precedes x_2." An object set $X(O_i)$ or simply X_i is composed of all objects in the order O_i. The length of O_i, i.e., $|X_i|$, is shortly denoted by L_i. An order of all objects, i.e., O_i s.t. $X(O_i)=X^*$, is called a complete order; otherwise, the order is incomplete. Rank, $r(O_i, x_j)$ or simply r_{ij}, is the cardinal number that indicates the position of the object x_j in the order O_i. For example, for $O_i = x_1 \succ x_3 \succ x_2$, $r(O_i, x_2)$ or r_{i2} is 3. Two orders, O_1 and O_2, are concordant if ordinal relations are consistent between any object pairs commonly contained in these two orders; otherwise, they are discordant. Formally, for two orders, O_1 and O_2, consider an object pair x_a and x_b such that $x_a, x_b \in X_1 \cap X_2, x_a \neq x_b$. We say that the orders O_1 and O_2 are concordant w.r.t. x_a and x_b if the two objects are placed in the same order, i.e., $(r_{1a} - r_{1b})(r_{2a} - r_{2b}) \geq 0$; otherwise, they are discordant. Further, O_1 and O_2 are concordant if O_1 and O_2 are concordant w.r.t. all object pairs such that $x_a, x_b \in X_1 \cap X_2, x_a \neq x_b$.

A pair set $\mathrm{Pair}(O_i)$ is composed of all the object pairs $x_a \succ x_b$, such that x_a precedes x_b in the order O_i. For example, from the order $O_1 = x_3 \succ x_2 \succ x_1$, three object pairs, $x_3 \succ x_2$, $x_3 \succ x_1$, and $x_2 \succ x_1$, are extracted. For a set of orders S, the $\mathrm{Pair}(S)$ is composed of all pairs in $\mathrm{Pair}(O_i)$ of $O_i \in S$. Note that if the same object pairs are contained in numbers of $\mathrm{Pair}(O_i)$, these pairs are multiply added into the $\mathrm{Pair}(S)$. For example, if the same object pairs $x_1 \succ x_2$ are extracted from O_5 and O_7 in S, both two ordered pairs $x_1 \succ x_2$ are multiply included in $\mathrm{Pair}(S)$.

The task of clustering orders is as follows. A set of sample orders, $S = \{O_1, O_2, \ldots, O_N\}$, $N \equiv |S|$, is given. Note that sample orders may be incomplete, i.e., $X_i \neq X_j$, $i \neq j$. In addition, O_i and O_j can be discordant. The aim of clustering is to divide the S into a partition. The partition, $\pi = \{C_1, C_2, \ldots, C_K\}$, $K = |\pi|$, is a set of all clusters. Clusters are mutually disjoint and exhaustive, i.e., $C_k \cap C_l = \emptyset, \forall k, l, k \neq l$ and $S = C_1 \cup C_2 \cup \cdots \cup C_K$. Partitions are generated such that orders in the same cluster are similar (internal cohesion), and those in different clusters are dissimilar (external isolation).

15.2.1 Similarity between Two Orders

Clusters are defined as a collection of *similar* orders; thus, the similarity measures between two orders are required. *Spearman's ρ* [9, 4] is one such measure, signifying the correlation between ranks of objects. The ρ between two orders, O_1 and O_2, consisting of the same objects (i.e., $X \equiv X(O_1) = X(O_2)$) is defined as:

$$\rho = \frac{\sum_{x_j \in X} (r_{1j} - \bar{r}_1)(r_{2j} - \bar{r}_2)}{\sqrt{\sum_{x_j \in X} (r_{1j} - \bar{r}_1)^2} \sqrt{\sum_{x_j \in X} (r_{2j} - \bar{r}_2)^2}},$$

where $\bar{r}_i = (1/L) \sum_{x_j \in X} r_{ij}$, $L = |X|$. If no tie in rank is allowed, this can be calculated by the simple formula:

$$\rho = 1 - \frac{6 \sum_{x_j \in X} (r_{1j} - r_{2j})^2}{L^3 - L}. \tag{15.1}$$

The ρ becomes 1 if the two orders are concordant, and -1 if one order is the reverse of the other order. Observing Equation (15.1), this similarity depends only on the term

$$d_S(O_1, O_2) = \sum_{x_j \in X} (r_{1j} - r_{2j})^2. \tag{15.2}$$

This is called **Spearman's distance**. If two or more objects are tied, we give the same *midrank* to these objects [4]. For example, consider an order $x_5 \succ x_2 \sim x_3$ ("$\sim$" denotes a tie in rank), in which x_2 and x_3 are ranked at the 2nd or 3rd positions. In this case, the midrank 2.5 is assigned to both objects.

Another widely used measure of the similarity of orders is *Kendall's τ*. Intuitively, this is defined as the number of concordant object pairs subtracted by that of discordant pairs, and then it is normalized. Formally, Kendall's τ is defined as

$$\tau = \frac{1}{L(L-1)/2} \sum_{x_a \succ x_b \in \text{Pair}(O_1)} \text{sgn}\big((r_{1a} - r_{1b})(r_{2a} - r_{2b})\big), \tag{15.3}$$

where $\text{sgn}(x)$ is a sign function that takes 1 if $x > 0$, 0 if $x = 0$, and -1 otherwise. Many other types of similarities between orders have been proposed (see [4, chapter 2]), but the above two are widely used and have been well studied.

In this paper, we adopt Spearman's ρ rather than Kendall's τ because of the following reasons: First, these two measures have similar properties. Both measures of similarities between two random orders asymptotically follow normal distribution as the length of the orders grows. Additionally, these are highly correlated, because the difference between the two measures is bounded by Daniels' inequality [9]:

$$-1 \leq \frac{3(L+2)}{L-2} \tau - \frac{2(L+1)}{L-2} \rho \leq 1.$$

Second, Spearman's ρ can be calculated more quickly. All of the object pairs have to be checked to derive Kendall's τ, so $O(L^2)$ time is required. In the case of Spearman's ρ, the most time consuming task is sorting objects to decide their ranks; thus, the time complexity is $O(L \log L)$. Further, the central orders under Spearman distance is tractable, but the derivation under Kendall's distance is NP-hard [10].

For the clustering task, distance or dissimilarity is more useful than similarity. We defined a dissimilarity between two orders based on ρ:

$$d_\rho(O_1, O_2) = 1 - \rho(O_1, O_2). \tag{15.4}$$

Since the range of ρ is $[-1, 1]$, this dissimilarity ranges $[0, 2]$. This dissimilarity becomes 0 if the two orders are concordant.

15.3 Methods

Here, we describe exiting clustering methods and our new clustering method.

15.3.1 Hierarchical Clustering Methods

In the literature of psychometrics, questionnaire data obtained by a ranking method have been processed by traditional clustering techniques [4]. First, for all pairs of orders in S, the dissimilarities in Section 15.2.1 are calculated, and a dissimilarity matrix for S is obtained. Next, this matrix can be clustered by standard hierarchical clustering methods, such as the group average method. In these survey researches, the size of the processed data set is rather small ($N < 1000$, $L^* < 10$, $L_i < 10$). Therefore, hierarchical clustering methods could cluster order sets, even though the time complexity of these methods is $O(N^2 \log(N))$ under non-Euclidean metric [7] and is costly. However, these method cannot be applied to a large-scale data, due to their computational cost.

Additionally, when the number of objects, L^*, is large, it is hard for respondents to sort all objects in X^*. Therefore, sample orders are generally incomplete, i.e., $X(O_i) \subset X^*$, the dissimilarities cannot be calculated because the dissimilarity measures are defined between two orders consisting of the same objects. One way to deal with incomplete orders is to introduce the notion of an *Incomplete Order Set* (IOS)[1] [4], which is defined as a set of all possible complete orders that are concordant with the given incomplete order. Given the incomplete order O that consists of the object set X, an IOS is defined as

$$\mathrm{ios}(O) = \{O_i^* | O_i^* \text{ is concordant with } O, X(O_i^*) = X^*\}.$$

This idea is not fit for large-scale data sets because the size of the set is $(L^*!/L!)$, which grows exponentially in accordance with L^*. Additionally, there are some difficulties in defining the distances between the two sets of orders. One possible definition is to adopt the arithmetic mean of the distances between orders in each of the two sets. However, this is not distance because $d(\mathrm{ios}_a, \mathrm{ios}_a)$ may not be 0. Therefore, more complicated distance, i.e., Hausdorff distance, has to be adopted.

Since the above IOS cannot be derived for a large-scale data set, we adopted the following heuristics in this paper. In such cases, the dissimilarity between the orders is determined based on the the objects included in both. Take, for example, the following two orders:

$$O_1 = x_1 \succ x_3 \succ x_4 \succ x_6, \qquad O_2 = x_5 \succ x_4 \succ x_3 \succ x_2 \succ x_6.$$

---

[1] In [4], this notion is referred by the term *incomplete ranking*, but we have adopted IOS to insist that this is a set of orders.

From these orders, all objects that are not included in both orders are eliminated. The generated orders become:

$$O_1' = x_3 \succ x_4 \succ x_6, \qquad\qquad O_2' = x_4 \succ x_3 \succ x_6.$$

The ranks of objects in these orders are:

$$r(O_1', x_3){=}1,\ r(O_1', x_4){=}2,\ r(O_1', x_6){=}3;$$
$$r(O_2', x_3){=}2,\ r(O_2', x_4){=}1,\ r(O_2', x_6){=}3.$$

Consequently, the Spearman's ρ becomes

$$\rho = 1 - \frac{6\big((1{-}2)^2 + (2{-}1)^2 + (3{-}3)^2\big)}{3^3 - 3} = 0.5.$$

If no common objects exists between the two orders, $\rho = 0$ (i.e., no correlation).

15.3.2 *k*-o'means-TMSE (Thurstone Minimum Square Error)

In [8], we proposed a k-o'means algorithm as a clustering method designed to process orders. To differentiate our new algorithm described in detail later, we call it by a ***k*-o'means-TMSE** algorithm.

A k-o'means-TMSE in Figure 15.1 is similar to the well-known k-means algorithm [11]. Specifically, an initial cluster is refined by the iterative process of estimating new cluster centers and the re-assigning of samples. This process is repeated until no changes in the cluster assignment is detected or the pre-defined iteration time is reached. However, different notions of dissimilarity and cluster centers have been used to handle orders. For the dissimilarity $d(\bar{O}_k, O_i)$, equation (15.4) was used in step 4. As a cluster center in step 3, we used the following notion of a *central order* [4]. Given a set of

---

Algorithm *k*-o'means(S, K, $maxIter$)
$S = \{O_1, \dots, O_N\}$: a set of orders
K: the number of clusters
$maxIter$: the limit of iteration times

1) S is randomly partitioned into a set of clusters: $\pi = \{C_1, \dots, C_K\}$,
     $\pi' := \pi, t := 0$.
2) $t := t + 1$, if $t > maxIter$ **goto** step 6.
3) **for each** cluster $C_k \in \pi$,
        derive the corresponding central order $\bar{O}_k$.
4) **for each** order O_i in S,
        assign it to the cluster: $\arg\min_{C_k} d(\bar{O}_k, O_i)$.
5) **if** $\pi = \pi'$ **then goto** step 6; **else** $\pi' := \pi$, **goto** step 2.
6) **output** π.

Fig. 15.1. *k*-o'means algorithm

orders C_k and a dissimilarity measure between orders $d(O_a, O_b)$, a central order $\bar{O}_k$ is defined as the order that minimizes the sum of dissimilarities:

$$\bar{O}_k = \arg\min_O \sum_{O_i \in C_k} d(O, O_i). \qquad (15.5)$$

Note that the order $\bar{O}_k$ consists of all the objects in C_k, i.e., $X_{C_k} = \cup_{O_i \in C_k} X(O_i)$. The dissimilarity $d(\bar{O}_k, O_i)$ is calculated over common objects as in Section 15.3.1. However, because $X_i \subseteq X(\bar{O}_k)$, the dissimilarity can always be calculated over X_i. Unfortunately, the optimal central order is not tractable except for a special cases. For example, if using a Kendall distance, the derivation of central orders is NP-hard even if all sample orders are complete [10].

Therefore, many approximation methods have been developed. However, to use as a sub-routine in a k-o'means algorithm, the following two constraints must be satisfied. First, the method must deal with incomplete orders that consist of objects randomly sampled from X^*. In [12], they proposed a method to derive a central order of top k lists, which are special kinds of incomplete orders. Top k list is an order that consists of the most preferred k objects, and the objects that are not among the top k list are implicitly ranked lower than these k objects. That is to say, the top k objects of a hidden complete order are observed. In our case, objects are randomly sampled, and such a restriction is not allowed. Second, the method should be executed without using iterative optimization techniques. Since central orders are derived K times in each loop of the k-o'means algorithm, the derivation method of central orders would seriously affect efficiency if it adopts the iterative optimization.

To our knowledge, the method satisfying these two constraints is the following one to derive the minimum square error solution under a generative model of Thurstone's law of comparative judgment [13]. Because we used this method to derive central orders, we call this clustering algorithm by the k-o'means-TMSE (Thurstone Minimum Square Error) algorithm. We describe this method for deriving central orders. First, the probability $\Pr[x_a \succ x_b]$ is estimated. The pair set of $\mathrm{Pair}(C_k)$ in Section 15.2 is generated from C_k in step 3 of k-o'means-TMSE. Next, we calculate the probabilities for every pair of objects in C_k:

$$\Pr[x_a \succ x_b] = \frac{|x_a \succ x_b| + 0.5}{|x_a \succ x_b| + |x_b \succ x_a| + 1},$$

where $|x_a \succ x_b|$ is the number of the object pairs, $x_a \succ x_b$, in the $\mathrm{Pair}(C_k)$. These probabilities are applied to a model of Thurstone's law of comparative judgment. This model assumes that scores are assigned to each object x_l, and an order is derived by sorting according to these scores. Scores follow a normal distribution; i.e., $N(\mu_l, \sigma)$, where μ_l is the mean score of the object x_l, and σ is a common constant standard deviation. Based on this model, the probability that object x_a precedes the x_b is

$$\Pr[x_a \succ x_b] = \int_{-\infty}^{\infty} \phi\left(\frac{t - \mu_a}{\sigma}\right) \int_{-\infty}^{t} \phi\left(\frac{u - \mu_b}{\sigma}\right) du \, dt$$

$$= \Phi\left(\frac{\mu_a - \mu_b}{\sqrt{2}\sigma}\right), \qquad (15.6)$$

where $\phi(\cdot)$ is a normal distribution density function, and $\Phi(\cdot)$ is a normal cumulative distribution function. Under the minimum square error criterion of this model [14], μ'_l, which is a linearly transformed image of μ_l, is analytically derived as

$$\mu'_l = \frac{1}{|X_{C_k}|} \sum_{x \in X_{C_k}} \Phi^{-1}\big(\Pr[x_l \succ x]\big), \qquad (15.7)$$

where $X_{C_k} = \bigcup_{O_i \in C_k} X_i$. The value of μ'_l is derived for each object in X_{C_k}. Finally, the central order $\bar{O}_k$ can be derived by sorting according to the corresponding μ'_l. Because the resultant partition by k-o'means-TMSE is dependent on the initial cluster, this algorithm is run multiple times, randomly changing the initial cluster; then, the partition minimizing the following total error is selected:

$$\sum_{C_k \in \pi} \sum_{O_i \in C_k} d(O_i, \bar{O}_k). \qquad (15.8)$$

This k-o'means-TMSE could successfully find the cluster structure in a set of incomplete orders due to the following reason: Because the dissimilarity in Section 15.3.1 was measured between two orders, the precision of the dissimilarities was unstable. On the other hand, in the case of k-o'means-TMSE, central orders are calculated based on the $|C_k|$ orders. $|C_k|$ is generally much larger than two, and much more information is available; thus, the central order can be stably calculated. The dissimilarity between the central orders and each sample order can be stably measured, too, because all of objects in a sample order always exist in the corresponding central order and so the full information in the sample orders can be considered.

However, the k-o'means-TMSE is not so efficient in terms of time and memory complexity. Time or memory complexity in N and K is linear, and these are efficient. However, complexity in terms of L^* is quadratic, and further, the constant factor is rather large due to the calculation of the inverse function of a normal distribution. Due to this inefficiency, this algorithm cannot be used if L^* is large. To overcome this inefficiency, we propose a new method in the next section.

15.3.3 k-o'means-EBC (Expected Borda Count)

To improve efficiency in computation time and memory requirement, though we used the k-o'means framework in Figure 15.1 and the dissimilarity measure d_ρ of equation (15.4) in step 4 of Figure 15.1, we employed other types of derivation procedures for the central orders.

Below, we describe this derivation method for a central order $\bar{O}_k$ of a cluster C_k in step 3 of Figure 15.1. We call this the **Expected Borda Count**(EBC) method, and our new clustering method is called a **k-o'means-EBC** algorithm. The Borda Count method is used to derive central orders from complete orders; we modified this so as to make it applicable to incomplete orders. The Borda Count method [15] was originally developed for determining the order of candidates in an election from a set of ranking votes. A set of complete orders, C_k, is given. First, for each object x_j in X^*, the vote count is calculated:

$$\text{vote}(x_j) = \sum_{O_i \in C_k} \left(L^* - r_{ij} + 1\right).$$

Then, a central order is derived by sorting objects $x_j \in X^*$ in descending order of $\text{vote}(x_j)$. Clearly, this method is equivalent to sorting the objects in ascending order of the following mean ranks:

$$\bar{r}_j = \frac{1}{|C_k|} \sum_{O_i \in C_k} r_{ij}. \tag{15.9}$$

If all sample orders are complete and Spearman's distance is used, it is known that the central order derived by the above Borda Count optimally minimizes Equation (15.5) [4, theorem 2.2].

Because all sample orders are complete, Spearman's distance is proportional to the distance d_ρ. Therefore, even in the case that d_ρ is used as dissimilarity, the optimal central order can be derived by this Borda Count method. This optimal central order can also be considered as a maximum likelihood estimator of the Mallows-θ model [16]. The Mallows-θ model is a distribution model of the complete order O, and is defined as

$$\Pr[O; O_0, \theta] \propto \exp(\theta d_S(O_0, O)), \tag{15.10}$$

where the parameters θ and O_0 are called a dispersion parameter and a modal order, respectively.

Unfortunately, this original Borda Count method cannot be applied to incomplete orders. To cope with incomplete orders, we must show the facts known in the order statistics literature. First, we assume that there is hidden complete order O_h^* which is randomly generated. A sample order $O_i \in C_k$ is generated by selecting objects from this O_h^* uniformly at random. That is to say, from a universal object set X^*, L_i objects are sampled without replacement; then, O_i is generated by sorting these objects so as to be concordant with O_h^*. Now we are given O_i generated through this process. In this case, the complete order O_h^* follows the distribution.

$$\Pr[O_h^*|O_i] = \begin{cases} \frac{L_i!}{L^*!} & \text{if } O_h^* \text{ and } O_i \text{ are concordant,} \\ 0 & \text{otherwise.} \end{cases} \tag{15.11}$$

Based on the theory of order statics from a without-replacement sample [17, section 3.7], if an object x_j is contained in X_i, the conditional expectation of ranks of the object x_j in the order O_h^* given O_i is

$$\mathrm{E}[r_j^*|O_i] = r_{ij} \frac{L^*+1}{L_i+1}, \quad \text{if } x_j \in X_i, \tag{15.12}$$

where the expectation is calculated over all possible complete orders, O_h^*, and $r_j^* \equiv r(O_h^*, x_j)$. If an object x_j is not contained in X_i, the object is at any rank in the hidden complete order uniformly at random; thus, an expectation of ranks is

$$\mathrm{E}[r_j^*|O_i] = \frac{1}{2}(L^*+1), \quad \text{if } x_j \notin X_i. \tag{15.13}$$

Next, we turn to the case where a set of orders, C_k, consists of orders independently generated through the above process. Each $O_i \in C_k$ is first converted to a set of all complete orders; thus, the total number of complete orders is $L^*!|C_k|$. For each complete order, we assign weights that follow equation (15.11). By the Borda Count method, an optimal central order for these weighted complete orders can be calculated. The mean rank of x_j (equation (15.9)) for these weighted complete orders is

$$E[\bar{r}_j] = \frac{1}{|C_k|} \sum_{O_i \in C_k} \sum_{O_h^* \in S(L^*)} \Pr[O_h^*|O_i] r(O_h^*, x_j)$$

$$= \frac{1}{|C_k|} \sum_{O_i \in C_k} E[r_j^*|O_i], \qquad (15.14)$$

where $S(L^*)^2$ is a set of all complete orders. A central order is derived by sorting objects $x_j \in X_{C_k}$ in ascending order of the corresponding $E[\bar{r}_j]$. Since objects are sorted according to the means of expectation of ranks, we call this method an **Expected Borda Count** (EBC).

A central order derived by an EBC method is optimal if the distance $d(O_i, \bar{O}_k)$ is measured by

$$\sum_{O_h^* \in S(L^*)} \Pr[O_h^*|O_i] d_S(O_h^*, \bar{O}_k). \qquad (15.15)$$

Hence, in step 4 of Figure 15.1, not d_ρ, but this equation (15.15) should be used. However, it is intractable to compute equation (15.15), because its computational complexity is $O(L^*(L^*!/L_i!))$. Therefore, we adopt d_ρ, and it empirically performed well, as is shown later. Furthermore, if all sample orders are complete, d_ρ is compatible with equation (15.15). Note that we also tried

$$d(\bar{O}, O_i) \sum_{x_j \in X^*} \left(r(\bar{O}_k, x_j) - E[r_j^*|O_i] \right)^2,$$

but empirically, it performed poorly.

The time complexity of a k-o'means-EBC is

$$O\left(K \max(N\bar{L} \log(\bar{L}), L^* \log L^*) \right), \qquad (15.16)$$

where $\bar{L}$ is the mean of L_i over S. First, in step 3 of Figure 15.1, the K central orders are derived. For each cluster, $O((N/K)\bar{L})$ time is required for the means of expected ranks and $O(L^* \log L^*)$ time for sorting objects. Hence, the total time required for deriving K central orders is $O(\max(N\bar{L}, KL^* \log L^*))$. Second, in step 4, N orders are classified into K clusters. Because $O(L \log \bar{L})$ time is required for calculating one dissimilarity, $O(N\bar{L} \log(\bar{L})K)$ time is required in total. The number of iterations is constant. Consequently, the total complexity becomes equation (15.16).

Note that the uniformity assumption of missing objects might look too strong. However, in the case of a questionnaire survey by ranking methods, the objects to be ranked by respondents can be controlled by surveyors.

---

[2] $S(L^*)$ is equivalent to a permutation group of order L^*.

Further, if all the sample orders are first converted into the expected rank vectors, $\langle \mathrm{E}[r_1^*|O_i], \ldots, \mathrm{E}[r_{L^*}^*|O_i] \rangle$, then an original k-means algorithm is applied to these vectors. One might suppose that this k-means is equivalent to our k-o'means-EBC, but this is not the case. A k-means is different from this k-o'means-EBC in terms of the derivation of centers; In the k-means case, the mean vectors of the expected ranks are directly used as cluster centers; in a k-o'means case, these means are sorted and converted to rank values. Therefore, in the k-means case, the centers that correspond to the same central orders are simultaneously kept during clustering. For example, two mean rank vectors $\langle 1.2, 1.5, 4.0 \rangle$ and $\langle 1, 5, 10 \rangle$, correspond to the same central order $x_1 \succ x_2 \succ x_3$, but these two vectors are not differentiated. On the other hand, in a k-o'means-EBC algorithm, they are considered as equivalent, and thus we suppose that the k-o'means-EBC algorithm can find the cluster structure reflecting the ordinal similarities among data.

15.4 Experiments on Artificial Data

We applied the algorithms in Section 15.3 to two types of data: artificially generated data and real questionnaire survey data. In the the former experiment, we examined the characteristics of each algorithm. In the latter experiment of the next section, we analyzed a questionnaire survey data on preferences in sushi.

15.4.1 Evaluation Criteria

The evaluation criteria for partitions was as follows. The same object set was divided into two different partitions: a true partition π^* and an estimated one $\hat{\pi}$. To measure the difference of $\hat{\pi}$ from π^*, we adopted the *ratio of information loss* (RIL) [18], which is also called the uncertainty coefficient in numerical taxonomy literature. The RIL is the ratio of the information that is not acquired to the total information required for estimating a correct partition. This criterion is defined based on the contingency table for indicator functions [11]. The indicator function $I((x_a, x_b), \pi)$ is 1 if an object pair (x_a, x_b) are in the same cluster; otherwise, it is 0. The contingency table is a 2×2 matrix consisting of elements, a_{st}, that are the number of object pairs satisfying the condition $I((x_a, x_b), \pi^*)=s$ and $I((x_a, x_b), \hat{\pi})=t$, among all the possible object pairs. RIL is defined as

$$\mathrm{RIL} = \frac{\sum_{s=0}^{1} \sum_{t=0}^{1} \frac{a_{st}}{a_{..}} \log_2 \frac{a_{.t}}{a_{st}}}{\sum_{s=0}^{1} \frac{a_{s.}}{a_{..}} \log_2 \frac{a_{..}}{a_{s.}}}, \tag{15.17}$$

where $a_{.t} = \sum_s a_{st}$, $a_{s.} = \sum_t a_{st}$, and $a_{..} = \sum_{s,t} a_{st}$. The range of the RIL is $[0, 1]$; it becomes 0 if two partitions are identical.

15.4.2 Data Generation Process

Test data were generated in the following two steps: In the first step, we generated the K orders to be used as central orders. One permutation (we called it a *pivot*) consisting of all objects in X^* was generated. The other $K - 1$ centers were generated by

Table 15.1. Parameters of experimental data

---

1) the number of sample orders: $N = 1000$
2) the length of the orders: $L_i = 10$
3) the total number of objects: $L^* = 10, 100$
4) the number of clusters: $K = \{2, 5, 10\}$
5) the inter-cluster isolation: $\{0.5, 0.2, 0.1, 0.001\}$
6) the intra-cluster cohesion: $\{1.0, 0.999, 0.99, 0.9\}$

---

transforming this pivot. Two adjacent objects in the pivot were randomly selected and exchanged. This exchange was repeated at specified times. By changing the number of exchanges, the inter-cluster isolation could be controlled.

In the second step, for each cluster, constituent orders were generated. From the central order, L_i objects were randomly selected. These objects were sorted so as to be concordant with the central order. Again, two adjacent object pairs were randomly exchanged. By changing the number of times that objects were exchanged, the intra-cluster cohesion could be controlled. Note that the sizes of clusters are equal.

The parameters of the data generator are summarized in Table 15.1. The differences between orders cannot be statistically tested if L_i is too short; on the other respondents cannot sort too many objects. Therefore, we set the order length to $L_i = 10$. **Param 1–2** are common for all the data. The total number of objects (**Param 3**) is set to 10 or 100. All the sample orders are complete if $L^* = 10$, and these are examined in Section 15.4.3. We examine the incomplete case ($L^* = 100$) in Section 15.4.4. **Param 4** was the number of clusters. It is difficult to partition if this number is large, since the sizes of the clusters then decrease. **Param 5** was the inter-cluster isolation that could be tuned by the number of times that objects are exchanged in the first step of the data generation process. This isolation is measured by the probability that the ρ between a pivot and another central order is smaller than that between a pivot and a random order. The larger the isolation, the more easily clusters are separated. **Param 6** was the the intra-cluster cohesion indicating the number of times that objects are exchanged in the second step of the data generation process. This cohesion is measured by the probability that the ρ between the central order and a sample one is larger than that between the central order and a random one. The larger the cohesion, the more easily a cluster could be detected.

For each setting, we generated 100 sample sets. For each sample set, we ran the algorithms five times using different initial partitions; then the best partition in terms of Equation (15.8) was selected. Below, we show the means of RIL over these sets.

15.4.3 Complete Order Case

We analyzed the characteristics of the methods in Section 15.3 by applying these to artificial data of complete orders. The two k-o'means methods were abbreviated to TMSE and EBC, respectively. Additionally, a group average hierarchical clustering method using dissimilarity as described in Section 15.2.1 was tested, and we denoted

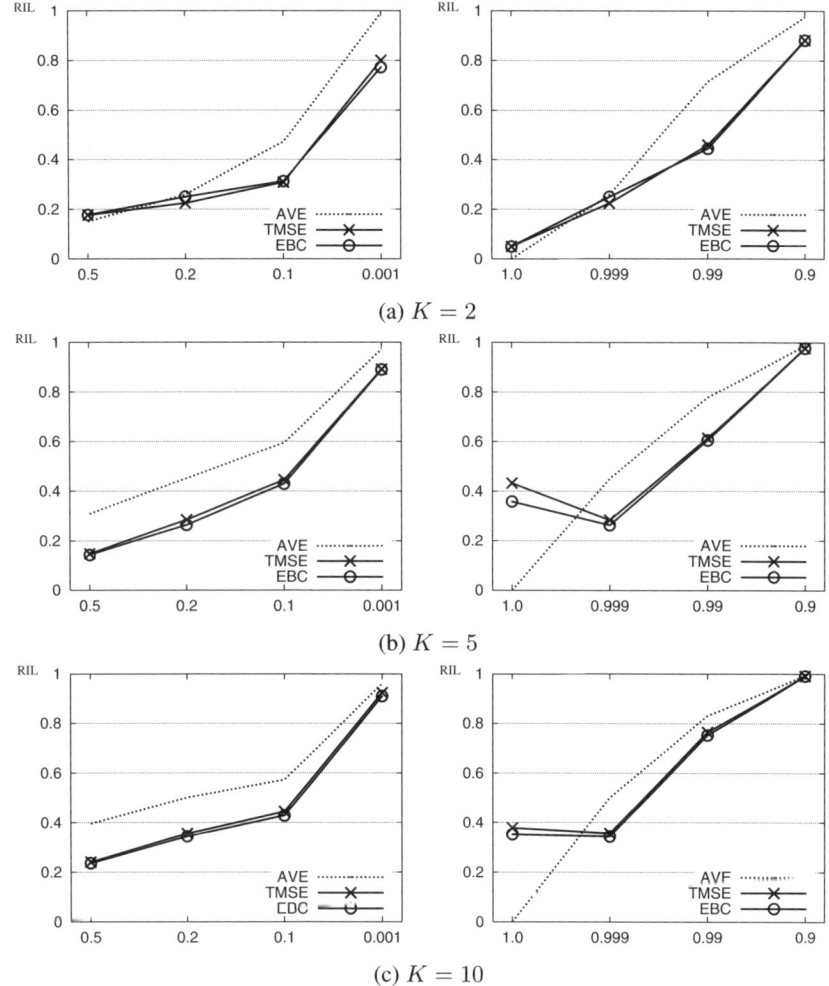

(a) $K = 2$

(b) $K = 5$

(c) $K = 10$

Fig. 15.2. Experimental results on artificial data of complete orders
NOTE: The left charts show the variation of RIL in the inter-cluster isolation when the intra-cluster cohesion is fixed to 0.999. The right charts show the variation of RIL in the intra-cluster cohesion when the inter-cluster isolation is fixed to 0.2.

this result by **AVE**. The experimental results on artificial data of complete orders (i.e, $L^* = 10$) are shown in Figure 15.2. In Figures 15.2(a), (b), and (c), the means of RIL are shown in cases of $K = 2$, 5, and 10, respectively. The left three charts show the variation of RIL in the inter-cluster isolation when the intra-cluster cohesion is fixed to 0.999. The right three charts show the variation of RIL in the intra-cluster cohesion when the inter-cluster isolation is fixed to 0.2.

As expected, the more inappropriate clusters were obtained when the inter-cluster isolation or the intra-cluster cohesion decreased and the number of clusters increased.

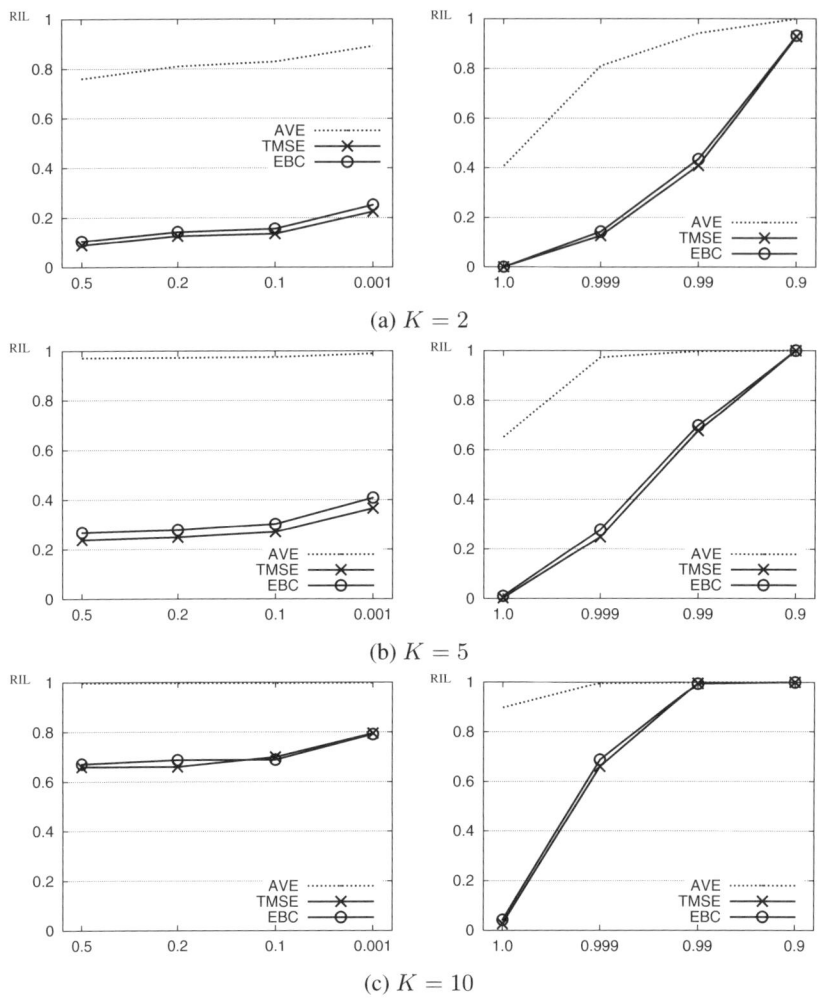

(a) $K = 2$

(b) $K = 5$

(c) $K = 10$

Fig. 15.3. Experimental results on artificial data of incomplete orders

NOTE: See note in Figure 15.2.

We begin with the variation of estimation performance according to the decrease of intra-cluster cohesion. If the cohesion is 1, sample orders are exactly concordant with their corresponding true central orders. In this trivial case, the **AVE** method succeeds almost perfectly in recovering the embedded cluster structure. Because the dissimilarities between sample orders are 0 if and only if they are in the same cluster, this method could lead to perfect clusters. Though both the **EBC** and **TMSE** methods found almost perfect clusters in the $K=2$ case, the performance gradually worsened when K increased. In a k-o'means clustering, a central order is chosen from the finite set, $\mathcal{S}(L^*)$. This is contrasted to the fact that a domain of centers is an infinite set in the clustering of real value vectors. Hence, the central orders of two clusters happen to agree, and one of these clusters is diminished during execution of the k-o'means. As the increase of K, clusters are merged with higher probability. For example, in the **EBC** case, when $K = 2$ and $K = 5$, clusters are merged in 7% and 35% of the trials, respectively. Such occurrence of merging degrades the ability of recovering clusters. As the cohesion increases, the performance of **AVE** became more drastically worse than the other two methods. Furthermore, in terms of the inter-cluster isolation, the performance of **AVE** became drastically worse as K increased, except for the trivial case in which the cohesion was 1. In the **AVE** method, the determination to merge clusters is based on local information, that is, a pair of clusters. Hence, the chance that orders belonging to different clusters would happen to be merged increases when orders are broadly distributed. When comparing **EBC** and **TMSE**, these two methods are almost completely the same.

15.4.4   Incomplete Order Case

We move to the experiments on artificial data of incomplete orders (i.e, $L^* = 100$). The results are shown in Figure 15.3. The meanings of the charts are the same as in Figure 15.2.

   TMSE was slightly better than **EBC** when $K = 2$ and $K = 5$ cases; but **EBC** overcame **TMSE** when $K = 10$. **AVE** was clearly the worst. We suppose that this advantage of the k-o'means is due to the fact that the dissimilarities between order pairs could not be measured precisely if the number of objects commonly included in these two orders is few. Furthermore, the time complexity of **AVE** is $O(N^2 \log N)$, while the k-o'means algorithms are computationally more inexpensive as in Equation (15.16). When comparing **TMSE** and **EBC**, **TMSE** would be slightly better. However, in terms of time complexity, **TMSE**'s $O(NL^* \max(L^*, K))$ is much worse than **EBC**'s $O(K \max(N\bar{L} \log(\bar{L}), L^* \log L^*)$ if L^* is large. In addition, while the required memory for **TMSE** is $O(L^{*2})$, **EBC** demands far less $O(KL^*)$. Therefore, it is reasonable to conclude that k-o'means-EBC is an efficient and effective method for clustering orders.

15.5   Experiments on Real Data

We applied our two k-o'means to questionnaire survey data, and proposed a method to interpret the acquired clusters of orders.

15.5.1   Data Sets

Since the notion of true clusters is meaningless for real data sets, we used the k-o'means as tools for exploratory analysis of a questionnaire survey of preference in sushi (a Japanese food). This data set was collected by the procedure in our previous works [19, 8]. In this data set, $N = 5000$, $L_i = 10$, and $L^* = 100$; in the survey, the probability distribution of sampling objects was not uniform as in equation (15.11). We designed it so that the more frequently supplied sushi in restaurants were more frequently shown to respondents. Objects were selected independently with probabilities ranging from 3.2% to 0.13%. Therefore, the assumption of the uniformity of the sampling distribution, introduced by the EBC method, was violated. The best result in terms of Equation (15.8) ware selected from 10 trials. The number of clusters, K, was set to 2. Note that responses of both authors were clustered into Cluster 1.

15.5.2 Qualitative Analysis of Order Clusters

In [8], we proposed a technique to interpret the acquired clusters based on the relation between attributes of objects and central orders. We applied this method to clusters derived by the EBC and TMSE methods. Table 15.2 shows Spearman's ρ between central orders of each cluster and an order of objects sorted according to the specific object attributes. For example, the third row presents the ρ between the central order and the sorted object sequence according to their price. Based on these correlations, we were able to learn what kind of object attributes affected the preferences of the respondents in each cluster. We will comment next on each of the object attributes.

Almost the same observations were obtained by both EBC and TMSE. The attribute A1 shows whether the object tasted heavy (i.e., high in fat) or light (i.e., low in fat). The positive correlation indicate a preference for heavy testing. The cluster 2 respondents preferred heavy-tasting sushi. The attribute A2 shows how frequently the respondent eats the sushi. The positive correlation indicates a preference for the sushi that the respondent infrequently eats. Respondents in both clusters preferred the sushi they usually eat. No clear difference was observed between clusters. The attribute A3 is the prices of the objects. The positive correlation indicates a preference for economical sushi. The cluster 2 respondents preferred more expensive sushi. The attribute A4 shows how frequently the objects are supplied at sushi shops. The positive correlation indicates a preference for the objects that fewer shops supply. Though the correlation of cluster 1 was rather larger, the difference was not very clear. Roughly speaking, the members of cluster 2 preferred more heavy-tasting and expensive sushi than those of cluster 1.

In this paper, we propose a new technique based on the changes in object ranks. First, a central order of all the sample orders was calculated, and was denoted by $\bar{O}^*$. Next, for each cluster, the central orders were also calculated, and were denoted by $\bar{O}_k$. Then, for each object x_j in X^*, the difference of ranks,

$$\mathrm{rankup}(x_j) = r(\bar{O}^*, x_j) - r(\bar{O}_k, x_j), \qquad (15.18)$$

Table 15.2. Relations between clusters and attributes of objects

Attribute	Cluster 1		Cluster 2	
	EBC	TMSE	EBC	TMSE
A1	0.0999	0.0349	0.3656	0.2634
A2	−0.5662	−0.7852	−0.4228	−0.6840
A3	−0.0012	−0.0724	−0.4965	−0.6403
A4	−0.1241	−0.4555	−0.1435	−0.5838

Table 15.3. The top 10 ranked up and the worst 10 ranked down sushi

#	Cluster 1		Cluster 2	
	2313		2687	
1	egg ♣	+74	ark shell ♡	+63
2	cucumber roll ♣	+62	crab liver ♠	+39
3	fermented bean roll ♣	+38	turban shell ♡	+26
4	octopus	+36	sea bass	+23
5	deep-fried tofu ♣	+33	abalone ♡	+22
6	salad ♣	+29	*tsubu* shell	+16
7	pickled plum & perilla leaf roll ♣	+28	angler liver ♠	+16
8	fermented bean ♣	+26	sea urchin ♠	+15
9	perilla leaf roll ♣	+24	clam ♡	+13
10	raw beef	+21	hardtail ◇	+13
	⋮		⋮	
91	flying fish ◇	-10	chili cod roe roll ♣	-15
92	young yellowtail ◇	-12	pickled plum roll ♣	-15
93	*battera* ◇	-13	shrimp	-17
94	sea bass	-14	tuna roll ♣	-19
95	amberjack ◇	-37	egg ♣	-19
96	hardtail ◇	-41	salad roll ♣	-27
97	fluke fin	-46	deep-fried tofu ♣	-30
98	abalone ♡	-63	salad ♣	-32
99	sea urchin ♠	-84	octopus	-57
100	salmon roe	-85	squid	-82

NOTE: Sushi in each cluster derived by k-o'means-EBC were sorted in descending order of rankup(x_j) (Equation (15.18)). In top row labeled "#", the sizes of clusters were listed. The upper half of the tables show the ranked up sushi, and the bottom half show the ranked down sushi. Just to the right of each sushi name, the rankup(x_j) values are shown.

was derived. We say that x_j is ranked up if rankup(x_j) is positive, and that it is ranked down if rankup(x_j) is negative. If the object x_j was ranked up, it was ranked higher in cluster center $\bar{O}_k$ than in the entire center $\bar{O}^*$. By observing the sushi whose the absolute values of rankup(x_j) were large, we investigated the characteristics of each cluster. Table 15.3 list the most 10 ranked up and the most 10 ranked down sushi in clusters derived by k-o'means-EBC. That is to say, we show the objects whose rankup(x_j) were the 1st to 10th largest, and were the 1st to 10th smallest. The upper half of the tables shows the ranked up sushi, and the bottom half shows the ranked down sushi. In the top row labeled "#", the sizes of the clusters are listed. Sushi names that we were not able to translate into English were written using their original Japanese names in *italics*. Just to the right of each sushi name, the rankup(x_j) values are shown.

We interpreted this table qualitatively. In this table, the mark ♠ indicates objects whose internal organs, such as liver or sweetbread, are eaten. The sushi marked by ◇ are so-called *blue fish*, and those marked by ♡ are clams or shells. These sushi were rather substantial and oily, as revealed in the A1 row of Table 15.2. However, we could not conclude that the respondents in cluster 2 preferred simply oily sushi. For example, sushi categorized as a *red fish meat*, e.g., fatty tuna, were not listed in the table, because the preference of sushi in this category were similar in both clusters. We can say that the respondents in cluster 2 preferred rather oily sushi, especially blue fish, clam/shell, or liver. The sushi marked by ♣ are very economical. Though these sushi were fairly ranked up in cluster 1, this would not indicate a preference for economical sushi. These would be ranked up because these respondents had sushi that they disliked more than these inexpensive types of sushi. Therefore, to interpret the acquired cluster of orders, not only should the values of equation (15.18) be observed, but also the kind of objects that were ranked up or ranked down.

15.6   Conclusions

We developed a new algorithm for clustering orders called the k-o'means-EBC method. This algorithm is far more efficient in computation and memory usage than k-o'means-TMSE. Therefore, this new algorithm can be applied even if the number of objects L^* is large. In the experiments on artificial data, our k-o'means outperformed the traditional hierarchical clustering. For artificial data, the prediction ability of k-o'means-TMSE is almost equal to that of k-o'means-EBC. Therefore, by taking computational cost into account, it could be concluded that the k-o'means-EBC method was superior to the k-o'means-TMSE for clustering orders. Additionally, we advocated the method to interpret the acquired ordinal clusters.

We plan to improve this method in the following ways. During clustering orders, undesired merges of clusters more frequently occur than in clustering of real value vectors. To overcome this defect, it is necessary to improve the initial clusters. For applying ordinal clustering to DNA microarray data, the curse of dimensionality must be solved. We want to develop a dimension reduction technique for orders like PCA. In the case of an Euclidean space, there are many points far from one point. However, in a case of a space of orders (a permutation group), the order most distant from one order is unique, i.e., the reverse order. Therefore, there are biases for central orders to become exact reversals of themselves. We also would like to lessen this bias.

Acknowledgments. This work is supported by the grants-in-aid 14658106 and 16700157 of the Japan society for the promotion of science.

References

1. Luaces, O., Bayón, G.F., Quevedo, J.R., Díez, J., del Coz, J.J., Bahamonde, A.: Analyzing sensory data using non-linear preference learning with feature subset selection. In: Boulicaut, J.-F., Esposito, F., Giannotti, F., Pedreschi, D. (eds.) ECML 2004. LNCS (LNAI), vol. 3201, pp. 286–297. Springer, Heidelberg (2004)
2. Fujibuchi, W., Kiseleva, L., Horton, P.: Searching for similar gene expression profiles across platforms. In: Proc. of the 16th Int'l Conf. on Genome Informatics, p. 143 (2005)
3. Everitt, B.S.: Cluster Analysis, 3rd edn. Edward Arnold (1993)
4. Marden, J.I.: Analyzing and Modeling Rank Data. Monographs on Statistics and Applied Probability, vol. 64. Chapman & Hall, Boca Raton (1995)

5. Branting, L.K., Broos, P.S.: Automated acquisition of user preference. Int'l Journal of Human-Computer Studies 46, 55–77 (1997)
6. Joachims, T.: Optimizing search engines using clickthrough data. In: Proc. of The 8th Int'l Conf. on Knowledge Discovery and Data Mining, pp. 133–142 (2002)
7. Olson, C.F.: Parallel algorithms for hierarchical clustering. Parallel Computing 21, 1313–1325 (1995)
8. Kamishima, T., Fujiki, J.: Clustering orders. In: Grieser, G., Tanaka, Y., Yamamoto, A. (eds.) DS 2003. LNCS (LNAI), vol. 2843, pp. 194–207. Springer, Heidelberg (2003)
9. Kendall, M., Gibbons, J.D.: Rank Correlation Methods, 5th edn. Oxford University Press, Oxford (1990)
10. Dwork, C., Kumar, R., Naor, M., Sivakumar, D.: Rank aggregation methods for the Web. In: Proc. of The 10th Int'l Conf. on World Wide Web, pp. 613–622 (2001)
11. Jain, A.K., Dubes, R.C.: Algorithms for Clustering Data. Prentice Hall, Englewood Cliffs (1988)
12. Fligner, M.A., Verducci, J.S.: Distance based ranking models. Journal of The Royal Statistical Society (B) 48(3), 359–369 (1986)
13. Thurstone, L.L.: A law of comparative judgment. Psychological Review 34, 273–286 (1927)
14. Mosteller, F.: Remarks on the method of paired comparisons: I — the least squares solution assuming equal standard deviations and equal correlations. Psychometrika 16(1), 3–9 (1951)
15. de Borda, J.C.: On elections by ballot (1784). In: McLean, I., Urken, A.B. (eds.) Classics of Social Choice, pp. 81–89. The University of Michigan Press (1995)
16. Mallows, C.L.: Non-null ranking models. I. Biometrika 44, 114–130 (1957)
17. Arnold, B.C., Balakrishnan, N., Nagaraja, H.N.: A First Course in Order Statistics. John Wiley & Sons, Inc., Chichester (1992)
18. Kamishima, T., Motoyoshi, F.: Learning from cluster examples. Machine Learning 53, 199–233 (2003)
19. Kamishima, T.: Nantonac collaborative filtering: Recommendation based on order responses. In: Proc. of The 9th Int'l Conf. on Knowledge Discovery and Data Mining, pp. 583–588 (2003)

16

Exploring Validity Indices
for Clustering Textual Data

Ahmad El Sayed, Hakim Hacid, and Djamel Zighed

University of Lyon
ERIC Laboratory- 5, avenue Pierre Mendès-France
69676 Bron cedex - France
{asayed, hhacid, dzighed}@eric.univ-lyon2.fr

Abstract. The goal of any clustering algorithm producing flat partitions of data, is to find both the optimal clustering solution and the optimal number of clusters. One natural way to reach this goal without the need for parameters, is to involve a validity index in a clustering process, which can lead to an objective selection of the optimal number of clusters. In this chapter, we provide two main contributions. Firstly, since validity indices have been mostly studied in a two or three-dimensionnal datasets, we have chosen to evaluate them in a real-world applications, document and word clustering. Secondly, we propose a new context-aware method that aims at enhancing the validity indices usage as stopping criteria in agglomerative algorithms. Experimental results show that the method is a step-forward in using, with more reliability, validity indices as stopping criteria.

16.1 Introduction

Due to the exponentially growing volume of textual data, clustering methods are gaining increasing attention in text applications, where they can play an essential role in offering more intelligence and efficiency to operations. By textual data, one can refer to characters, n-grams, words, chunks, sentences, documents, etc. In this chapter, we focus on document and word clustering. Both tasks can be extremely useful in a wide range of applications. On the one hand, document clustering plays a key role especially in Information Retrieval (IR), by improving systems' precision and recall [1], by enabling a search without typing through the *scatter/gather* method [15], and by enabling an easier information access by groups [37], or by exploratory browsing [18]. On the other hand, word clustering seeks applications like knowledge acquisition from text [6], query expansion in IR [26], and word sense disambiguation [34].

A well-known and inherent issue in cluster analysis is to require a *minimal input parameters* [13]. Yet, most clustering methods still require the predefinition of a number of parameters usually unknown by the user, such as the desired number of clusters. In practice, this is an ill-posed problem since the final partitions will depend on subjectively chosen parameters that do not necessarily fit the dataset. This can lead to discover spurious patterns not really existing, or to fail to discover the true patterns.

D.A. Zighed et al. (Eds.): Mining Complex Data, SCI 165, pp. 281–300.
springerlink.com   © Springer-Verlag Berlin Heidelberg 2009

The goal of any clustering algorithm seeking flat partitions of data is to find the optimal clustering solution and the optimal number of clusters k. One natural way to reach this without the need for parameters, is to evaluate the quality of different clustering solutions along different number of clusters, in order to finally choose the solution giving the best results. In cluster analysis, the procedure of evaluating the results is known as *cluster validation* [16], and the indices that aims at comparing different solutions with different parameters are known as *relative validity indices* [12].

Relative indices can be involved in clustering methods in two different ways:

- They can be involved as external indicators about the quality of the clustering solution(s) provided by an algorithm.
- They can be involved as criterion functions driving the entire clustering process. In this case, algorithms are often called *incremental* [8] .

Involving a validity index as a criterion function leads usually to a very high complexity. Indeed, this complexity depends on each validity index, and on the type of algorithm that tends to optimize it. For instance, in agglomerative algorithms, at each level of the process, instead of merging two clusters by means of a 'classic' criterion function (e.g., by average-linkage), a validity index VI is calculated evaluating the different solutions that could be obtained following the different merging possibilities. The pair of clusters that will be chosen for merging is the one that optimizes VI. Thus, at a level having k clusters, the complexity of evaluating the different solutions is $O(k(k-1)/2)$;

In the literature, an ambiguity remains unresolved about the usage of validity indices. What is sure, is that they are highly effective, since they seem to yield the only way to bypass the need for input parameters in most clustering algorithms. However, despite their central role, the real utility of validity indices was neither fully exploited nor enough investigated in the literature, especially when dealing with high-dimensional data like text. Along those lines, we provide two main contributions that broadly aim at exploring and exploiting relative indices in clustering algorithms. The applications of interest are document and word clustering. More particularly, our contributions try to bring answers to the key questions below.

- How "good" are relative indices at detecting the optimal clustering solution? Are they better used as criterion functions or simply as external indicators? In case of criterion functions, which index is most likely to drive an agglomerative algorithm to an 'optimal partition'?
- How reliably can we use validity indices as stopping criteria (to terminate the process) in agglomerative algorithm? To which extent can we enhance their reliability for such usage?

The remainder of this chapter is organized as follows. After an overview on cluster validity indices in the next section, we describe in Section 16.3 our new context-aware method. Relative indices are evaluated and compared in Section 16.4. Then, we evaluate our context-aware method in Section 16.5. We conclude in Section 16.6 by summarizing and drawing some future trends.

16.2   Cluster Validity Indices

There exist three kinds of cluster validity indices VI, namely internal, external, and relative indices [16]: (1) External indices evaluate a clustering solution by comparing it to an *a-priori* specified structure that reflects the desired result over the dataset (e.g., *FScore measure, entropy, Jaccard Coefficient, Rand Statistic*). (2) Internal indices assess the intrinsic adequacy between the data structure and the imposed solution basing on quantities and features extracted from the dataset itself (e.g., *CPCC, Hubert τ statistic*). (3) Relative indices compare a clustering solution to another one obtained with different parameters. This can help choosing the parameters that best fit the dataset. Relative indices tend to maximize the intra-cluster compactness and the inter-cluster separation (e.g., *DB, Dunn indices, C1..C4, S_Dbw*).

Since our concern is broadly to find the optimal solution across different k values, our focus in the following will be on relative validity indices. We can distinguish two categories of them, depending on whether or not they scale with the numbers of clusters [12]:

16.2.1   Relative Indices Scaling with the Number of Clusters

Some relative validity indices have the nature to follow systematically the trend of the number of clusters k, which means that, as k increases, their values will keep either increasing or decreasing. Thus, the definition of the optimal k cannot rely on the maximum/minimum value of a VI. It is usually chosen by inspection taking the plot having the more significant local change (jump or drop) in the values of VI, appearing like a "knee" or an "elbow". The intuition is that quick jumps/drops are expected when we are still behind the optimal k, and slower jumps/drops are expected once reaching the optimal k. However, given the many variations in the values of VI, it is often difficult and unclear in practice how to identify the right "knee" in the curve. To overcome this shortcoming, two approaches are widely used: The gap statistics [36] and the stability approach [2, 20].

Among indices in this category, we can find: CH [35], $Diff$ [19], the Hubert τ statistic [31]. Another set of indices (i.e., $I1, I2, E1, H1, H2$) are developed by Zhao [38] specifically for document clustering purposes.

16.2.2   Relative Indices not Scaling with the Number of Clusters

Indices under this category do not systematically follow the trend of k. In this case, the optimal k is more easily chosen as the point on the plot maximizing/minimizing VI. Among indices developed for generic clustering purposes, we can cite: *Dunn [9], the modified Dunn (m-dunn) [3], Davies-Bouldin (DB) [7], RMSSDT, SPR, RS, CD [33], SD, S_Dbw [12], SF [29]*. Another bunch of indices (i.e., *C1, C2, C3, C4*) were developed by Raskutti [27] for document clustering purposes.

Due to their facility of interpretation, we choose to focus on indices not scaling with k. The other reason is related to our primary goal which is to enable the usage of relative indices as stopping criteria. This is definitely a hard task that will get much harder if the optimal k must be selected using relatively sophisticated techniques like the "knee" detection, the gap statistics, or the stability approach.

A known drawback when involving validity indices in clustering is the computational cost that quickly becomes prohibitive when scaling to large and high-dimensional datasets. The main reason is that pairwise similarities between the dataset elements/clusters have to be calculated. As a preliminary attempt to reduce complexity, we propose a new validity index, $H3$, that we define as follows[1].

$$H3 = \frac{\sum_{i=1}^{k} n_i \cdot \sum_{j=1}^{n_i} sim(e_j, S_i)}{(\sum_{i=1}^{k} sim(S_i, S))/k}$$

where sim denotes the similarity between two objects, S_i denotes the centroid of cluster C_i containing n_i elements, e_j denotes a data element, and S denotes the collection's centroid which is the average vector of all clusters' centroids.

$H3$ is significantly less expensive than other indices. The reason is that $H3$ deals with centroids to calculate the inter-cluster separation and the intra-cluster compactness. It has a linear complexity of $O(k + n)$. $H3$ is inspired from the $H1$ and $H2$ indices proposed by Zhao [38]. The difference is that $H3$ does not follow the trend of k after having removed its sensitivity to k in an ad-hoc manner. As a matter of fact, the intra-cluster similarity decreases as k decreases, thus the quality of clustering continuously deteriorates from an intra-cluster point of view. We consider that an optimal partition is reached, when the average of inter-cluster similarities, that tends to improve while grouping similar objects, is no more able to overwhelm the intra-cluster deterioration.

16.3   A Method for Enhancing Relative Indices Usage as Stopping Criteria

In this section, we propose a method that aims to reduce the complexity of clustering algorithms when involving relative indices as criterion functions. One potential option is to use these indices to terminate the clustering process at a point where the "optimal" solution is reached, which allows to discard all the remaining unnecessary part of the process.

16.3.1   Problem Definition

The classical usage of relative validity indices for determining the k yielding the optimal clustering solution comes *a-posteriori*, after evaluating the different

---

[1] Evaluation of the $H3$ index is out of scope in this chapter.

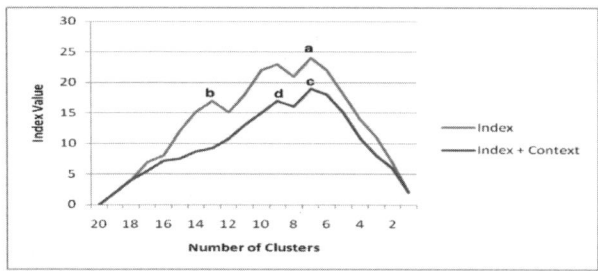

Fig. 16.1. A validity index values obtained along different number of clusters

solutions provided by a clustering algorithm through all the possible k values [24, 27, 12, 39]. Consider in Figure 16.1 the plot (that we call *index*) of a validity index against the different k values. Using an agglomerative algorithm, once reaching the flat optimal solution at $k = a$, all the remaining actions (until $k = 1$) are obviously a time waste because we will end up by considering the solution provided at a. Hence, finding a relevant stopping criterion is primordial. In probabilistic clustering algorithms (mixture models), many stopping criteria are defined quantifying the degree to which a model fits a dataset; among the most known criteria, we can find the Bayesian Information Criterion (BIC) [10], and the Minimum Description Length (MDL) criterion [28]. In non-probabilistic algorithms - which are our concern in this chapter - stopping criteria rely in most cases on input user parameters. For instance, in agglomerative algorithms, these parameters can be a predefined number of clusters, a minimum similarity between clusters, a maximum similarity gap between successive levels, etc. This kind of stopping criteria has serious limitations since users often ignore the parameters that best fit the datasets [13].

A promising approach to address this issue is to make use of relative indices in order to develop an incremental agglomerative algorithm [8] able to stop once reaching the "right" optimal solution in terms of a VI at $k = a$. Thus, an intuitive approach is to let the clustering process go on while optimizing a VI in a stepwise fashion, and to stop once reaching a point ($k = b$) where no further (significant) improvement[2] can be made with any (merging) action [25]. However, such an ad-hoc approach suffers from ignoring, at the specific level b, whether it has truly reached the optimal solution (i.e., $a = b$), or a better solution will come afterward if it accepts a quality decrease at b (which is the case in Figure 16.1). The major problem is that validity indices are using too much local information to take a global decision, e.g., stopping the process. As one could notice, addressing the described issue is a tough and challenging task. Along those lines, we developed a method that aims at enhancing agglomerative algorithms with context-aware decisions taken along with validity indices.

---

[2] While a significant improvement is required with indices scaling with k, a slight improvement is enough for indices not scaling with k.

16.3.2   Context-Aware Method

Context-Aware Clustering. The notion of context was introduced to cluster analysis in [17, 22]. Context was typically involved to provide a more reliable similarity calculation between objects by taking into account their relative nearest neighbor objects. Our goal by involving context in clustering is totally different; actually, we aim to involve context in order to enhance the ability of validity indices to be used as stopping criteria where a first drop (FD) in the quality of a clustering solution can more relevantly indicate that the optimal solution has been reached. Reconsidering Figure 16.1, the goal is to enhance the classic *index* curve with context-awareness, in order to obtain another temperate curve (*index + context* curve) where the FD (at d) approaches as much as possible the optimal solution (at c).

To achieve this goal, the idea is to provide clustering algorithms with a wider view on the dataset partition, which will enable them to take decisions while having in "mind" an "idea" on what could happen next if a specific action is undertaken. As we are seeking the hierarchical agglomerative algorithm, the method applies the following heuristic at each level j of the process:

1. Consider the M closest pairs of clusters;
2. estimate VI after trying to merge each of the M pairs;
3. among the mergings that improve VI (over the previous VI at $j-1$), assess the merging with the lowest *Context Risk* (CR);
4. if no merging improves VI, merge the pair that optimizes (maximizes/minimizes) VI.

Consider the bi-dimensional dataset provided in Figure 16.2 where each point represents a cluster centroid[3]. Suppose that at this stage we have two merging options: (1) merging C_i and C_j into a new cluster C_p, and (2) merging C_m and C_n into a new cluster C_q. Suppose that C_p optimizes a validity index VI, while C_q simply improves it. Before taking any definitive decision, the method examines the context of C_p and C_q in terms of their K Nearest Neighbors K-NN (i.e., surrounding clusters)[4]. If the context of C_p tells that merging C_i and C_j could lead to a global quality degradation in terms of VI in next iterations, the method chooses to create rather C_q improving VI at a minimal context risk.

Applying this method surely implies a more temperate and a slower improvement in VI (as shown at the curve *index + context* in Figure 16.1), but has the advantage of continuously pushing, as much as possible, risky merging actions entailing possible future degradations for later processing. We argue that taking the "safest" action at each level leads an expected degradation to occur as late as possible during the process. Thus, a first drop (FD) is likely to occur closer to the optimal solution, which will offer the possibility to the algorithm to consider more relevantly FD at $(d + 1)$ as a stopping criteria, and the solution provided at d as the optimal clustering solution.

---

[3] S_i is the centroid of cluster C_i.

[4] We set $K = 10$ in our experiments.

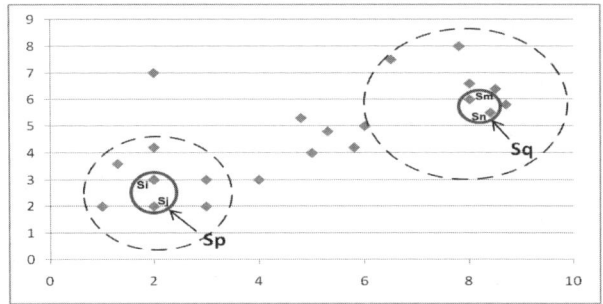

Fig. 16.2. An illustration of a bi-dimensional dataset

Note that calculating VI for all the possible mergings between k clusters will lead to a high complexity of $O(k^2)$ at *each level* of the process. We overcome this by considering, at a given level, only the M closest pairs of clusters[5], since they form the most potential candidates to improve VI.

Context Space Composition. For each new cluster candidate C_p, a Context Risk CR expresses how risky can be assessing C_p for the overall clustering quality in the expected upcoming mergings given the context of C_p. Consider the two new clusters candidates C_p and C_q depicted by their centroids respectively in Figures 16.3 and 16.4 with five context clusters each $(C_1...C_5)$. We assume that C_p, with its K-NN $(K = 5)$ neither too close nor too distant from its centroid, is more risky than C_q, with its K-NN either too close or too distant from its centroid. Therefore, we define a *context space* as the space including the K-NN of a new cluster candidate C_p. Then, as shown in Figure 16.3, we decompose the context space of C_p into three layers that we define below.

Intra layer. Clusters within this layer reduce CR as they should not lead to a quick drop in VI. For this, they have to be *close* enough to C_p, therefore, likely to be merged with C_p in next iterations without causing a significant degradation (comparing to the previous mergings) in the global intra-cluster compactness. As a matter of fact, the clusters are getting larger over mergings, and thus the intra-cluster is continuously deteriorating. At a level k where FD did not occurred yet, we suppose that all the previous mergings that caused degradations in the intra-cluster are acceptable. This layer is delimited by the thresholds $t0 = 0$ and $t1 = \delta(C_p)$. We define $\delta(C_p)$ as the radius of the new cluster candidate C_p augmented by the standard deviation of radius values obtained following the previous mergings. A radius is the maximum distance between the centroid of C_p and an element within C_p.

$$\delta(C_p) = radius(C_p) + Std(radius(C_n...C_{k-1}))$$

---

[5] We set $M = 10$ in our experiments.

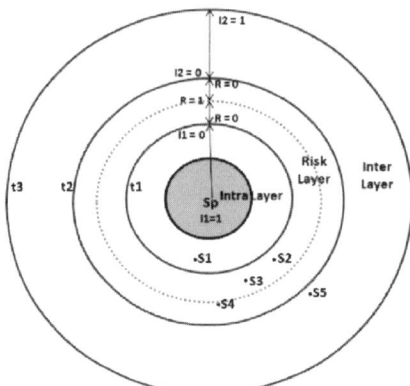

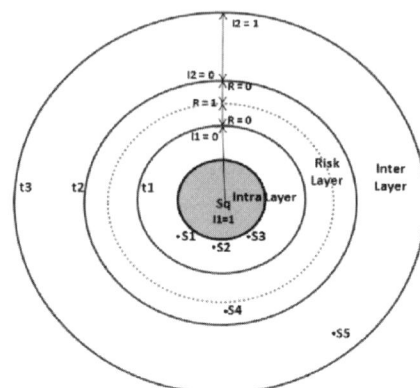

Fig. 16.3. The three-layers context space of the new risky cluster candidate C_p

Fig. 16.4. The three-layers context space of the new non-risky cluster candidate C_q

Inter layer. Clusters within this layer also reduce CR as they should not lead to a quick drop in VI. For this, they have to be *distant* enough from C_p, therefore, not likely to be merged with C_p in next iterations. Further, keeping them outside would contribute to improve (or at least not to deteriorate) the global inter-cluster separation. This layer is delimited by a first threshold $t2 = \Delta(\text{K}-NN(C_p))$. We define $\Delta(\text{K}-NN(C_p))$ as the average pairwise inter-cluster distance between the K-NN of C_p, reduced by the standard deviation of its homologous values obtained following the previous mergings. Getting the average separation between clusters surrounding C_p, will give a hint on the minimum required inter-distance to improve the local inter-cluster separation around C_p, which will most likely improve the global inter-cluster separation.

$$\Delta(K - NN(C_p)) = AvgInter(C_p) - Std(AvgInter(C_n..C_{k-1}))$$

$$AvgInter(C_p) = \frac{\sum_{i=1}^{K} \sum_{j=1}^{K} dist(C_i, C_j)}{K.(K-1)/2} \quad i \neq j$$

We decided to set the same margin for the intra and inter layers in order to have balanced scores in both layers. Subsequently, we define the other inter-layer threshold $t3 = t2 + t1$.

Risk layer. Clusters within this layer increase CR because we consider that they could lead to a fast drop in the global clustering quality, whether on the inter-cluster or intra-cluster level. Actually, these clusters, if merged with C_p in next iterations, would contribute to a significant degradation in the intra-cluster compactness since they are not enough close to C_p. Further, clusters within this layer, if not merged with C_p in next iterations, would not contribute to any significant amelioration in the inter-cluster separation since they are not enough distant from C_p. This layer is delimited by the thresholds $t1$ and $t2$ previously defined.

Context Risk Calculation. In order to calculate CR for a candidate cluster C_p, we use the following formula:

$$CR(C_p) = \frac{1}{K}(\sum_{i=1}^{n1} R(C_i, C_p) - \sum_{j=1}^{n2} I1(C_j, C_p) - \sum_{h=1}^{n3} I2(C_h, C_p))$$

where $R(C_i, C_p)$, $I1(C_j, C_p)$, $I2(C_h, C_p)$ denote the score given for a cluster C_p situated respectively in the risk layer, intra layer, and inter layer. All the scores are distributed along a $[0,1]$ range according to their distances with the centroid of C_p (See Figures 16.3 and 16.4). $n1$, $n2$, $n3$ denote the number of clusters situated respectively in the risk, intra and inter layers. Consequently, CR varies between -1 (for a minimal risk) and 1 (for a maximal risk). For a contextual cluster C_x having a distance d_x with S_p, its score is calculated with respect to the following conditions:

$$\left\{ \begin{array}{l} if \quad d_x < t_1 \quad then \quad I1(C_x, C_p) = \frac{t_1 - d_x}{t1} \\ else \quad if \quad d_x > t_2 \quad then \quad I2(C_x, C_p) = \frac{d_x - t_2}{t1} \\ else \quad if \quad d_x > t_3 \quad then \quad I2(C_x, C_p) = 1 \\ else \quad if \quad t_1 \leq d_x < (t1 + t2)/2 \quad then \quad R(C_x, C_p) = \frac{d_x - t_1}{t1/2} \\ else \quad if \quad (t1 + t2)/2 \leq d_x \leq t2 \quad then \quad R(C_x, C_p) = \frac{t_2 - d_x}{t1/2} \end{array} \right\}$$

In terms of complexity, CR cannot be considered as computationally expensive. In fact, given K nearest neighbors, and p clusters at a specific iteration, we compute CR for M candidate clusters resulted from the merging of the M closest pairs of clusters. Therefore, at a given iteration, the added complexity to a clustering algorithm is $O(M(p-1+K+K(K-1)/2))$. Furthermore, we argue that the parameters M and K have "second order" effect on the results. In other words, they are not 'critical' parameters, and their choices depend solely on the extent to which we are able to augment the complexity of the algorithm.

16.4 Experimental Study on Relative Validity Indices

16.4.1 Motivation

Since their high computational cost, relative indices are most often studied in small-dimensional datasets including small numbers of clusters [12, 24, 9, 7, 21, 4, 29]. The principal motivation was the ability to visually inspect data in two/three dimensions, which help to easily predefine the optimal number of clusters k. Consequently, indices can be compared according to their ability to detect the optimal k. Even though such experiments can provide a preliminary idea about the performance of each validity index, they do not reflect the reality of their performance in real-world applications, where data is often multidimensional and include a large number of clusters.

In this section, we investigate, through an experimental study, the utility of different relative indices for an agglomerative algorithm in two different contexts: Document and word clustering. In such contexts, many questions remain unaddressed in the literature.

- How "good" are relative indices at evaluating partitions? Is it worthwhile involving relative indices as criterion functions in an agglomerative algorithm? Or involving them as external indicators could be enough (leading to comparable results with much lower complexity)? The goal here is of course to find the best trade-off between effectiveness and efficiency among the different approaches.
- Consider the case where relative indices are involved as criterion functions. A key question arises: Which index will most likely guide the agglomerative algorithm to the optimal clustering solution and to the optimal number of clusters in each application?

Indices must indeed be evaluated according to their ability to identify both, the optimal clustering solution and the optimal number of clusters. Note that these two goals do not necessarily overlap; actually, since algorithms are error-prone, an optimal solution can lie under a number of clusters different from the "real" optimal number of clusters. Moreover, an algorithm can provide poor solutions at the "real" optimal number of clusters. For this reason, we have chosen to separate between the two concepts.

The rest of this section is organized as follows: We start by describing the two benchmarks used for our experiments. Then, we present our evaluation methodology and the obtained results for both benchmarks in the following two subsections. We end up this section by discussing the results.

16.4.2   Benchmark for Document Clustering

The benchmark used for document clustering is constituted of two datasets, whose general characteristics are summarized in Table 16.1. These are two distinct collections (no common document) extracted from the Reuters corpus[6]. Basically, the Reuters corpus contains over 800,000 manually categorized newswire stories (documents), each of which consisting of few hundred up to several thousand words. Each document has been manually categorized into one or multiple topics, such as *"Economics, Markets, Corporate/Industrial"*. For our experiments, documents are preprocessed by applying the classical techniques of Natural Language Processing (NLP) provided by Gate[7]: Tokenization, stop-word removal, POS tagging, words lemmatization. The Vector-Space Model is used to represent each document d by a vector v in a multidimensional space, where each dimension represents a word expressed by its $tf.idf$ score [30]. Finally, a similarity is calculated between a pair of documents by means of the *cosines coefficient* between their feature vectors.

---

[6] Reuters corpus, volume 1 (RCV 1), English language, release date: 2000-11-03.
[7] http://www.gate.ac.uk/

Table 16.1. Summary of datasets used for our study on document clustering

Dataset	# of documents	# of topics
DS1	500	48
DS2	1000	65

16.4.3 Benchmark for Word Clustering

The benchmark used for word clustering is constituted of two datasets, whose general characteristics are summarized in Table 16.2. These are two distinct datasets (no common words) extracted from the SemCor corpus [32]. SemCor is a collection of 352 texts where each token is annotated with POS, lemma, and a sense (synset from WordNet [23]). A polysemous word can be associated to many senses in the corpus, as long as it can occur in many different contexts. Words are represented with respect to the Distributional Hypothesis [14]. Thus, a word w is represented by a vector v in a multidimensional space, where each dimension is a local context word w' expressed by its PMI score (Pointwise Mutual Information) with w [5]. A 2-word window size is considered for context words. We turned to language-independent practices representing a context by a plain word concatenated with its relative positions to the target word (e.g., *before-tea*, *after-coffee*). Finally, a semantic similarity is calculated between a pair of words by means of the *cosines coefficient* between their feature vectors.

Table 16.2. Summary of datasets used for our study on word clustering

Dataset	# of words	# of senses
DS3	500	52
DS4	1000	78

16.4.4 Evaluation Methodology

The advantage of using the outlined corpora (i.e., Reuters, SemCor) is that objects have been manually pre-classified by experts. Having this at hand, one can define such artificial structures as the ideal "Gold Standard" structures for a clustering algorithm. Indeed, we have to point out that these structures reflect only a certain level of granularity that could be too specific or too generic for the data. Thus, we cannot claim that the predefined partition for each dataset is the *only* correct partition, but it is indeed a correct one that we could reliably consider as a "Gold Standard".

Subsequently, an external validity index (e.g., *FScore*) is used to evaluate the partition provided by an algorithm against the predefined partition. Moreover, one effective way for evaluating relative indices is to compare their behaviors with those of external indices which we suppose bear the optimal behaviors since they are based on structures set *a-priori* by experts.

Our experiments include 12 algorithms after having run the agglomerative algorithm separately along with each of the 8 relative validity indices (i.e., *C1, C2, C3, C4, H3, DB, Dunn, Dunn-like*), and the 4 classic criterion functions (i.e., *single-linkage, complete-linkage, average-linkage, mean-linkage*). Then, each solution provided at each level of the clustering process is evaluated by means of the target relative index (predicted quality) and the *FScore* measure (real quality).

16.4.5   Experimental Results

Following the described methodology, we study the ability of each criterion function to lead the algorithm to reach the predefined structure, in terms of reaching both the optimal clustering solution, and the optimal number of clusters.

On Reaching the Optimal Clustering Solution. The optimal solution is defined as the solution that maximizes the *FScore* with the predefined solution. Approving the *FScore* output as the "Gold Standard" output at each iteration, we present in Figures 16.5 and 16.6 the indices results evaluated from three different angles:

– Their correlation with the *FScore*: By studying correlation between the predicted values and the real values, we can figure out to which extent a relative index can behave similarly to an external index. Correlation represents also how "good" are relative indices at evaluating solutions, independently of the clustering method.
– The optimal *FScore* reached across the different k values: It represents the optimal clustering quality that a VI can reach if it shares the same optimal k with *FScore*. Values express also to which extent, (merging) actions based on a given index can lead to correct/incorrect partitions among clusters.
– The *FScore* reached at the optimal value of VI: This is a good indicator of the overall solution quality that a VI can reach. By comparing these values to the previous values (i.e., optimal *FScore*), one can check to which extent the predicted optimal solution detected by a VI can approach the real optimal solution.

On Reaching the Optimal Number of Clusters. The optimal k is defined as the number of distinct classes (i.e., topics, senses) in each dataset. We present in Figures 16.7, 16.8 the indices results evaluated from two different angles:

– k at the optimal value of VI, which represents to which extent a VI, with its actual trend for determining the optimal k, is able to approach the real optimal k value;
– k at the optimal value of the *FScore*, which represents to which extend a VI, if it had the trend of *FScore* for determining the optimal k, is able to approach the real optimal k value.

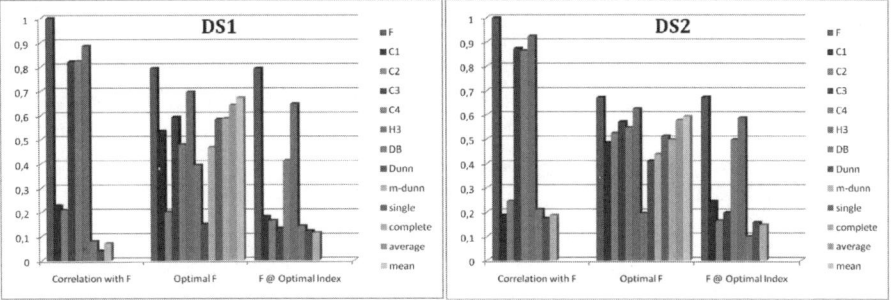

Fig. 16.5. Indices ability to identify the optimal clustering solution - application on DS1 and DS2 for document clustering

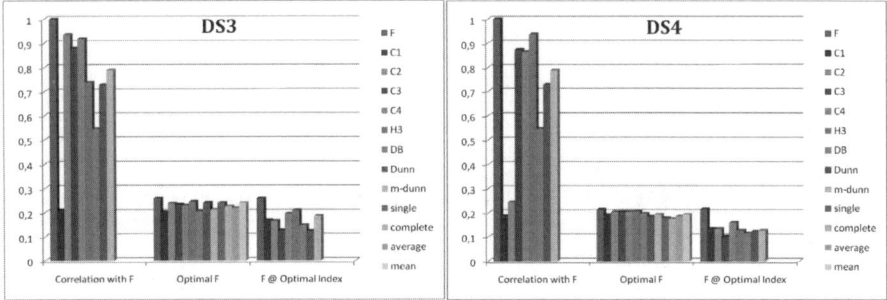

Fig. 16.6. Indices ability to identify the optimal clustering solution - application on DS3 and DS4 for word clustering

16.4.0 Discussion

After examining the different graphs in Figures 16.5, 16.6, 16.7, and 16.8, we can conclude the following remarks:

On involving indices as external indicators. An important first remark is that most indices perform "well" at evaluating solutions, especially when applied on words. By looking at Figures 16.5, 16.6, we can notice the high correlations that most indices have with the $FScore$, which means that they have comparable behaviors to an external index relying on a predefined structure. For word clustering, on the top of the list, we can find the $C2$ and the $H3$ indices with the outstanding correlations of 0.936 and 0.938 on DS3 and DS4 respectively, which are of course encouraging results. For document clustering, correlations were less satisfying. Nevertheless, we could obtain very high correlations with three indices, i.e., $H3$, $C3$, $C4$, but very low correlations, which could barely go beyond 0.2, were obtained with the other indices.

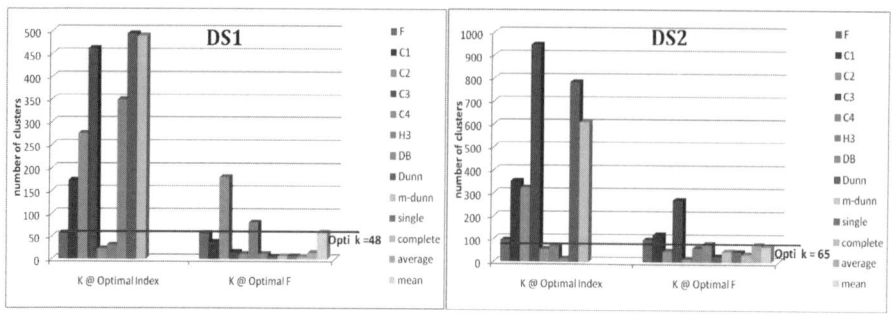

Fig. 16.7. Indices ability to identify the optimal number of clusters - application on DS1 and DS2 for document clustering

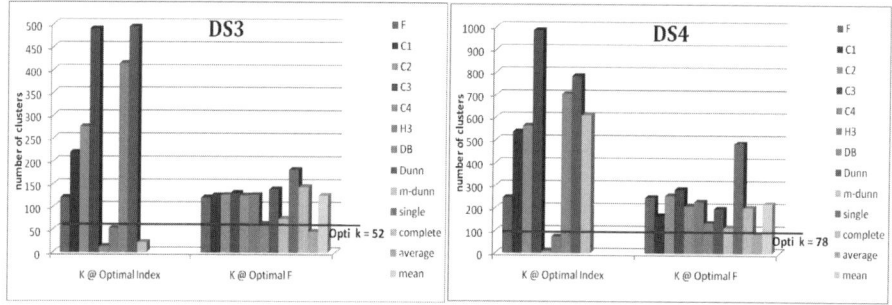

Fig. 16.8. Indices ability to identify the optimal number of clusters - application on DS3 and DS4 for word clusterin

On involving indices as criterion functions. Next, we compare the ability of validity indices - when involved as criterion functions - to drive an algorithm toward "optimal" partitions. That is, by looking at Figures 16.5, 16.6 (*"Optimal F"*), we can conclude that the $H3$ and $C3$ indices could lead the agglomerative algorithm to the optimal partitions in terms of *FScore* in all our datasets. For document clustering, the $H3$ index could lead to an *FScore* of 0.695 and 0.592 respectively in DS1 and DS2. These are interesting rates since they are somehow comparable to the "upper-bound" rates reached when involving the *FScore* index itself as criterion function. On the other hand, for word clustering, much lower *FScores* were noticed in DS3 and DS4. That is, a maximal *FScore* of only 0.245 was attained when using the $H3$ index. This wide gap in *FScore* values between both applications, i.e., word and document clustering, is certainly due to the relatively poor representation of words patterns. We argue that representing documents by their content terms is much more meaningful for similarity calculations than representing words by their context words. This is what makes word clustering look like a hard tasks, comparing to other clustering applications.

On indices rigidity over the number of clusters. In spite of their ability to reach "good" clustering solutions, relative indices provide relatively poor solutions at their top-ranked solutions. In other words, while they could reach high *FScores* at a specific k, they have hard-time detecting this optimal k, leading unfortunately to poor solutions comparing to what they could have reached. This can be explained by the somehow rigid trends over the number of clusters. Even if these indices are supposed to be completely insensitive to k, depending uniquely on the dataset, it is not always the case. This lead them to indicate optimal solutions at "wrong places". The gaps between the *FScores* are dramatic; for instance, while $C3$ could lead the algorithm to an *FScore* of 0.60 in DS4, it could only reach an *FScore* of 0.21 at its optimal value. The only exceptions to this are the $H3$ and $C4$ indices which seem to provide comparable *FScores*, since their high ability to detect the predefined optimal k.

The rigid trend over k, can be seen more clearly in Figures 16.7, 16.8. By focusing on the first colon "*K @ Optimal index*", we can notice that most indices are keeping similar relative trends over the optimal k along different datasets. In our datasets, they generally show trends to large optimal k comparing to the "real" optimal k which is remarkably smaller (*"K @ Optimal F*). These differences, between the predicted optimal k, and the "real" optimal k are naturally behind the wide gaps in *FScore* that we have mentioned above. This explain also why $H3$ and $C4$ are exceptions to the other indices. In fact, their high ability for reaching the optimal k is surely affecting their high ability for reaching the optimal clustering solutions.

Are indices better used as external indicators or criterion functions? As mentioned earlier, involving indices as criterion functions leads to much higher complexity than by involving them as external indicators. In fact, it is natural to believe that driving an algorithm by optimizing a validity index, would overcome the approach that drives an algorithm basing on a "blind" similarity between patterns (e.g., mean-linkage), without paying any consideration to the whole clustering quality. However, surprisingly, our results showed that the difference is not significant. For instance, consider the *'one-to-one'* maximal reached rates with each approach; the $H3$/*mean-linkage* criteria could lead to maximal *FScore* of 0.695/0.671, 0.624/0.593, 0.245/0.240, and 0.204/0.192 respectively in DS1, DS2, DS3, and DS4. Thus, slight improvements were noticed when involving the $H3$ index, but results are comparable, especially when involving the other indices. Depending on each task requirements, the open question remains: *Is it worthwhile to considerably increase the algorithm complexity to reach only a slight improvement in partitions quality?* In most cases, the answer will be no. However, for one seeking *definitely* the optimal partitions out of an algorithm, a complexity reduction is highly recommended. This is broadly the purpose of the method that we proposed in Section 16.3, which aims to reduce algorithms' complexity by using indices as stopping criteria. This method is evaluated in the next section.

16.5   Evaluating Our Context-Aware Method

In this section, we present an experimental study, which is an attempt to answer the following question: *How reliable can be the usage relative indices as stopping criteria in agglomerative clustering?*. Along those lines, we explore the added-value of enhancing a clustering process with context-awareness in order to enable validity indices usage as stopping criteria. We evaluate the proposed method on the four benchmarks described in Sections 16.4.2 and 16.4.3, for document and word clustering respectively. We excluded from the following experiments some indices that showed to be inappropriate for the context-aware method because they provide too unstable curves to be stabilized (e.g., *Dunn*, *m-Dunn*). At last, our experiments will be carried out on 5 indices, namely *DB [7]*, *C1*, *C2*, *C4 [27]*, *and H3*.

Therefore, the experiments include 10 algorithms after having run the agglomerative algorithm 2 times for each of the 5 relative indices (with and without context-awareness). Each solution provided at each level of the clustering process is evaluated by means of the target relative index (predicted quality) and the *FScore* (real quality).

As stressed earlier in this chapter, the goal is to approach, as much as possible, the solution provided before FD to the optimal solution. The optimal solution is defined as being the solution at k where a specific VI reached its maximum or minimum, depending on whether we tend to maximize or minimize VI.

16.5.1   On Approaching the Optimal Clustering Solution

We first study to which extent the context-aware method allows FD to approach the optimal clustering solution reached under a specific number of clusters k. Therefore, we demonstrate in Figures 16.9 and 16.10 the complete agglomerative clustering process ($k = n \to k = 1$) divided into three parts:

– **P1:** This part goes from the initial set ($k = n$) to the last point before FD. Thus, using a VI as a stopping criterion will lead the process to the last point of P1.
– **P2:** This part goes from FD to the optimal clustering solution. It represents the part that must be processed but will not if VI is used as a stopping criterion.
– **P3:** This part goes from the optimal solution until the root cluster ($k = 1$), which forms the unnecessary part that will be performed in vain if VI is not used as a stopping criterion.

By observing Figures 16.9 and 16.10, we can quickly notice the added-value of the context-aware method for both applications word and document clustering. On the first hand, it avoids a clustering algorithm from processing all the P3 parts which is a great time waste. On the other hand, it contributes to reduce P2, since in most cases, FD occurs remarkably closer to the optimal solution. This will surely enable us to consider more relevantly a solution before FD as the

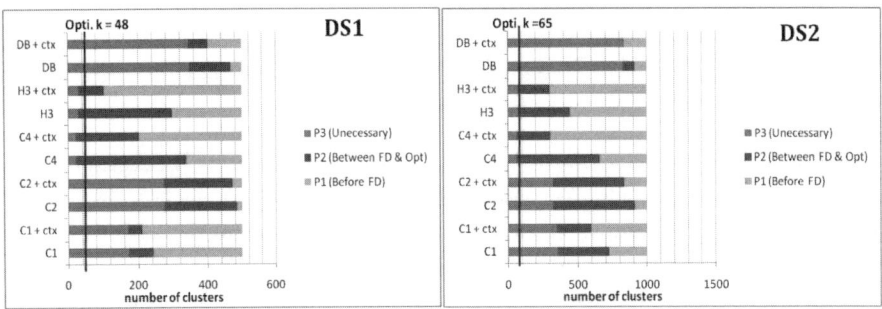

Fig. 16.9. The added-value of the context-aware method in approaching the optimal k - application on DS1 and DS2 for document clustering

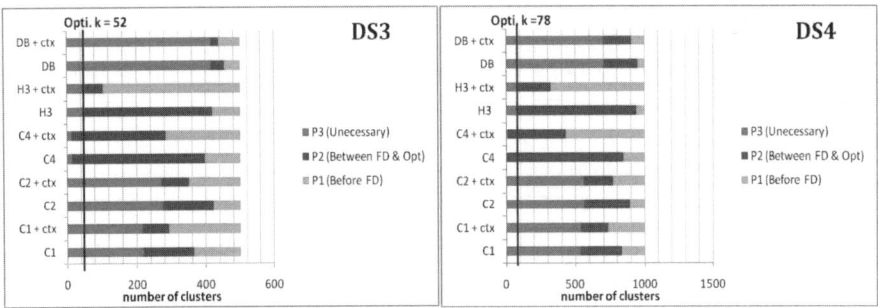

Fig. 16.10. The added-value of the context-aware method in approaching the optimal k - application on DS3 and DS4 for word clustering

optimal solution. The contribution is made clear when observing, for instance, the performance of the $C4$ index in DS2: While the optimal solution in terms of $C4$ is found at $k = 57$ with $F = 0.49$, using the index as stopping criterion entails a first drop to occur at $k = 655$ with $F = 0.21$. However, when adding context-awareness to the process, a first drop occurs closer at $k = 302$ with $F = 0.354$. Similarly, consider the $H3$ index in DS3: While the optimal solution is found at $k = 53$ with $F = 0.21$, using the index alone entails a first drop to occur at $k = 418$ with $F = 0.146$. However, when adding context-awareness to the process, a first drop occurred at $k = 102$ with $F = 0.197$, which is indeed a promising result.

16.5.2 On the Quality of the Optimal Solutions

Since a context-aware algorithm is no more taking the merging decisions that optimizes VI, one may imagine that the method, although approaching the optimal solution, can deteriorate the quality of this solution. However, results show the opposite. Actually, the histograms illustrating the $FScore$ at the optimal value of each VI, assess that with the context-aware method we can still have a comparable and sometimes better clustering quality than the standard method without involving any context-awareness. In average, using the method led the $FScore$ at the optimal value of VI to an improvement of 1.27% and 0.16% in DS2 and DS3, and to a deterioration of 0.21% and 0.14% in DS1 and DS4 respectively. This can be explained by the "safe" decisions taken at each step of the process. Although not improving VI at the highest speed, these context-aware decisions, by foreseeing the upcoming mergings, provide better clustering possibilities in future iterations, and thus competitive partitions quality at the end.

16.5.3   On the Quality of the Final Solutions

More informative than the quality of the optimal solutions, is the quality of the final provided solutions obtained when stopping the process before FD. Thus, these solutions, provided with/without using context-awareness, are also evaluated in terms of $FScore$. In average, using the context-aware method contributed to an $FScore$ improvement of 63.14%, 30.16%, 10.04%, and 19.53% in DS1, DS2, DS3, and DS4 respectively. We can notice that the largest improvements are noticed in the document clustering datasets. This is not surprising given the relatively poor representation of words patterns comparing to documents.

16.6   Conclusion and Future Trends

On the hand, we presented an experimental study that showed that indices perform generally "well" at evaluating solutions, especially when dealing with words. However, although they are supposed to be completely insensitive to the number of clusters k, they have showed some rigidity to k, leading to erroneous top-ranked solutions. In addition to that, we saw that these indices when involved as criterion functions yield slightly better results to the case where indices were simply used as external indicators.

On the other hand, we studied the feasibility of using relative indices as stopping criteria in agglomerative clustering algorithms. Experiments performed in two applications, document and word clustering, showed that indices used *alone* are not effective for such purpose. Thus, we presented a method that aims to smooth indices' plots by taking the "safest" decision at each level of a clustering process. We demonstrated that the method could remarkably enhance the usage of relative indices as stopping criteria.

An important drawback in most relative indices is their high computational cost. Yet, their utilization seems crucial in view of a parameter-free clustering. That is, an important trend is to develop methods that could accurately approximate their values on *reduced* and *representative* subsets of data. Among the few works that have been conducted in this direction, we can cite [11]. Such works, if tied with efficient clustering methods (e.g., CLIQUE, PROCLUS, Bisective k-means, frequent pattern-based methods), may enable an objective clustering on large and high-dimensional datasets.

References

1. Baeza-Yates, R.A., Ribeiro-Neto, B.A.: Modern Information Retrieval. ACM Press/Addison-Wesley (1999)
2. Ben-Hur, A., Elisseeff, A., Guyon, I.: A stability based method for discovering structure in clustered data. In: Pacific Symposium on Biocomputing, pp. 6–17 (2002)
3. Bezdek, J.C., Li, W., Attikiouzel, Y., Windham, M.P.: A geometric approach to cluster validity for normal mixtures. Soft Comput. 1(4), 166–179 (1997)
4. Chou, C.-H., Su, M.-C., Lai, E.: A new cluster validity measure and its application to image compression. Pattern Anal. Appl. 7(2), 205–220 (2004)
5. Christopher Manning, H.S.: Foundations of statistical natural language processing (1999)
6. Cimiano, P., Hotho, A., Staab, S.: Comparing conceptual, divise and agglomerative clustering for learning taxonomies from text. In: ECAI, pp. 435–439 (2004)
7. Davies, D.L., B.D.: A cluster separation measure. IEEE Transactions on Pattern Analysis and Machine Intelligence, 1(2) (1979)
8. Duda, R.O., Hart, P.E., Stork, D.G.: Pattern classification. John Willey & Sons (2001)
9. Dunn, J.C.: Well separated clusters and optimal fuzzy paritions. Journal Cybern. 4, 95–104 (1974)
10. Fraley, C., Raftery, A.E.: How many clusters? which clustering method? answers via model-based cluster analysis. Comput. J. 41(8), 578–588 (1998)
11. Greene, D., Cunningham, P.: Efficient prediction-based validation for document clustering. In: ECML, pp. 663–670 (2006)
12. Halkidi, M., Batistakis, Y., Vazirgiannis, M.: Clustering validity checking methods: Part ii. SIGMOD Record 31(3), 19–27 (2002)
13. Han, J., Kamber, M.: Data Mining: Concepts and Techniques. Morgan Kaufmann, San Francisco (2006)
14. Harris, Z.S.: Distributional structure (1985)
15. Hearst, M.A., Pedersen, J.O.. Reexamining the cluster hypothesis: Scatter/gather on retrieval results. In: SIGIR, pp. 76–84 (1996)
16. Jain, A.K., Dubes, R.C.: Algorithms for Clustering Data. Prentice-Hall, Englewood Cliffs (1988)
17. Jarvis, R.A., Patrick, E.A.: Clustering using a similarity measure based on shared near neighbors. IEEE Trans. Comput. 22(11), 1025–1034 (1973)
18. Koller, D., Sahami, M.: Hierarchically classifying documents using very few words. In: ICML, pp. 170–178 (1997)
19. Krzanowski, W.J., Lai, Y.T.: A criterion for determining the number of groups in a data set using sum-of-squares clustering. Biometrics 44, 23–34 (1988)
20. Lange, T., Roth, V., Braun, M.L., Buhmann, J.M.: Stability-based validation of clustering solutions. Neural Comput. 16(6), 1299–1323 (2004)
21. Maulik, U., Bandyopadhyay, S.: Performance evaluation of some clustering algorithms and validity indices. IEEE Trans. Pattern Anal. Mach. Intell. 24(12), 1650–1654 (2002)
22. Michalski, R., Stepp, R., Diday, E.: A recent advance in data analysis: Clustering objects into classes characterized by conjuctive concepts. Progress in Pattern Recognition 1 (1983)
23. Miller, G.A.: Wordnet: A lexical database for english. Commun. ACM 38(11), 39–41 (1995)

24. Milligan, G.W., Cooper, M.C.: An examination of procedures for determining the number of clusters in a data set. Psychometrika 50(2), 159–179 (1985)
25. Pedersen, T., Kulkarni, A.: Selecting the "right" number of senses based on clustering criterion functions. In: EACL (2006)
26. Qiu, Y., Frei, H.-P.: Concept based query expansion. In: SIGIR 1993: Proc. of the 16th annual Int. ACM SIGIR Conf. on Research and development in information retrieval, pp. 160–169. ACM, New York (1993)
27. Raskutti, B., Leckie, C.: An evaluation of criteria for measuring the quality of clusters. In: IJCAI, pp. 905–910 (1999)
28. Rissanen, J.: Stochastic complexity in statistical inquiry. World Scientific Publishing Co., Singapore (1989)
29. Saitta, S., Raphael, B., Smith, I.F.C.: A bounded index for cluster validity. In: MLDM, pp. 174–187 (2007)
30. Salton, G., Buckley, C.: Term weighting approaches in automatic text retrieval. Technical report, Ithaca, NY, USA (1987)
31. Sergios Theodoridis, K.K.: Pattern recognition. Academic Press, London (1999)
32. Shari Landes, R.I.T., Leacock, C.: Building semantic concordances, pp. 199–216 (1998)
33. Sharma, S.: Applied multivariate techniques. John Wiley and Sons, Chichester (1996)
34. Stokoe, C., Oakes, M.P., Tait, J.: Word sense disambiguation in information retrieval revisited. In: SIGIR, pp. 159–166 (2003)
35. Harabasz, C.T.: A dendrite method for cluster analysis. Communications in Statistics 3, 1–27 (1974)
36. Tibshirani, R., Walther, G., Hastie, T.: Estimating the number of clusters in a dataset via the gap statistic. Technical report, Dept. of Statistics, Stanford University (2000)
37. Zamir, O., Etzioni, O., Madani, O., Karp, R.M.: Fast and intuitive clustering of web documents. In: KDD, pp. 287–290 (1997)
38. Zhao, Y., Karypis, G.: Empirical and theoretical comparisons of selected criterion functions for document clustering. Machine Learning 55(3), 311–331 (2004)
39. Zhao, Y., Karypis, G., Fayyad, U.M.: Hierarchical clustering algorithms for document datasets. Data Min. Knowl. Discov. 10(2), 141–168 (2005)

Author Index

Printing: Krips bv, Meppel, The Netherlands
Binding: Stürtz, Würzburg, Germany